# Superstrings

# NATO ASI Series

**Advanced Science Institutes Series**

*A series presenting the results of activities sponsored by the NATO Science Committee, which aims at the dissemination of advanced scientific and technological knowledge, with a view to strengthening links between scientific communities.*

The series is published by an international board of publishers in conjunction with the NATO Scientific Affairs Division

| | | |
|---|---|---|
| A | **Life Sciences** | Plenum Publishing Corporation |
| B | **Physics** | New York and London |
| C | **Mathematical and Physical Sciences** | Kluwer Academic Publishers Dordrecht, Boston, and London |
| D | **Behavioral and Social Sciences** | |
| E | **Applied Sciences** | |
| F | **Computer and Systems Sciences** | Springer-Verlag |
| G | **Ecological Sciences** | Berlin, Heidelberg, New York, London, |
| H | **Cell Biology** | Paris, and Tokyo |

***Recent Volumes in this Series***

*Volume 170*—Physics and Applications of Quantum Wells and Superlattices
edited by E. E. Mendez and K. von Klitzing

*Volume 171*—Atomic and Molecular Processes with Short Intense Laser Pulses
edited by André D. Bandrauk

*Volume 172*—Chemical Physics of Intercalation
edited by A. P. Legrand and S. Flandrois

*Volume 173*—Particle Physics: *Cargèse 1987*
edited by Maurice Lévy, Jean-Louis Basdevant, Maurice Jacob, David Speiser, Jacques Weyers, and Raymond Gastmans

*Volume 174*—Physicochemical Hydrodynamics: Interfacial Phenomena
edited by Manuel G. Velarde

*Volume 175*—Superstrings
edited by Peter G. O. Freund and K. T. Mahanthappa

*Volume 176*—Nonlinear Evolution and Chaotic Phenomena
edited by Giovanni Gallavotti, Marcello Anile, and Paul F. Zwiefel

*Series B: Physics*

# Superstrings

Edited by

Peter G. O. Freund

University of Chicago
Chicago, Illinois

and

K. T. Mahanthappa

University of Colorado
Boulder, Colorado

Plenum Press
New York and London
Published in cooperation with NATO Scientific Affairs Division

Proceedings of a NATO Advanced Research Workshop on
Superstrings,
held July 27–August 1, 1987,
in Boulder, Colorado

Library of Congress Cataloging in Publication Data

NATO Advanced Research Workshop on Superstrings (1987: Boulder, Colo.)
Supersprings.

(NATO ASI series. Series B, Physics; v. 175)
"Proceedings of a NATO Advanced Research Workshop on Superstrings, held July 27–August 1, 1987, in Boulder, Colorado."—T.p. verso.
"Published in cooperation with NATO Scientific Affairs Division."
Includes bibliographical references and index.
1. Superstring theories—Congresses. I. Freund, Peter G. O. (Peter George Oliver), date. II. Mahanthappa, K. T. III. North Atlantic Treaty Organization. Scientific Affairs Division. IV. Title. V. Series.
QC794.6.S85N37 1987 539.7′21 88-6032

ISBN-13: 978-1-4612-8293-8 e-ISBN-13: 978-1-4613-1015-0
DOI: 10.1007/ 978-1-4613-1015-0



Softcover reprint of the hardcover 1st edition 1988

A Division of Plenum Publishing Corporation
233 Spring Street, New York, N.Y. 10013

PREFACE

The Advancea Kesearch Workshop on Superstrings was held on the campus of the University of Colorado at Boulder from July 27th through August 1, 1987.

Since the work of Green and Schwartz in the summer of 1984, string theories have elicited tremendous amount of interest from both theoretical physicists and mathematicians. The objective of the Workshop was to bring together practitioners in the field to discuss the progress and problems, and possible directions of future research.

There were ten talks of one hour each and twenty three talks of one-half hour each. The talks covered new formulations and technical developments. There were intense discussions both during and at the end of the lectures; further discussions continued during lunch and dinner. These proceedings contain all talks given at the Workshop except those by Victor Kac, Darwin Chang and Doron Gepner.

The Workshop was sponsored by the North Atlantic Treaty Organization, which provided generous financial support enabling many young physicists from the U.S. and abroad to participate in the Workshop.

Additional co-sponsors were the U.S. Department of Energy and the University of Colorado. The former offered further financial assistance and the latter furnished clerical and technical services and its campus facilities for the purpose of the organization and running of the Workshop.

The International Organizing Committee consisted of John Ellis, Francois Englert, Peter G.O. Freund (co-director), K. T. Mahanthappa (co-director) and Abdus Salam.

Jon Bjorkman, Vasilios (Bill) Koures and Richard Loft assisted in the day-to-day functioning of the Workshop. Prameela Mahanthappa and Kathy Oliver organized the social programs. Without the assistance of Linda Frueh it would not have been possible to organize and run the Workshop so smoothly.

We thank the lecturers for their stimulating contributions and for their cooperation in the preparation of this volume. We are grateful to Linda Frueh for her editorial assitance.

*Peter G.O. Freund*
*K. T. Mahanthappa*

Chicago and Boulder
November, 1987

CONTENTS

CHAPTER IV: COMPACTIFICATION AND PHENOMENOLOGY



CHAPTER V: POLYAKOV STRINGS



CHAPTER VI: LOOPS



CHAPTER VII: ANOMALIES AND CHIRAL BOSONIZATION

CHAPTER VIII: SUPERMEMBRANES AND OTHER ALTERNATIVES

# Chapter I
# Mathematical Techniques

# MODULAR INVARIANCE AND

# INFINITE-DIMENSIONAL ALGEBRAS

Peter Goddard

Department of Applied Mathematics and Theoretical Physics,
University of Cambridge, Silver Street
Cambridge, CB3 9EW, U.K.

## INTRODUCTION

This lecture is concerned with certain infinite-dimensional algebras, namely affine Kac-Moody algebras and the Virasoro algebra, their representations and the modular transformation properties of the characters of these representations. These algebras and their properties are of central importance in the study of conformally invariant two-dimensional systems, and hence in analysing the critical behaviour of suitable two-dimensional statistical systems, and in string theories where they are of relevance because of the two-dimensional nature of the string world-sheet. For a recent review of these algebras in relation to their applications in physics see [1]. Our objective is to show how the properties of representations of the Virasoro algebra can be understood in terms of the corresponding properties of affine Kac-Moody representations [2–4].

The Virasoro algebra [5], $\hat{v}$, has a basis consisting of $L_n$, $n \in \mathbf{Z}$, and a central element, $c$, satisfying the commutation relations,

$$[L_m, L_n] = (m-n)L_{m+n} + \frac{c}{12}m(m^2-1)\delta_{m,-n}, \qquad (1a)$$

$$[L_n, c] = 0. \qquad (1b)$$

In physical applications the representations of $\hat{v}$ which are relevant are the unitary ones, i.e. those which satisfy the hermiticity condition

$$L_n^\dagger = L_{-n}. \qquad (2)$$

In an irreducible unitary representation the central element $c$ takes a fixed real value.

A further typical physical requirement is that the spectrum of $L_0$ be bounded below (because it corresponds to an energy or dilatation operator). Then, since

$$L_0 L_n = L_n(L_0 - n), \qquad (3)$$

the $L_n$, $n > 0$, act as lowering operators for the eigenvalues of $L_0$. Thus, if $|h\rangle$ is an eigenvector of $L_0$ corresponding to the lowest eigenvalue $h$,

$$L_0|h\rangle = h|h\rangle, \qquad (4a)$$

$$L_n|h\rangle = 0, \qquad n > 0. \qquad (4b)$$

By a perverse inversion of nomenclature, such a state $|h\rangle$ is called a ***highest weight state.*** It is not difficult to see that the whole of an irreducible representation is built up from such a highest weight state by the action of the algebra, and that the representation space is spanned by states of the form

$$(L_{-1})^{n_1}(L_{-2})^{n_2}\dots(L_{-r})^{n_r}|h\rangle. \tag{5}$$

Such a representation is called an irreducible ***highest weight representation.*** Using the commutation relations (1), the hermiticity condition (2), and the highest weight condition (4), the scalar product of any two states of the form (5) can be calculated in terms of $c$ and $h$. Thus the representations are characterised by the pair of numbers $(c, h)$ and one then wishes to know for which values of $(c, h)$ there do indeed exist unitary representations. (Here we are using unitary in the sense of (2) holding with respect to a positive definite scalar product.) We shall give the answer to this question and describe the construction of such representations but, before we do this, we give some motivation for this study from the way the Virasoro algebra occurs in two-dimensional conformal field theory and the study of critical phenomena.

## CONFORMAL INVARIANCE

At a critical point of a suitable two-dimensional statistical system scaling behaviour occurs which can be described by a two-dimensional conformally invariant Euclidean quantum field theory. The measurable critical exponents in such a theory are linear combinations of the scaling dimensions of appropriate fields. The conformal symmetry means that the states of the field theory form representations of the Lie algebra of the conformal group (or more accurately a central extension of it). In two dimensions this algebra consists of two commuting copies of the Virasoro algebra. For systems possessing the "reflection positivity" property these representations have to be unitary in the sense described above. Many systems of interest have this property and we shall restrict attention to them. The value of the central element $c$ is common to both copies of the algebra and is a characteristic of the theory under consideration. The scaling dimensions of fields are related to the values of the highest weights $h$. Thus the list of unitary representations places important restrictions on the possible scaling dimensions in the theory. In the last couple of years it has been realised that the possible combinations of irreducible representations that can occur in a theory are further restricted in suitable physical contexts by the requirement of modular invariance.

The power of conformal invariance in two-dimensional quantum field theory was made evident by the work of Polyakov [6] and Belavin, Polyakov and Zamolodchikov [7]. The constraints following from modular invariance have emphasised and analysed by Cardy [8], by Gepner and Witten [9] and by Cappelli, Itzykson and Zuber [10].

Using the complex plane to parameterize two-dimensional Euclidean space, a basis for the infinitessimal conformal transformations is provided by the transformations

$$z \mapsto z + \varepsilon z^{n+1}, \tag{6}$$

with $\varepsilon$ a complex number. The corresponding generator can be denoted by $\varepsilon L_n + \varepsilon^* \bar{L}_n$. Here $\varepsilon^*$ is the complex conjugate of $\varepsilon$ but $\bar{L}_n$ is independent of $L_n$. In a conformally invariant theory the stress-energy tensor is traceless and conserved and thus of the form

$$\Theta^{\mu\nu} dx^\mu dx^\nu = \Theta(z)dz^2 + \bar{\Theta}(z^*)(dz^*)^2. \tag{7}$$

The generators of conformal transformations are given by the moments of

$$\Theta(z) = \sum_{n\in\mathbf{Z}} L_n z^{-n-2}, \quad \text{and} \quad \bar{\Theta}(z^*) = \sum_{n\in\mathbf{Z}} \bar{L}_n (z^*)^{-n-2}. \tag{8}$$

The operators $L_n$ and $\bar{L}_n$ satisfy two commuting Virasoro algebras with the same central term $c$.

In such a field theory there are certain ***primary fields*** $\varphi$ which have simple conformal transformation properties, specified by commutation relations

$$[L_n, \varphi] = z^{n+1}\frac{\partial\varphi}{\partial z} + h(n+1)z^n\varphi, \tag{9a}$$

$$[\bar{L}_n, \varphi] = (z^*)^{n+1}\frac{\partial\varphi}{\partial z^*} + \bar{h}(n+1)(z^*)^n\varphi. \tag{9b}$$

The vacuum state $|0\rangle$ satisfies the conditions

$$L_n|0\rangle = 0, \qquad n \geq -1. \tag{10}$$

From this it follows that $\varphi(0)$ creates from the vacuum a highest weight state for the Virasoro algebras $L_n$ and $\bar{L}_n$,

$$L_0\varphi(0)|0\rangle = h\varphi(0)|0\rangle, \tag{11a}$$

$$L_n\varphi(0)|0\rangle = 0, \qquad n > 0, \tag{11b}$$

with similar equations obtained by replacing $L_n$ by $\bar{L}_n$ and $h$ by $\bar{h}$, provided that the limit of $\varphi(z)|0\rangle$ as $z \to 0$ exists, which we assume. These equations are obtained by letting equations (9) act on the vacuum and then sending $z \to 0$. We assume that all the states which are highest weight states with respect to both $L_n$ and $\bar{L}_n$ are obtained in this way, i.e. there is a one-to-one correspondence between such highest weight states and primary fields.

The conditions (9–11) can be used to determine two and three point functions up to a multiplicative constant. For the two point function we have

$$\langle 0|\varphi(z_1)\varphi(z_2)|0\rangle = r^{-2(h+\bar{h})}e^{-2i\theta(h-\bar{h})}\langle 0|\varphi(1)\varphi(0)|0\rangle \tag{12}$$

where

$$z_2 - z_1 = re^{i\theta}. \tag{13}$$

This shows that $h + \bar{h}$ is the scaling dimension of the field $\varphi$ and $h - \bar{h}$ is its spin.

## VIRASORO REPRESENTATIONS

Returning to the question of for which values of $(c, h)$ do there exist unitary representations of $\hat{v}$, to analyse this, we consider the matrix $\mathcal{M}_N(c, h)$ formed by the scalar products of the states (15) for which $L_0$ has eigenvalue $h + N$, i.e. those for which $\sum jn_j = N$. This is a $\pi_N \times \pi_N$ dimensional matrix, where

$$\sum_{N=0}^{\infty} \pi_N q^N = \prod_{n=1}^{\infty}(1 - q^n) \tag{14}$$

These matrices need to be positive semi-definite for each positive N. A remarkable formula for det $\mathcal{M}_N(c, h)$ was given by Kac [11,12], and this enabled Friedan, Qiu and

Shenker [13] to find necessary conditions on $(c, h)$ for unitarity; we shall see, by explicit construction, that these conditions are also sufficient.

To see something of the way these conditions arise, consider

$$\begin{aligned}\|L_{-n}|h\rangle\|^2 &= \langle h|[L_n, L_{-n}]|h\rangle, \qquad \text{if } n > 0,\\ &= \{2nh + \frac{c}{12}n(n^2-1)\}\||h\rangle\|^2,\end{aligned} \tag{15}$$

which must be non-negative for all positive $n$. From this it immediately follows that

$$c \geq 0, \qquad h \geq 0. \tag{16}$$

It is not difficult to estabish using Kac's formula that all the values

$$c \geq 1, \qquad h \geq 0, \tag{17}$$

correspond to unitary representations. The point $(c, h) = (0, 0)$ corresponds to the trivial representation, i.e. $L_n \equiv 0$. It is tempting to suppose that this is the only unitary representation apart from the continuum of representations (17), but actually there are three representations well-known from the spinning string theory of Ramond-Neveu-Schwarz [14–17], corresponding to a single periodic (Ramond) or anti-periodic (Neveu-Schwarz) fermion field. These correspond to $c = \frac{1}{2}$ and $h = 0$ or $\frac{1}{2}$ (NS case) and $h = \frac{1}{16}$ (R case). In fact, $c = 0$ and $\frac{1}{2}$ are the first two terms in a discrete series of unitary representations.

The result established by Friedan, Qiu and Shenker [13] is that for a unitary highest weight representation it is necessary that *either*

$$c \geq 1, \qquad h \geq 0, \tag{18a}$$

*or*

$$c = 1 - \frac{6}{(m+2)(m+3)}, \qquad m = 0, 1, \ldots \tag{18b}$$

$$h = \frac{[(m+3)p - (m+2)q]^2 - 1}{4(m+2)(m+3)}, \tag{18c}$$

where $p = 1, 2, \ldots m+1$ and $q = 1, 2, \ldots p$. These representations have been identified with two-dimensional statistical systems as indicated in the following table:

| | | | |
|---|---|---|---|
| $m = 0$ | $c = 0$ | $h = 0.$ | |
| $m = 1$ | $c = \frac{1}{2}$ | $h = 0, \frac{1}{16}, \frac{1}{2}.$ | Ising model |
| $m = 2$ | $c = \frac{7}{10}$ | $h = 0, \frac{3}{80}, \frac{1}{10}, \frac{7}{16}, \frac{3}{5}, \frac{3}{2}.$ | Tricritical Ising model |
| $m = 3$ | $c = \frac{4}{5}$ | $h = 0, \frac{1}{40}, \frac{1}{15}, \frac{1}{8}, \frac{2}{5}, \frac{21}{40}, \frac{2}{3}, \frac{7}{5}, \frac{13}{8}, 3.$ | 3-state Potts model |

In the conformal field theory, associated with the critical behaviour of a two-dimensional statistical system, with a value of $c < 1$, the primary fields are labelled by pairs of values $(h, \bar{h})$, where both $h$ and $\bar{h}$ are chosen from the list of possible values (18c). For example, in the Ising model, which has $c = \frac{1}{2}$, there are three primary fields, namely the following:

| | | |
|---|---|---|
| $(0, 0)$ | 1 | identity operator; |
| $(\frac{1}{2}, \frac{1}{2})$ | $\varepsilon$ | magnetisation; |

$(\frac{1}{16}, \frac{1}{16})$ $\sigma$ energy density.

In this case the primary fields are labelled by pairs $(h, \bar{h}) = (\eta, \eta)$, where $\eta$ ranges over the set of values given by (18c). In general it not necessary that $h = \bar{h}$ but the set of pairs that can in practice occur as labelling the primary fields in the conformal field theory associated with a statistical system is in fact severely constrained for $c < 1$ and there actually only a few possibilities.

## MODULAR INVARIANCE

What produces these constraints is the requirement of *modular invariance* [8]. If $Z(\ell, \ell')$ denotes the partition function for a statistical system at a critical point defined on an $\ell \times \ell'$ rectangle with periodic boundary conditions, that is effectively on a torus, then evidently

$$Z(\ell, \ell') = Z(\ell', \ell). \tag{19}$$

We can obtain a continuum theory on the torus by letting $\ell, \ell' \to \infty$ with the ratio of $\ell$ to $\ell'$ kept fixed. In such a limit

$$Z \sim e^{-f\ell\ell'} I, \tag{20}$$

where $I$ is a function of the ratio. We can generalise this to more general tori by considering a theory defined on a parallelogram. If we think of our original rectangle as one with vertices at the origin, $\ell$ and $i\ell'$ in the complex plane, we can obtain this generalisation by letting $i\ell'$ move off the imaginary axis. By letting $\ell \to \infty$ with the ratio $\tau = i\ell'/\ell$ fixed, we obtain a function $I(\tau)$. It is easy to see that (a) interchanging $\ell$ and $\ell'$ and (b) replacing $i\ell'$ by $i\ell' + \ell$ do not change the periodic theory, i.e yield indentical tori. As a result

$$I(\tau) = I(-1/\tau) \tag{21a}$$

and

$$I(\tau) = I(1 + \tau). \tag{21b}$$

The transformations $\tau \mapsto -1/\tau$ and $\tau \mapsto \tau + 1$ generate a discrete group, the *modular group*,

$$\tau \mapsto \tau' = \frac{a\tau + b}{c\tau + d}, \tag{22}$$

where $a, b, c$ and $d$ are integers and $ad - bc = 1$, which maps the upper half complex $\tau$-plane into itself. It follows that $I(\tau)$ is invariant under the whole of the modular group (22).

The function $I(\tau)$ can be expressed in terms of the representations of the Virasoro algebras $L_n$ and $\bar{L}_n$, or equivalently the primary fields that occur in the theory. In fact

$$I(\tau) = q^{-\frac{c}{24}} (q^*)^{-\frac{c}{24}} \operatorname{trace}\{q^{L_0} (q^*)^{\bar{L}_0}\} \tag{23a}$$

$$= \sum_j \chi_{c,h_j}(\tau) \chi_{c,\bar{h}_j}(\tau)^* \tag{23b}$$

where

$$q = e^{2\pi i \tau}, \tag{24}$$

and $\chi_{c,h}$ denotes the character of the representation $(c, h)$ of the Virasoro algebra,

$$\chi_{c,h}(\tau) = q^{-\frac{c}{24}} \operatorname{trace}\{q^{L_0}\}. \tag{25}$$

The sum in (24) is over the pairs $(h_j, \bar{h}_j)$ labelling the primary fields in the theory. Thus the modular invariance requirements (21) impose constraints on the combinations of primary fields that can occur in a theory.

KAC-MOODY REPRESENTATIONS

We shall be able to see how to form such modular invariant sesquilinear combinations of Virasoro characters for representations with $c < 1$ once we have described the construction of such representations. As a preliminary to doing this, we first review some elementary properties of affine Kac-Moody algebras. Given a compact Lie algebra $g$, with a basis consisting of $t^a$, $1 \leq a \leq \dim g$, satisfying the commutation relations

$$[t^a, t^b] = i\, f^{abc}\, t^c, \tag{26}$$

where the structure constants $f^{abc}$ are totally antisymmetric, the affine Kac-Moody algebra $\hat{g}$ associated with $g$ is

$$[T^a_m, T^b_n] = i\, f^{abc} T^c_{m+n} + k\, m\, \delta_{m,-n}\, \delta^{ab}, \tag{27a}$$

where $k$ is a central element,

$$[T^a_m, k] = 0. \tag{27b}$$

A representation of $\hat{g}$ is said to be *unitary* if the hermiticity conditions

$$(T^a_n)^\dagger = T^a_{-n}, \qquad k^\dagger = k, \tag{28}$$

are satisfied. Since $k$ is an hermitian generator commuting with all the other generators of the algebra, like $c$ in the Virasoro algebra, it takes a definite value in any irreducible unitary representation.

Our construction of representations of the Virasoro algebra is based on the fact that it is always possible to extend any representation of $\hat{g}$ so that we also have a representation of $\hat{v}$ which interrelates with that of $\hat{g}$ so that we have a representation of a semi-direct product, defined by (1) and (27) together with

$$[L_m, T^a_n] = -nT^a_{m+n}. \tag{29}$$

The representations of $\hat{g}$ which are physically interesting are those which are unitary and for which the spectrum of $L_0$ is bounded below. Such representations are also called *unitary highest weight representations.* [We do not need to bother about the precise form of $L_0$; it is enough for the purposes of defining highest weight representations that it satisfy (29).]

It is not surprising that affine Kac-Moody algebras and the Virasoro algebra are related. If $G$ denotes a compact Lie group with Lie algebra $g$, then $\hat{g}$ is the algebra of the group of maps from the unit circle, $S^1 \to G$, with the group operation defined by pointwise multiplication (at least if we put $k = 0$). The Virasoro algebra, $\hat{v}$, is the algebra of maps $S^1 \to S^1$, with the group operation defined by composition. Thus, there is a natural action of $\hat{v}$ on $\hat{g}$, making them into a semi-direct product.

In a highest weight representation of $\hat{g}$ there must be states *highest weight states* $\Psi$, which will be those for which the eigenvalue of $L_0$ is least. Such states will satisfy

$$T^a_n \Psi = 0, \qquad \text{for } n > 0. \tag{30}$$

Acting on these states with the operators $T^a_{-n}$, $n > 0$, generates an invariant subspace, and hence the whole representation space if it is irreducible. In this case it is spanned by states of the form

$$T^{a_1}_{-n_1} T^{a_2}_{-n_2} \ldots T^{a_r}_{-n_r} \Psi. \tag{31}$$

The subalgebra $\{T^a_0\}$ of $\hat{g}$, which is isomorphic to $g$, maps the states $\Psi$ satisfying (30) into themselves. Thus these highest weight states form a representation of the finite-dimensional algebra $g$, which we call the *vacuum representation.* This must be an irreducible representation if we are to obtain an irreducible representation of $\hat{g}$. The algebra (27) and the hermiticity conditions (28) enables us to calculate all the scalar products of the states (31) provided that we know the value of $k$ and the representation of $\hat{g}$ formed by the vacuum states. The classification of unitary highest weight representation then amounts to deciding for which values of $k$ and for which vacuum representations the scalar products, calculated as described above, are really consistent with a positive definite scalar product.

We first of all answer this question for $su(2)$ and then discuss how it generalises. The affine algebra $s\hat{u}(2)$ is defined by

$$[T^a_m, T^b_n] = i\epsilon_{abc} T^c_{m+n} + km\delta^{ab}, \tag{32}$$

and its representations are labelled by $(k, l)$, where $l$ is the spin of the vacuum representation. From the representation theory of $su(2)$, we know that $2l \in \mathbf{Z}$. To see what further constraints there are, write $s\hat{u}(2)$ in terms of step operators,

$$[T^+_m, T^-_n] = 2T^3_{m+n} + 2mk\delta_{m,-n}, \tag{33}$$

where $T^\pm_m = T^1_m \pm iT^2_m$. Now, if $|\mu\rangle$ denotes a vacuum state of helicity $\mu$,

$$\begin{aligned} \|T^+_{-1}|\mu\rangle\|^2 &= \langle\mu|[T^-_1, T^+_{-1}]|\mu\rangle \\ &= 2(k-\mu)\||\mu\rangle\|^2, \end{aligned} \tag{34}$$

from which we deduce that $k \geq |\mu|$. Further, we see that $T^-_1$, $T^+_{-1}$, $T^3_0 + k$ constitutes an algebra isomorphic to $su(2)$. Hence, $2\mu + 2k \in \mathbf{Z}$ for any eigenvalue $\mu$ of $T^3_0$. Since we all ways have $2\mu \in \mathbf{Z}$, it follows that $2k$ is an integer, and as we have just seen, it must be non-negative. This integer is called the *level* and we will denote it in general by $x$. For a given level $x$, the spin of the vacuum $l$, must satisfy

$$2l \leq x. \tag{35}$$

Thus, for $x = 0$ there is only one possibility, $l = 0$, and it is easy to see that the representation is trivial. For $x = 1$ we can have $l = 0$ or $\frac{1}{2}$. The condition (35) is not only necessary but also sufficient and this can be demonstrated by constructing the corresponding unitary representations explicitly.

For a general affine Kac-Moody algebra $\hat{g}$, the level is defined by

$$x = \frac{2k}{\psi^2} \tag{36}$$

and is a non-negative integer again. Here $\psi$ denotes any long root of the algebra $g$, which for convenience we are assuming to be simple. By considering $su(2)$ subalgebras of $\hat{g}$ of the form

$$E^\beta_{-1}, E^{-\beta}_1, \frac{2}{\beta^2}(\beta \cdot H_0 + k), \tag{37}$$

where $H$ denotes the elements of a Cartan subalgebra of $g$ and $E^\beta$ a step operator, $\beta$ being a root of $g$, we see that any weight $\lambda$ of a vacuum representation (and so in particular its highest weight) must satisfy

$$\|\beta{\cdot}\lambda\| \leq k \qquad \text{for all roots } \lambda. \tag{38}$$

Again it is possible to show by construction that the condition (38) is not only necessary but also sufficient for $\lambda$ to the highest weight of the vacuum representation of a unitary highest weight representation of $\hat{g}$.

SUGAWARA CONSTRUCTION

The fact that a representation of $\hat{g}$ can be extended to a representation of the semi-direct product of $\hat{v}$ and $\hat{g}$ is demonstrated by the Sugawara construction [18], which represents the Virasoro algebra as bilinear quantities in the Kac-Moody generators, roughly

$$\Theta(z) \equiv \sum_{n\in\mathbf{Z}} L_n z^{-n-2} \sim T^a(z)T^a(z), \tag{39}$$

where a sum over $a$ is implied and

$$T^a(z) = \sum_n T^a_n z^{-n-1}. \tag{40}$$

To make this precise, we need to be careful about such matters as normal ordering, so that the expressions are well-defined. We define a normal ordering operation by

$$\begin{aligned} {}^\times_\times T^a_m T^a_n {}^\times_\times &= T^a_m T^a_n \qquad \text{if } n \geq 0, \\ &= T^a_n T^a_m \qquad \text{if } n \leq 0. \end{aligned} \tag{41}$$

Note that this definition is phrased entirely in terms of the operators $T^a_n$ and is not in any way dependent on the way they might be constructed out of any more basic oscillators. First, as a precise interpretation of (39), consider

$$\tilde{\mathcal{L}}^g_n = \tfrac{1}{2}\sum_m {}^\times_\times T^a_m T^a_{n-m} {}^\times_\times. \tag{42}$$

Calculation shows that this is not quite a representation of the Virasoro algebra. First, from eq.(27), one can deduce that

$$[\tilde{\mathcal{L}}^g_m, T^a_n] = -(k + \tfrac{1}{2}Q^g)nT^a_{m+n}, \tag{43}$$

where $Q^g$ is the quadratic Casimir operator of $g$ in the adjoint representation,

$$f^{abc}f^{abd} = Q^g\delta^{cd}. \tag{44}$$

Thus we are led to renormalise the definition (42) to give

$$\mathcal{L}^g_n = \frac{1}{2k + Q^g}\sum_m {}^\times_\times T^a_m T^a_{n-m} {}^\times_\times. \tag{45}$$

Then, it follows from eq.(43) that

$$[\mathcal{L}^g_m, \mathcal{L}^g_n] = (m-n)\mathcal{L}^g_{m+n} + \frac{c^g}{12}m(m^2-1)\delta_{m,-n}, \tag{46}$$

where

$$c^g = \frac{2k \dim g}{2k + Q^g} \tag{47a}$$

$$= \frac{x \dim g}{x + h^g}. \tag{47b}$$

We have used

$$h^g = \frac{Q^g}{\psi^2} \tag{48}$$

to denote the *dual Coxeter number* of $g$, which is always an integer. From (47) it is clear that the value of the central charge $c^g$ provided by the Sugawara construction is always rational but it is not in general integral or even half-integral as it would be for free bosons or free fermions. In fact, it is not difficult to see that

$$\text{rank} g \leq c^g \leq \dim g. \tag{49}$$

If $g$ is an abelian algebra, these inequalities become equalities.

COSET CONSTRUCTION

Since the Sugawara construction always gives a value of $c$ bigger than 1, to construct the representations in the discrete series (18b,c) we must use an elaboration of the Sugawara construction, due to Goddard, Kent and Olive [2.3]. This construction is based on an algebra and a subalgebra, $g \supset h$, and so can be associated with a coset space $g/h$. Given such a pair, we can use the Sugawara construction to define two Virasoro algebras, $\mathcal{L}_n^g$ and $\mathcal{L}_n^h$, with corresponding values of $c$, denoted $c^g$ and $c^h$, given by the general formula (47). It then follows that if $T_n^a$ is an element of $\hat{h}$, the relation (29) will hold in respect of both Virasoro algebras, i.e.

$$[\mathcal{L}_m^g, T_n^a] = -nT_{m+n}^a, \tag{50a}$$

$$[\mathcal{L}_m^h, T_n^a] = -nT_{m+n}^a, \tag{50b}$$

but this will not hold if $T_n^a$ is a member of $\hat{g}$ that is not in $\hat{h}$. From (50) it immediately follows that $\mathcal{L}_m^g - \mathcal{L}_m^h$ commutes with the whole of $\hat{h}$. We define

$$K_n = \mathcal{L}_n^g - \mathcal{L}_n^h. \tag{51}$$

Since $\mathcal{L}_m^h$ is constructed out of the elements of $\hat{h}$, it follows that it commutes with $K_n$. Then, since

$$\mathcal{L}_n^g = \mathcal{L}_n^h + K_n, \tag{52}$$

$$[\mathcal{L}_m^g, \mathcal{L}_n^g] = [\mathcal{L}_m^h, \mathcal{L}_n^h] + [K_m, K_n]. \tag{53}$$

From this it is easy to see that $K_m$ satisfies the Virasoro algebra,

$$[K_m, K_n] = (m-n)K_{m+n} + \frac{c^K}{12}m(m^2-1)\delta_{m,-n}, \tag{54}$$

where the value of the central term is given by

$$c^K = c^g - c^h. \tag{55}$$

Note that (52) expresses $\mathcal{L}_n^g$ as the sum of two commuting Virasoro algebras.

In the above we have implicitly assumed that both $g$ and $h$ are simple algebras. However in practice we wish to apply the construction in cases in which either or both are semi-simple. To extend it to cover these cases we only need to describe how the Sugawara construction applies in semi-simple cases. This is quite straightforward. If $g$ is the direct sum of simple algebras $g_j$,

$$g = g_1 \oplus g_2 \oplus \ldots \oplus g_M, \tag{56}$$

we apply the Sugawara construction to each of the simple components and add the results

$$\mathcal{L}_n^g = \mathcal{L}_n^{g_1} + \mathcal{L}_n^{g_2} + \ldots + \mathcal{L}_n^{g_M}. \tag{57}$$

This provides a Virasoro algebra with a value of $c$ given by

$$c^g = c^{g_1} + c^{g_2} + \ldots + c^{g_M}, \tag{58}$$

where each $c^{g_j}$ is given by (47)

$$c^{g_j} = \frac{x_j \dim g_j}{x_j + h^{g_j}}, \tag{59}$$

where $h^{g_j}$ is the dual Coxeter number of $g_j$ and $x_j$ is the level of the representation of $\hat{g}_j$. (To specify a highest weight unitary representation of $g$ we need to specify the vacuum representation and the level for each component $\hat{g}_j$.)

It is easy to use the construction (51) to produce the whole of the discrete series of Virasoro representations. To do this take $g = su(2) \oplus su(2)$ and $h$ to be the diagonal $su(2)$ subalgebra. For a level $x$ representation of $s\hat{u}(2)$,

$$c^{su(2)} = \frac{3x}{x+2}, \tag{60}$$

as $h^{su(2)} = 2$. If we take a representation of $\hat{g}$ formed by a level $m$ representation of the first $s\hat{u}(2)$ factor and a level 1 representation of the second, we obtain a level $m+1$ representation of $\hat{h}$. Then (55) becomes

$$\begin{aligned} c^K &= \frac{3m}{m+2} + 1 - \frac{3(m+1)}{(m+3)} \\ &= 1 - \frac{6}{(m+2)(m+3)}, \end{aligned} \tag{57}$$

giving the whole sequence (18b).

This construction is manifestly unitary. To demonstrate [3] that it yields all the values of $h$ given by (18c) it is necessary to consider in detail the decomposition of these representations of $\hat{g}$ into irreducible representations of the direct sum of algebras $\hat{h} \oplus \hat{v}$, where $\hat{v}$ is the Virasoro algebra $\{K_n\}$. It can be shown [4] that this decomposition is a finite one if and only if $c^K < 1$. To show that finite reducibility follows from this condition, consider the action of the equation

$$\mathcal{L}_0^g = \mathcal{L}_0^h + K_0 \tag{58}$$

on the highest weight states for $\hat{h} \oplus \hat{v}$ in a given representation of $\hat{g}$. On such a state the value of the right hand side is seen to be

$$\frac{Q_\lambda^h}{2k + Q^h} + h^K, \tag{59}$$

where $Q^h_\lambda$ is the value of the quadratic Casimir operator of $h$ in the appropriate vacuum representation and $h^K$ is one of the permitted values of $h$ for the given value of $c = c^K$. Each of these two quantities is bounded, given the representation of $\hat{g}$, because the level of the representation of $\hat{h}$ is determined by that of $\hat{g}$ and so the choice of vacuum representations of $\hat{h}$ is finite, and there are only finitely many values of $h$ available in (18c) for a given $c < 1$. A list of all the pairs $g \supset h$ for which $c^K < 1$ is provided in [4].

It follows that when $c^K < 1$ the character of an irreducible representation of $\hat{g}$ can be written as a finite sum of products of characters of $\hat{h}$ and $\hat{v}$. The character of a representation $(x, \lambda)$ of $\hat{g}$, where $x$ is the level and $\lambda$ the highest weight of the vacuum representation, is defined by

$$\chi^g_{x,\lambda}(\tau, \theta) = q^{-\frac{c}{24}} trace\{q^{L^g_0} e^{i\theta\cdot H_0}\}, \tag{60}$$

where $q$ is given by (24). If the representation $(x, \lambda)$ decomposes into the direct sum of representations $(y, \mu_j) \times (c^K, h_j)$,

$$(x, \lambda) = \bigoplus_j (y, \mu_j) \times (c^K, h_j). \tag{61}$$

This direct sum is finite given that $c^K < 1$. We have in consequence a relationship between characters

$$\chi^g_{x,\lambda}(\tau, \theta) = \sum_j \chi^h_{y,\mu_j}(\tau, \theta) \chi_{c^K, h_j}(\tau), \tag{62}$$

provided that $\theta \cdot H_0$ is in $\hat{h}$ as well as $\hat{g}$. We shall see that this relationship between characters can be very useful in studying the modular transformation properties of the characters of Virasoro representations.

The irreducible unitary highest weight representations of $s\hat{u}(2)$ are thus labelled by a pair $(m, l)$, where $2l \le m$, $l$ is the spin of the vacuum representation and $m$ is the level. Here $m \in \mathbf{Z}$ and $2l \in \mathbf{Z}$. Thus, in the case of $g = su(2) \oplus su(2)$, the irreducible representations of $\hat{g}$ can be labelled $(m, \frac{1}{2}(p-1)) \times (1, \delta)$, where $p \in \mathbf{Z}$, $1 \le p \le m+1$ and $\delta = 0$ or $\frac{1}{2}$. Taking $h$ to be diagonal $su(2)$, as in (57), the form that (61) assumes is [3]

$$(m, \tfrac{1}{2}(p-1)) \times (1, \delta) = \bigoplus_{q=1}^{m+2} n^\delta_{p,q}(m+1, \tfrac{1}{2}(q-1)) \times (c, h_{p,q}), \tag{63}$$

where $c, h_{p,q}$ are given by (18b) and (18c) and $n^\delta_{p,q} = 1$ if $p - q \in \mathbf{Z} + 2\delta$ and 0 otherwise. This demonstrates that all the representations in the discrete series are indeed unitary.

## MODULAR INVARIANTS

The modular transformation properties of the characters $\chi^g_{x,\lambda}(\tau, \theta)$ are calculable in a straightforward fashion using the Kac-Weyl character formula [19]. Putting the characters of the representations of $\hat{g}$ of a given level together to form a vector $\chi^g_x$. Under a modular transformation these characters transform into linear combinations of one another and so the transformation is described by a unitary matrix $M^g_x$,

$$\chi^g_x \mapsto M^g_x \chi^g_x. \tag{64}$$

From such matrix transformation rules, we can deduce the modular properties of Virasoro properties. From (62) we obtain a relation

$$\chi^g_x(\tau,\theta) = \chi^v(\tau)\chi^h_y(\tau,\theta), \tag{65}$$

where $\chi^v$ is a matrix whose elements are characters of $\hat{v}$ for $c = c^K$. Hence, from (64), under a modular transformation these characters also transform into linear combinations of themselves,

$$\chi^v \mapsto (M^g_x)^{-1}\chi^v M^h_y. \tag{66}$$

In this way we deduce from (63) a relation, first found by inspection by Gepner [20] between the modular transformation properties of Virasoro characters and $\hat{su}(2)$ characters.

Since the matrices $M^g_x$ are unitary, it follows that ${\chi^g_x}^\dagger \chi^g_x$ is invariant under modular transformations. In what follows we shall refer to this as the *trivial invariant* for level $x$ representations of $\hat{g}$. Other invariants can be described by matrices $K^g_x$ preserved by the unitary transformations $M^g_x$,

$$(M^g_x)^{-1}K^g_x M^g_x = K^g_x. \tag{67}$$

Given such invariants for level $x$ representations of $\hat{g}$ and level $y$ representations of $\hat{h}$, we can use (66) to form an invariant for Virasoro representations with $c = c^{\hat{K}}$, namely

$$trace\{{K^h_y}^\dagger {\chi^v}^\dagger K^g_x \chi^v\}. \tag{68}$$

Applying this to the construction of the discrete series that we have given, we can construct invariants for the value of $c$ given by (18b) from invariants for level $m$ and level $m+1$ representations of $\hat{su}(2)$ (together with the trivial invariant for level 1, which is the only invariant for that level).

The invariants which are of relevance in the conformal field theories are those in which the coefficients of the products of characters are positive integers. Cappelli, Itzykson and Zuber conjectured [10] that all such modular invariants for representations of the Virasoro algebra with $c < 1$ are obtained in this way from $\hat{su}(2)$ invariants with positive integral coefficients and they further conjectured a complete list of of such invariants. These conjectures have recently been proved [21,22]. The conjectured list of $\hat{su}(2)$ invariants consisted of those that could be obtained by taking positive integral linear combinations of a certain basic set labelled by the simple simply-laced algebras $A_n, n \geq 2; D_n, n \geq 4; E_6, E_7$, and $E_8$, the level being given by subtracting 2 from the dual Coxeter number of the labelling algebra.

Just as pairs $g \supset h$ can be used to obtain modular invariants for the Virasoro algebra, they can be used to obtain non-trivial invariants for $\hat{h}$ from trivial invariants for $\hat{g}$ when $c^K = 0$. In this case the representations of $\hat{g}$ are finitely reducible with respect to $\hat{h}$ and the matrix $\chi^v$ consists of constants; $(\chi^v)^\dagger\chi^v$ then defines an invariant for $\hat{h}$ which may be non-trivial. The $\hat{su}(2)$ invariants labelled by $E_6$ and $E_8$ can be obtained in this way from the pairs $so(5) \supset su(2)$ and $G_2 \supset su(2)$ [23,4].

To obtain the other $\hat{su}(2)$ invariants listed by Cappelli, Itzykson and Zuber we need a slight refinement of this procedure. If we have a pair $g \supset h$, where $h$ has the structure $su(2) \oplus h'$ and $c^K = 0$ for representations of $\hat{g}$ of a suitable level (which must in practice be 1), we can obtain a modular invariant for $\hat{su}(2)$ from trivial invariants for $g$ and $h'$. To get such a pair $g \supset su(2) \oplus h'$ associated with each simple simply-laced algebra we use the symmetric space theorem of [24]. This implies that if $g' \supset h$ is a symmetric space then, using the $N = \dim g' - \dim h$ dimensional representation of $h$ provided by the tangent space to the symmetric space, we obtain a pair $so(N) \supset h$ such that $c^K = 0$ for the level 1 representation of $s\hat{o}(N)$. If $g'$ is a simple simply-laced algebra and $\psi$ one of its roots, we can use

$$\sigma = e^{i\pi\psi \cdot H}, \tag{69}$$

where $H$ denotes a Cartan subalgebra of $g'$, to obtain a symmetric space $g' \supset h$ by defining

$$h = \{X \in g' : \sigma X \sigma^{-1} = X\}. \tag{70}$$

This construction is due to Nahm [25] who has shown that it leads from the simply-laced algebra $g'$ to a level $(h^{g'} - 2)$ $s\hat{u}(2)$ modular invariant which is a combination of the trivial invariant and the non-trivial $s\hat{u}(2)$ invariant Cappelli, Itzykson and Zuber [10] associated with $g'$.

It is hoped the arguments presented in this lecture have illustrated how representations of the Virasoro algebra, and in particular their modular properties, can be understood in terms of the representations of affine Kac-Moody algebras.

ACKNOWLEDGEMENTS

I am grateful to Peter Bowcock, Adrian Kent, Werner Nahm and David Olive for discussions on the subject matter of this lecture.

REFERENCES

1. P. Goddard and D. Olive, Int. J. Mod. Phys. **A1** (1986) 303.
2. P. Goddard, A. Kent and D. Olive, Phys. Lett. **152B** (1985) 88.
3. P. Goddard, A. Kent and D. Olive, Commun. Math. Phys. **103** (1986) 105.
4. P. Bowcock and P. Goddard, Nucl. Phys. **B285** **[FS19]** (1987) 651.
5. M.A. Virasoro, Phys. Rev. **D1** (1970) 2933.
6. A.M. Polyakov, JETP Lett. **12** (1970) 381.
7. A.A. Belavin, A.M. Polyakov and A.B. Zamolodchikov, Nucl. Phys. **B241** (1984) 333.
8. J.L. Cardy, Nucl. Phys. **B270** **[FS16]** (1986) 186.
9. D. Gepner and E. Witten, Nucl. Phys. **B278** (1987) 493.
10. A. Cappelli, C. Itzykson and J.B. Zuber, **B280** **[FS18]** (1987) 445.
11. V.G. Kac, *Lecture Notes in Physics* **94** (1979) 441.
12. B.L. Feigin and D.B. Fuks, Funct. Anal. App. **16** (1982) 114.
13. D. Friedan, Z. Qiu and S. Shenker, Phys. Rev. Lett. **52**
14. K. Bardacki and M.B. Halpern, Phys. Rev. **D3** (1971) 2493.
15. P. Ramond, Phys. Rev. **D3** (1971) 2415.
16. A. Neveu and J.H. Schwarz, Nucl. Phys. **B31** (1971) 86.

17. A. Neveu, J.H. Schwarz and C.B. Thorn, Phys. Lett. **35B** (1971) 529.

18. H. Sugawara, Phys. Rev. **170** (1968) 1659.

19. V.G. Kac, Funct. Anal. Appl. **8** (1974) 68.

20. D. Gepner, Nucl. Phys. **287** (1987) 111.

21. A. Cappelli, C. Itzykson and J.B. Zuber, Saclay preprint S. Ph-T/87-59 (1987).

22. A. Kato, Tokyo preprint UT-509 (1987).

23. P. Bouwknegt and W. Nahm, Phys. Lett. **184B** (1987) 359.

24. P. Goddard, W. Nahm and D. Olive, Phys. Lett. **160B** (1985) 111.

25. W. Nahm, Stony Brook preprint ITP-SB-87-7 (1987).

# OFF-SHELL BRST INVARIANT SUPERJACOBIANS [1]

**S. James Gates, Jr.** [2]

**Department of Physics and Astronomy**
**University of Maryland**
**College Park, MD 20742 USA**

**Abstract**

In this talk I review the off-shell superspace representations of the Jacobian factors arising in going to an orthonormal gauge in spinning and heterotic string theories.

In many references, including "the big green book"[1], we find that the $FP$ ghost action for the NSR $N = 1$ spinning string is given by the expression,

$$S^{(1)}_{FP,SUSY} = -i \int d^2\sigma \left[ b^{ab}\, \partial_a c_b + \beta^{\alpha a}\, \partial_a \gamma_\alpha \right] , \tag{1}$$

$c_a$ and $\gamma_\alpha$ are the diffeomorphic and supersymmetry ghosts and $b^{ab}$ and $\beta^{\alpha a}$ are anti-ghosts. An interesting question is, "How does this generalize to arbitrary supergravity background fields $\hat{e}_a{}^m$ , $\hat{\psi}_a{}^\alpha$ , $\hat{A}$ ?" This is important when we recall that the partition function for the spinning string includes a sum over the modular parameters for two-surfaces of arbitrary genus. The background supergravity fields can depend on these modular parameters.

A partial answer to the question was given by Freedman and Warner [2] who showed that (1) can be generalized in the presence of background supergravity fields

$$S_{FP,SUSY} = -i \int d^2\sigma\, \hat{e}^{-1} \left[ b^{ab} \hat{\nabla}_a c_b + \beta^{\alpha a} \hat{\nabla}_a \gamma_\alpha + \ldots \right] . \tag{2}$$

The commutator of the local supersymmetry variations was found to take the form

$$[\delta_Q(\epsilon_1) \, , \, \delta_Q(\epsilon_2)] = 2\, \delta_P(\xi^a) + 2\, \delta_{LL}(\lambda) + \delta_\Delta(\xi^a) \, ,$$

$$\xi^a \equiv i(\epsilon_1 \gamma^a \epsilon_2) \; , \quad \lambda \equiv \hat{A} \left( \epsilon_1 \gamma^3 \epsilon_2 \right) \; ,$$

---

[1] Invited talk given at the NATO Workshop on Superstrings, University of Colorado at Boulder , July 27 - August 1, 1987.

[2] Research supported by National Science Foundation Grant PHY 86-19077.

$$\delta_\Delta\left(\xi^b\right)c_a = \xi^b\left(\hat\nabla_a c_b + \hat\nabla_b c_a - \eta_{ab}\hat\nabla\cdot c\right),$$
$$\delta_\Delta\left(\xi^c\right)b^{ab} = \xi^a\hat\nabla_c b^{cb} + \xi^b\hat\nabla_c b^{ca} - \eta^{ab}\xi^c\hat\nabla^d b_{cd}\,. \tag{3}$$

Here $\delta_P$ and $\delta_{LL}$ are variations with respect to translations and local Lorentz transformations. As can be seen the commutator algebra does **not** close unless we include the term involving $\delta_\Delta$. This is typical behavior in a supersymmetric system when a set of auxiliary fields is absent. Ordinarily in on-shell supergravity systems such nonclosure is proportional to a fermionic equation of motion. Here, however, since the diffeomorphic ghost and anti-ghost obey first order equations of motion, the nonclosure is manifested on them. In fact $\delta_\Delta\left(\xi^b\right)c_a$ and $\delta_\Delta\left(\xi^c\right)b^{ab}$ are proportional to the equations of motion of the antighost and ghost, respectively. It is now simple to show that $\left(Q_{BRST}\right)^2 \neq 0$ unless $\delta_\Delta\left(\xi^b\right)c_a = \delta_\Delta\left(\xi^c\right)b^{ab} = 0$.

This raises the question of how to realize $\left(Q_{BRST}\right)^2 = 0$ without reference to the equations of motion? By way of comparison, this situation is not present for bosonic strings. So we must confront this problem by working off-shell in the supersymmetric case. Since this only arises off-shell, it seems reasonable to expect to solve the problem by including auxiliary fields. We need to find these.

For this purpose it is useful to recall the purely bosonic action where $S^{(1)}_{FP} = -i\int d^2\sigma\, b^{ab}\partial_a c_b$. Even through there is not an anti-BRST symmetry of the gauged fixed action, it is suggestive to introduce an anti-BRST generator and write this in the form $S^{(1)}_{FP} = i\int d^2\sigma\left(\bar Q_{BRST}\tilde h^{ab}\right)\cdot\left(Q_{BRST}\tilde h_{ab}\right)$, $\tilde h_{ab} \equiv h_{ab} - \frac12\eta_{ab}h_c{}^c$ and $\bar Q_{BRST}\tilde h_{ab} = ib_{ab}$, $Q_{BRST}\tilde h_{ab} = i\left(\partial_a c_b + \partial_b c_a - \eta_{ab}\partial\cdot c\right)$. So formally at least, the Fadeev-Popov action takes the usual form. This representation shows that **formally** the anti-ghost may be regarded as arising from the application of $\bar Q_{BRST}$ to the conformal graviton $\tilde h_{ab}$. Thus to solve the problem of missing auxiliary fields we only need to know the supersymmetric analog of $\tilde h_{ab}$. However, an analysis [3] of the solution to the off-shell $D = 2$, $N = 1$ constraints [4] has already provided the answer to this question. The superconformal $D = 2$, $N = 1$ prepotential is a pure "spin-$\frac32$" superfield $H_\alpha{}^b$ with gauge variation $\delta H_\alpha{}^b = (\gamma_c\gamma^b)_\alpha{}^\beta D_\beta K^c$. Application of the anti-BRST operator to this superfield implies that the superfield version of the anti-ghost-ghost superfield action must be of the form

$$S_{NL,WZ} = \frac12\int d^2\sigma\, d^2\zeta\, \hat E^{-1}\, B^{\alpha a}\,\hat\nabla_\alpha C_a\,, \tag{4}$$

where we have used the notational conventions of ref. [5]. It has been shown that (4) correctly reproduces (1) and possesses an appropriate set of auxiliary fields. To see the appearance of these auxiliary fields, it is most convenient set the background supergravity superfields to zero. In this limit, (4) takes the form

$$S_{NL,WZ} = \frac12\int d^2\sigma\, d^2\zeta\, B^{\alpha a}\, D_\alpha C_a\,,$$
$$= -i\int d^2\sigma\left[b^{ab}\partial_a c_b + \beta^{\alpha a}\partial_a\gamma_\alpha + \rho^{\alpha a}\chi_{\alpha a} + v^a y_a\right], \tag{5}$$

where the extra auxiliary fields are manifest by comparison with (1). So equation (4) is the off-shell extension of the often quoted result. This is not the end of the story of super ghosts. At this stage we have achieved the goal of having $(Q_{BRST})^2 = 0$ without the use of the equations of motion.

Now let us add the superfield NSR string action [4]

$$S_{NSR} = -\frac{1}{4} \int d^2\sigma \, d^2\zeta \, \eta_{\underline{ab}} \, (D^\alpha X^{\underline{a}}) \, (D_\alpha X^{\underline{b}}) \ , \tag{6}$$

together with (5). (Note that the absence of supergravity variables in these two actions may be regarded as choosing an orthonormal gauge for these variables.) The BRST invariance of the total action $S_{NSR} + S_{NL,WZ}$ is a necessary requirement for the unitarity of the quantized theory. Alternately, this invariance would signal that we have properly represented the Jacobian factor arising by going to the orthonormal gauge. After defining $Q_{BRST} \, X^{\underline{a}}$ by following the usual procedure, we find that $Q_{BRST} \, (S_{NSR} + S_{NL,WZ}) \neq 0$! We might be tempted to conclude that somehow the superfield formalism is the cause of this problem. This is, of course, nonsense. In fact, the source of the problem is that, off-shell, the purely bosonic expression $-i \int d^2\sigma \, b^{ab} \partial_a c_b$ is **not** the proper representation of the Jacobian arising by going to orthonormal gauge. This is most easily seen by considering a toroidally compactified bosonic string theory. For our later convenience, we will use light-cone coordinates from here on. Thus we find

$$S_{FP}^{(1)} \equiv -i \int d^2\sigma \, b^{ab} \partial_a c_b = -i \int d^2\sigma \left( b_{--}{}^{++} \Delta_{++}{}^{--} + b_{++}{}^{--} \Delta_{--}{}^{++} \right) \ ,$$

$$\Delta_{++}{}^{--} \equiv \partial_{++} c^{--} \ , \quad \Delta_{--}{}^{++} \equiv \partial_{--} c^{++} \ . \tag{7}$$

Similarly the bosonic string coordinate action becomes

$$S_{string} = -\frac{1}{2} \int d^2\sigma \, \eta_{ab} \, (\partial_{++} X^a) \left( \partial_{--} X^b \right) \ . \tag{8}$$

In order to induce a toroidal compactification we introduce $2N$ Majorana-Weyl spinors $\vec{\lambda}_-$ and $2N$ Majorana spinors $\vec{\chi}_+$, whose action is just

$$S_M = -i \frac{1}{2} \int d^2\sigma \left( \vec{\lambda}_- \cdot \partial_{++} \vec{\lambda}_- - \vec{\chi}_+ \cdot \partial_{--} \vec{\chi}_+ \right) \ . \tag{9}$$

Now it is possible to show that BRST variations exist so that $Q_{BRST} \, (S_{string} + S_{FP}^{(1)}) = 0$ without using equations of motion. However it is **not** possible to arrange $Q_{BRST} \left( S_{string} + S_M + S_{FP}^{(1)} \right) = 0$ independent of equations of motion. This is the nonsupersymmetric analog of what was found earlier.

The source of the problem is a subtle point concerning the complete set of ghosts and anti-ghosts which properly represent the Jacobian factor. As we noted earlier, the symmetric traceless anti-ghost, $b_{ab}$, may be thought to arise by applying an anti-BRST operator to the conformal graviton $\tilde{h}_{ab}$. But the full graviton has two additional parts, the Weyl and Lorentz compensators, $\psi$ and $\ell$ respectively. Application of the anti-BRST operator to these give rise to two additional anti-ghosts $\hat{b}$ and $b$. Thus, the proper representation of the Jacobian arising in going to orthonormal gauge is given by

$$S_{FP} = -i\int d^2\sigma\ \left[b_{--}{}^{++}\,\Delta_{++}{}^{--} + b_{++}{}^{--}\Delta_{--}{}^{++} + b\Delta + \hat{b}\hat{\Delta}\right] \ ,$$

$$\Delta \equiv \partial_{++}\,c^{++} - \partial_{--}\,c^{--} - 2c\ ,$$

$$\hat{\Delta} \equiv \partial_{++}\,c^{++} + \partial_{--}\,c^{--} - 2\hat{c} \quad , \tag{10}$$

where $\hat{c}$ and $c$ are the BRST ghosts associated with local Weyl and Lorentz gauge parameters. Since the gauge variations of $\psi$ and $\ell$ are algebraic, it is usually said [6] that their elimination is trivial. In the case of the uncompactified bosonic string this is true. But for the compactified bosonic string this is false! BRST invariance of the action along with $(Q_{BRST})^2 = 0$ (without use of equations of motion) requires the extra anti-ghosts. It is simple to find BRST variations such that $Q_{BRST}\ (S_{string} + S_M + S_{FP}) = 0$ without using equations of motion. Explicitly we find

$$Q_{BRST}X^{\underline{a}} = -\,ic^a\partial_a X^{\underline{a}},$$

$$Q_{BRST}c^a = -\,ic^b\partial_b c^a\ ,\ Q_{BRST}c = -ic^b\partial_b c\ ,\ Q_{BRST}\hat{c} = -ic^b\partial_b\hat{c}\ ,$$

$$Q_{BRST}\vec{\chi}_+ = -\,ic^a\partial_a\vec{\chi}_+\ -\ i\tfrac{1}{2}c\vec{\chi}_+\ -\ i\tfrac{1}{2}\hat{c}\vec{\chi}_+\ ,$$

$$Q_{BRST}\vec{\lambda}_- = -\,ic^a\partial_a\vec{\lambda}_-\ +\ i\tfrac{1}{2}c\vec{\lambda}_-\ -\ i\tfrac{1}{2}\hat{c}\vec{\lambda}_-\ ,$$

$$Q_{BRST}b_{++}{}^{--} = -\,i\partial_a(c^a b_{++}{}^{--}) + i(\ \partial_{++}c^{++} - \partial_{--}c^{--}\ )b_{++}{}^{--}$$
$$+\,(\partial_{--}X^{\underline{a}})(\partial_{--}X_{\underline{a}}) + i\tfrac{1}{2}(\vec{\chi}_+\cdot\partial_{++}\vec{\chi}_+) - i2k\ b\Delta_{++}{}^{--}\quad ,$$

$$Q_{BRST}b_{--}{}^{++} = -\,i\partial_a(c^a b_{--}{}^{++}) - i(\ \partial_{++}c^{++} - \partial_{--}c^{--}\ )b_{--}{}^{++}$$
$$+\,(\partial_{++}X^{\underline{a}})(\partial_{++}X_{\underline{a}}) - i\tfrac{1}{2}(\vec{\lambda}_-\cdot\partial_{--}\vec{\lambda}_-) + i2(1-k)b\Delta_{--}{}^{++}\quad ,$$

$$Q_{BRST}b = -\,i\partial_a(c^a b) + \tfrac{1}{2}(\mathcal{L}_{(+M)} - \mathcal{L}_{(-M)})\quad ,$$
$$Q_{BRST}\hat{b} = -\,i\partial_a(c^a\hat{b}) + \tfrac{1}{2}(\mathcal{L}_{(+M)} + \mathcal{L}_{(-M)})\quad , \tag{11}$$

where $\mathcal{L}_{(+M)} = -i\frac{1}{2}\vec{\chi}_+\cdot\partial_{--}\vec{\chi}_+$, $\mathcal{L}_{(-M)} = i\frac{1}{2}\vec{\lambda}_-\cdot\partial_{++}\vec{\lambda}_-$ and $k$ is any real number. Given the explicit form of the BRST transformation laws of the fields, it is possible to write a functional differential representation of the BRST operator and calculate its fourier coefficients,

$$Q_{BRST} = (Q_{BRST}X^{\underline{a}})\frac{\delta}{\delta X^{\underline{a}}}\ +\ (Q_{BRST}\vec{\chi}_+)\cdot\frac{\delta}{\delta\vec{\chi}_+}\ +\ (Q_{BRST}\vec{\lambda}_-)\cdot\frac{\delta}{\delta\vec{\lambda}_-}\ +\ \dots$$
$$= \sum_n U_{(n)}e^{in\sigma} \tag{12}$$

where the dots stand for similar terms for all of the other fields in (11).

Before returning to the supersymmetric case, we note that background gravitation can be coupled to the action (10) above by making the substitutions $\int d^2\sigma \to \int d^2\sigma\,\hat{e}^{-1}$ ,

$\partial_{++} \to \hat{\nabla}_{++}$, $\partial_{--} \to \hat{\nabla}_{--}$. The resulting action is invariant under the background scale transformations of the form,

$$\delta \hat{e}_a{}^m = S\hat{e}_a{}^m \quad , \quad \delta c^{++} = -Sc^{++} \quad , \quad \delta c^{--} = -Sc^{--} \quad ,$$
$$\delta b_{ab} = 2Sb_{ab} \quad , \quad \delta b = 2Sb \quad , \quad \delta \hat{b} = 2S\hat{b} \quad ,$$
$$\delta c = \tfrac{1}{2}(\, c^{++}\hat{\nabla}_{++}S \; - \; c^{--}\hat{\nabla}_{--}S \,) \; ,$$
$$\delta \hat{c} = \tfrac{1}{2}(\, c^{++}\hat{\nabla}_{++}S \; + \; c^{--}\hat{\nabla}_{--}S \,) \; , \tag{13}$$

where $S$ is the parameter of the scale transformation. By using Noether's theorem we are able to derive the complete ghost number current which takes, in covariant notation, the form $j_a^{(GN)} = -i(b_{ab} + \epsilon_{ab}b + \eta_{ab}\hat{b})c^b$. The "extra" terms in (10) in comparison with (7) may be formally written as $i \int d^2\sigma \left(\bar{Q}_{BRST}\psi\right)(Q_{BRST}\psi) +$ $\left(\bar{Q}_{BRST}\ell\right)(Q_{BRST}\ell)$.

So to solve our supersymmetric problem, we must generalize (10). However, this is now a trivial extension of our nonsupersymmetric discussion above. We note that in (10) we have simply "restored" the compensating parts of the graviton and applied the formal BRST and anti-BRST operators to them. Thus for the supersymmetric case the same logic dictates that we should restore the superfield supergravity compensators (given in ref.[3,5]) analogously. This immediately leads to

$$\begin{aligned} S_{GH} = -\tfrac{1}{4}\int d^2\sigma d^2\zeta \hat{E}^{-1}\{\; & 2B^{\alpha a}\hat{\nabla}_\alpha C_a + iB_\alpha[(\gamma^a)^{\alpha\beta}\hat{\nabla}_\beta C_a - 4C^\alpha] \\ & + B[(\gamma^3)^{\alpha\beta}\hat{\nabla}_\beta C_\alpha - iC] \; + \; \hat{B}[\hat{\nabla}_\alpha C^\alpha - i\hat{C}]\;\} \; , \end{aligned} \tag{14}$$

as the supersymmetric generalization in the presence of background supergravity superfields. This action possesses the supersymmetric generalization of (13). It is background scale invariant under

$$\begin{aligned} &\delta_S \ddot{E} = S\ddot{E} \quad , \qquad \delta_S \ddot{\nabla}_\alpha = \tfrac{1}{2}S\ddot{\nabla}_\alpha + (\gamma^3)_\alpha{}^\beta(\hat{\nabla}_\beta S)\mathcal{M} \quad , \\ &\delta_S B^{\alpha a} = \tfrac{3}{2}SB^{\alpha a} \quad , \qquad \delta_S B^\alpha = \tfrac{3}{2}SB^\alpha \quad , \qquad \delta_S B = SB \quad , \qquad \delta_S \hat{B} = S\hat{B} \quad , \\ &\delta_S C_a = -SC_a \quad , \qquad \delta_S C_\alpha = -\tfrac{1}{2}SC_\alpha - \tfrac{1}{2}(\gamma_a)_\alpha{}^\beta(\hat{\nabla}_\beta S)C^a \quad , \\ &\delta_S C = i(\gamma^3)^{\alpha\beta}(\hat{\nabla}_\alpha S)C_\beta + \epsilon^{ab}(\hat{\nabla}_a S)C_b - i\tfrac{1}{2}(\gamma^3\gamma^a)^{\alpha\beta}(\hat{\nabla}_\alpha S)(\hat{\nabla}_\beta C_a) \quad , \\ &\delta_S \hat{C} = -i(\hat{\nabla}_\alpha S)C^\alpha - (\hat{\nabla}_a S)C^a - i\tfrac{1}{2}(\gamma^a)^{\alpha\beta}(\hat{\nabla}_\alpha S)(\hat{\nabla}_\beta C_a) \quad . \end{aligned} \tag{15}$$

The anti-ghosts $B^{\alpha a}$, $B^\alpha$, $B$, and $\hat{B}$ arise by applying the anti-BRST operator to the superfield supergravity prepotentials $H^{\alpha a}$, $H^\alpha$, $L$, and $\Psi$. Similarly the coefficients multiplying each anti-ghost superfield represents the effect of applying the BRST operator to the corresponding prepotential. (These factors are derivable by simply considering the linearized theory and following the usual BRST prescription.) The first term in (14) contains the first two terms of (10). The two terms on the second line in (14) produce the third and fourth terms of (10) along with supersymmetric completion terms. The second term in (14) has no nonsupersymmetric analog. This is due to the fact that superfield supergravity possesses a local supersymmetry compensator (called $H^\alpha$ in ref. [3,5]) along with local Lorentz and Weyl compensators. The presence of this term is crucial for off-shell BRST invariance of the NSR superfield action. This is most apparent if we calculate the effect of the BRST operator on $\mathcal{L}_{NSR}$. We find

$$\begin{aligned} Q_{BRST}\mathcal{L}_{NSR} &= -i[\partial_a(C^a\mathcal{L}_{NSR})] - [D_\alpha((C^\alpha - 2\Delta^\alpha)\mathcal{L}_{NSR})] \\ &\quad + \tfrac{1}{4}(D^\alpha X^{\underline{a}})(\partial^a X_{\underline{a}})(\gamma^b\gamma_a)_\alpha{}^\beta \Delta_{\beta b} \\ &\quad - [\, (D_\alpha X^{\underline{a}})(D^2 X_{\underline{a}}) \,]\Delta^\alpha \ , \end{aligned} \tag{16}$$

where $\Delta_{\alpha a} \equiv D_\alpha C_a$ , $\Delta^\alpha \equiv C^\alpha - \frac{1}{4}(\gamma_a)^{\alpha\beta}D_\beta C^a$ . We might be tempted to set $\Delta^\alpha$ to zero as a definition but this would imply that $(Q_{BRST})^2 = 0$ only if some dynamical equation of motion is satisfied. (This same phenomenon can be seen in the nonsupersymmetric compactified string if we attempted to set either $\Delta$ or $\hat{\Delta}$ to zero.) If we now add (6) and (14) and demand that the sum is BRST invariant, we are led to the following superfield BRST transformation laws.

$$\begin{aligned} Q_{BRST}X^{\underline{a}} &= -\mathcal{C}\cdot DX^{\underline{a}} \equiv -(iC^a\partial_a + C^\alpha D_\alpha)X^{\underline{a}} \ , \\ Q_{BRST}C^b &= -\mathcal{C}\cdot DC^b - (\gamma^b)_{\alpha\beta}C^\alpha C^\beta \ , \qquad Q_{BRST}C^\beta = -\mathcal{C}\cdot DC^\beta \ , \\ Q_{BRST}C &= -\mathcal{C}\cdot DC \ , \quad Q_{BRST}\hat{C} = -\mathcal{C}\cdot D\hat{C} \ , \\ Q_{BRST}B^{\alpha a} &= -\mathcal{C}\cdot DB^{\alpha a} - B^{\alpha a}[i\partial_b C^b + D_\beta C^\beta] - i(\gamma^a\gamma^c)_\beta{}^\alpha B^{\beta b}\partial_c C_b \\ &\quad - i\tfrac{1}{8}(\gamma^a\gamma^b)_\gamma{}^\alpha B_\delta[C^{\delta\gamma}D^2C_b + i2(\gamma^c)^{\delta\gamma}\partial_b C_c] - i\tfrac{1}{4}(\gamma^a\gamma^c)_\beta{}^\alpha \hat{B}\partial_c C^\beta \\ &\quad - i\tfrac{3}{64}(\gamma^3\gamma^c\gamma^a\gamma^b)_\beta{}^\alpha B\partial_b D_\beta C_c + \tfrac{1}{64}(\gamma^3\gamma^a\gamma^b)^{\beta\alpha}D_\beta B D^2 C_b \ , \\ Q_{BRST}B_\alpha &= -\mathcal{C}\cdot DB_\alpha - B_\alpha[i\partial_b C^b + D_\beta C^\beta] \\ &\quad + B_\delta[\tfrac{1}{2}(\gamma^a)^{\delta\gamma}(\gamma_a)_{\alpha\epsilon}D_\gamma\Delta^\epsilon + \tfrac{1}{4}(\gamma^a)_\alpha{}^\delta D^2 C_a - i\tfrac{1}{4}(\gamma^b\gamma^a)_\alpha{}^\delta\partial_b C_a] \\ &\quad + D_\gamma B_\delta[\tfrac{1}{8}(\gamma^a)^{\delta\gamma}(\gamma_a\gamma^b)_\alpha{}^\epsilon D_\epsilon C_b - \tfrac{1}{4}(\gamma^a)^{\delta\gamma}D_\alpha C_a] \\ &\quad + B^{\beta a}[(\gamma^b)_{\alpha\beta}\partial_b C_a - i(\gamma_a)_{\gamma\alpha}D_\beta C^\gamma] \\ &\quad - (\gamma^3)_\alpha{}^\beta B[\tfrac{1}{4}(\gamma^a)_{\beta\delta}\partial_a\Delta^\delta - i\tfrac{1}{4}D^2\Delta_\beta + \tfrac{1}{8}D_\beta\partial_a C^a] \\ &\quad + i\tfrac{1}{4}(\gamma^3)_\alpha{}^\beta D_\delta B[D_\beta\Delta^\delta + \tfrac{1}{4}(\gamma_b)^{\delta\epsilon}D_\beta D_\epsilon C^b] \\ &\quad - \tfrac{1}{4}(\gamma^3)_\alpha{}^\beta\partial_b B[D_\beta C^b + 2(\gamma^b)_{\beta\delta}C^\delta] - 2(\gamma^a)_{\alpha\beta}\hat{B}(\partial_a C^\beta) \ , \\ Q_{BRST}B &= -\mathcal{C}\cdot DB + B[i\partial_a C^a + D_\alpha C^\alpha] \ , \\ Q_{BRST}\hat{B} &= -\mathcal{C}\cdot D\hat{B} + \hat{B}[i\partial_a C^a + D_\alpha C^\alpha] \ . \end{aligned} \tag{17}$$

By simply reinterpreting equation (12) as a superfield equation and using the results above, we can write a superfunctional differential representatation of the BRST operator for the NSR theory. Application of the Noether method to (14) leads to the following expression for the ghost number current.

$$J_\alpha{}^{(GN)} = 2B_\alpha{}^a C_a - i(\gamma^a)_\alpha{}^\beta B_\beta C_a + \hat{B}C_\alpha + (\gamma^3)_\alpha{}^\beta BC_\beta \ . \tag{18}$$

The first two terms in the off-shell ghost number current allow us to couple the spacetime dilaton field $\Phi(X)$ in a manner that preserves the background scale invariance

$$S_{Dilaton} = -\tfrac{1}{4}\int d^2\sigma d^2\zeta \hat{E}^{-1}[\ \hat{\nabla}^\alpha\Phi(X)\ ][\ k_1 B_\alpha{}^a C_a - ik_2(\gamma^a)_\alpha{}^\beta B_\beta C_a\ ]\ , \tag{19}$$

where $k_1$ and $k_2$ are constants.

The proper off-shell treatment of the heterotic string [7] follows much the same pattern. It can be obtained by either truncation of the (1,1) superspace construction above or by directly using unidexterous superspace [8]. In the following we will use the latter method. The general solution of (1,0) supergravity involves five prepotentials $H_+{}^{--}$, $H_{--}{}^{++}$, $\Psi$, $H^+$, and $L$. Application of an anti-BRST operator to these yield a corresponding set of anti-ghosts $B_{--}{}^{++}$, $B_+{}^{--}$, $\hat{B}_-$, $B$, and $B_-$. (Note the Lorentz structures of the corresponding anti-ghosts are different due to the fact the (1,0) measure transforms as a spinor.) The (1,0) version of equation (10) becomes

$$\begin{aligned} S_{GH}{}' \ =\ \int d^2\sigma d\zeta^- \hat{E}^{-1}[\ & -iB_+{}^{--}\hat{\nabla}_{--}C^{++}\ +\ B_{--}{}^{++}\hat{\nabla}_+C^{--} \\ & + B(\hat{\nabla}_+C^{++} + 2C^+) \\ & + iB_-(\hat{\nabla}_{++}C^{++} - \hat{\nabla}_{--}C^{--} - 2C)\ ] \\ & + i\hat{B}_-(\hat{\nabla}_{++}C^{++} + \hat{\nabla}_{--}C^{--} - 2\hat{C})\ ]\ , \end{aligned} \tag{20}$$

and possesses the expected background superscale invariance with respect to

$$\begin{aligned} &\delta_S\hat{\nabla}_+ = \tfrac{1}{2}S\hat{\nabla}_+ + (\hat{\nabla}_+S)\mathcal{M}\ , \qquad \delta_S\hat{\nabla}_{--} = S\hat{\nabla}_{--} - (\hat{\nabla}_{--}S)\mathcal{M}\ , \\ &\delta_S\hat{\nabla}_{++} = S\hat{\nabla}_{++} - i(\hat{\nabla}_+S)\hat{\nabla}_+ + (\hat{\nabla}_{++}S)\mathcal{M}\ , \qquad \delta_S\hat{E} = \tfrac{3}{2}S\hat{E}\ , \\ &\delta_S B_+{}^{--} = \tfrac{3}{2}SB_+{}^{--}\ , \quad \delta_S B_{--}{}^{++} = 2SB_{--}{}^{++}\ , \quad \delta_S C_a = -SC_a\ , \\ &\delta_S B = 2SB\ , \qquad \delta_S B_- = \tfrac{3}{2}SB_-\ , \qquad \delta_S\hat{B}_- = \tfrac{3}{2}S\hat{B}_-\ , \\ &\delta_S C^+ - -\tfrac{1}{2}SC^+ + (\hat{\nabla}_+S)C^{++}\ , \\ &\delta_S C = -\tfrac{1}{2}i(\hat{\nabla}_+S)(\hat{\nabla}_+C^{++}) - (\hat{\nabla}_{++}S)C^{++} + (\hat{\nabla}_{--}S)C^{--}\ , \\ &\delta_S\hat{C} = -\tfrac{1}{2}i(\hat{\nabla}_+S)(\hat{\nabla}_+C^{++}) - (\hat{\nabla}_{++}S)C^{++} - (\hat{\nabla}_{--}S)C^{--}\ . \end{aligned} \tag{21}$$

The superfield action for the heterotic string

$$S_H = i\tfrac{1}{2}\int d^2\sigma d\zeta^-[\ D_+X^a\partial_{--}X_a\ +\ i\eta_-{}^{\hat{I}}D_+\eta_-{}^{\hat{I}}\ ]\ , \tag{22}$$

implies that the BRST variation of this lagrangian becomes

$$\begin{aligned} Q_{BRST}\mathcal{L}_H = &- i[\partial_a(C^a\mathcal{L}_H)] - [D_+((C^+ - \Delta^+)\mathcal{L}_H)] \\ &- \tfrac{1}{2}(\partial_{--}X^{\underline{a}})(\partial_{--}X_{\underline{a}})\Delta_+{}^{--} - \tfrac{1}{2}(D_+X^{\underline{a}})(\partial_{++}X_{\underline{a}})\Delta_{--}{}^{++} \\ &+ i\tfrac{1}{2}[\ (D_+X^{\underline{a}})\partial_{--}(D_+X_{\underline{a}})\ ]\Delta^+ \\ &+ i\tfrac{1}{2}(\eta_-{}^{\hat{I}}\partial_{--}\eta_-{}^{\hat{I}})\Delta_+{}^{--} + \tfrac{1}{2}(D_+\eta_-{}^{\hat{I}})(D_+\eta_-{}^{\hat{I}})\Delta^+ \\ &- i\tfrac{1}{2}\mathcal{L}_{-MSM}(\Delta - \hat{\Delta})\ , \end{aligned} \tag{23}$$

where $\mathcal{L}_{-MSM} \equiv -\frac{1}{2}\eta_-{}^{\hat{I}}D_+\eta_-{}^{\hat{I}}$, $\Delta_{--}{}^{++} \equiv \partial_{--}C^{++}$, $\Delta_+{}^{--} \equiv D_+C^{--}$, and $\Delta^+ \equiv D_+C^{++} + 2C^+$. Upon taking the flat background limit of (20) and adding it to (22), we find that the sum is invariant under the following BRST variations.

$$Q_{BRST}X^{\underline{a}} = -\mathcal{C}\cdot DX^{\underline{a}} \quad , \quad Q_{BRST}\eta_-{}^{\hat{I}} = -\mathcal{C}\cdot D\eta_-{}^{\hat{I}} + i\tfrac{1}{2}C\eta_-{}^{\hat{I}} - i\tfrac{1}{2}\hat{C}\eta_-{}^{\hat{I}} \ ,$$

$$\begin{aligned}
Q_{BRST}B_+{}^{--} &= -\mathcal{C}\cdot DB_+{}^{--} - B_+{}^{--}[i2\partial_{++}C^{++} + D_+C^+] \\
&\quad - i2B_-(\partial_{++}C^{--}) + i\tfrac{1}{2}(D_+X^{\underline{a}})(\partial_{++}X_{\underline{a}}) \ , \\
Q_{BRST}B_{--}{}^{++} &= -\mathcal{C}\cdot DB_{--}{}^{++} + B_{--}{}^{++}[i\partial_aC^a + 2D_+C^+] \\
&\quad + iB(\partial_{--}C^{++}) - i(B_- - \hat{B}_-)(\partial_{--}C^+) \\
&\quad + \tfrac{1}{2}[\,(\partial_{--}X^{\underline{a}})(\partial_{--}X_{\underline{a}}) - i(\eta_-{}^{\hat{I}}\partial_{--}\eta_-{}^{\hat{I}})\,] \\
Q_{BRST}B &= -\mathcal{C}\cdot DB + B[i2\partial_{++}C^{++} + i\partial_{--}C^{--} + 2D_+C^+] \\
&\quad - iB_+{}^{--}(\partial_{--}C^+) + iB_{--}{}^{++}(\partial_{++}C^{--}) \\
&\quad + i(B_- + \hat{B}_-)(\partial_{++}C^+) \\
&\quad - i\tfrac{1}{2}[\,(D_+X^{\underline{a}})\partial_{--}(D_+X_{\underline{a}}) - i(D_+\eta_-{}^{\hat{I}})(D_+\eta_-{}^{\hat{I}})\,] \\
Q_{BRST}B_- &= -\mathcal{C}\cdot DB_- - B_-[i\partial_aC^a + D_+C^+] + \tfrac{1}{2}\mathcal{L}_{-MSM} \ , \\
Q_{BRST}\hat{B}_- &= -\mathcal{C}\cdot D\hat{B}_- - \hat{B}_-[i\partial_aC^a + D_+C^+] - \tfrac{1}{2}\mathcal{L}_{-MSM} \ ,
\end{aligned}$$

$$Q_{BRST}C^+ = -[iC^a\partial_a + C^+D_+]C^+ \equiv -\mathcal{C}\cdot DC^+ \ ,$$

$$Q_{BRST}C^{++} = -\mathcal{C}\cdot DC^{++} - C^+C^+ \quad , \quad Q_{BRST}C^{--} = -\mathcal{C}\cdot DC^{--} \ ,$$

$$Q_{BRST}C = -\mathcal{C}\cdot DC \quad , \quad Q_{BRST}\hat{C} = -\mathcal{C}\cdot D\hat{C} \ . \qquad (24)$$

The ghost action in (20) also permits the derivation of the (1,0) ghost number current which has components

$$\begin{aligned}
J_{-++}{}^{(GN)} &= -\,i[\,B_+{}^{--}C^{++} + (B_- - \hat{B}_-)C^{--}\,] \ , \\
J_{---}{}^{(GN)} &= i(\hat{B}_- - B_-)C^{++} \ , \qquad (25)\\
J_{--}{}^{(GN)} &= -\,[\,B_{--}{}^{++}C^{--} + BC^{++}\,] \ .
\end{aligned}$$

As in the case of the NSR string, these ghost number currents permit the dilaton to be coupled in a background scale invariant manner via the action

$$S_{Dilaton} = \tfrac{1}{2}\int d^2\sigma d\zeta^- \hat{E}^{-1}[\,k_1(\hat{\nabla}_+\Phi)J_{--}{}^{(GN)} + k_2(\hat{\nabla}_{--}\Phi)J_{-++}{}^{(GN)}\,] \ . \qquad (26)$$

Elsewhere [9], we have verified that the superspace formulation of the bosonized description of the heterotic string is treated in the same fashion. We have also checked that the world sheet supersymmetric description [10] of NSR and heterotic string toroidal compactification is consistent with the classical superfield BRST symmetries described above.

**Acknowledgement**

I wish to acknowledge the collaboration of R. Brooks in carrying out this research.

## REFERENCES


[1] M. B. Green, J. H. Schwarz, and E. Witten, *Superstring Theory* , Camb. Univ. Press, 1987, Camb. U. K.

[2] D. Z. Freedman and N. P. Warner, *Phys. Rev.* **D34** (1986) 3084.

[3] S. J. Gates, Jr. and H. Nishino, *Class. Quantum Grav.* **3** (1986) 391.

[4] P. Di Vecchia and P. Howe, *Phys. Lett.* **65B** (1976) 471; P. Howe, *Phys. Lett.* **70B** (1977) 453; ibid.*J. Phys. A: Math. Gen.* **12** (1979) 393; M. Brown and S. J. Gates, Jr., *Ann. of Phys.* **122** (1979) 443.

[5] R. Brooks and S. J. Gates, Jr., *Nucl. Phys.* **B287** (1987) 699.

[6] P. van Nieuwenhuizen, , SUNY Stony Brook preprint # ITP-SB-87-9 (1987).

[7] D.J. Gross, J.A. Harvey, E. Martinec and R. Rohm, *Phys. Rev. Lett.* **54** (1985) 502.

[8] R. Brooks, F. Muhammad, and S. J. Gates, Jr., *Nucl. Phys.* **B268** (1986) 599.

[9] R. Brooks and S. J. Gates, Jr., M.I.T. Preprint # CTP-1485, to appear in Nucl. Phys.

[10] S. J. Gates, Jr., R. Brooks, and F. Muhammad *Phys.Lett.* **194B** (1987) 35.

# Chapter II
# String Field Theory

# A GEOMETRY INDEPENDENT FIELD THEORY OF STRING

Keiji Kikkawa

Department of Physics
Osaka University
Toyonaka 560, Japan

The equation of motion that follows from the purely cubic action model is shown to provide a set of equations for background fields which define a geometry of the space-time. Using a sigma model as a supplementary tool, a systematic method for solving the string field theory in curved space is presented.

One of the most important issues in string theories is to understand how the space-time with the critical dimension 26 or 10 can be compactified. In a conformal field theory approach[1,2] one begins with writing down a $\sigma$-model type action with background fields and then look for conditions over the background to make the theory conformal invariant when it is quantized. The conditions were first derived by requiring the $\beta$-functions for background field couplings being zero[1], and later by requiring the nilpotency[2] of BRST charge $Q_B$.

These conditions provide us with a set of equations of motion for background fields. The $\sigma$-model action, into which one of the solutions to the equations is substituted, defines then a string theory in the background fields. Although this approach is supposed to describe a correct string behavior in the curved space provided that the backward reaction from quantum strings is negligible, the formalism is not self-contained. Because, depending on a choice of the background solutions, one has to begin with a different $\sigma$-model action associated with the choice. The theory which we want is a theory that covers all solutions to the background field equation so that, for instance, the quantum transition from one geometry to others is able to be treated within the framework.

The other possible approach is to begin with the field theory of string.[3,4,5] If one has a field theoretical action of string, one may find a spontaneously compactified space-time as a solution to the equation of motion. This is a gravity counterpart of the spontaneous symmetry breakdown in usual local field theories. However, field theoretic formalisms so far proposed are all geometry dependent from the outset. In Kyoto-CERN[4] theory, for example, the action is given by

$$S = \Phi \cdot Q \ \Phi + \frac{2}{3} g \Phi \cdot ( \Phi * \Phi ) \qquad (1)$$

where $\Phi$ is the string field for a closed string, a functional of string coordinate $X^\mu(\sigma)$ for $0 \leq \sigma \leq \pi$ and ghost fields $C(\sigma)$ and $\bar{C}(\sigma)$. The BRST operator[6] Q defines the kinetic energy term of string. It is the operator Q that explicitly depends on the background geometry. In both Witten's and Kyoto-CERN formalisms, Q is given in a flat Minkowski space $M_{26}$ with the critical dimension 26. It may be possible to reach some compactified and curved space-time configurations by applying a non-perturbative method to (1). Although interesting developments[7] are observed recently, whether these approaches cover all possibilities of compactifications is not known. What we want in this direction is to find a geometry independent action of string field theory.

Recently a number of people[8] pointed out that the cubic interaction term in (1) is geometry independent and hence that the purely cubic action without the kinetic term may define a meaningful field theory. This idea has been subsequently realized by Hata, Itoh, Kunitomo, Kugo and Ogawa (HIKKO)[9] and also by Horowitz, Lykken, Rohm and Strominger (HLRS)[10]. According to them the ordinary action with a kinetic energy term in a geometry, such as (1), follows by making expansion around a certain classical solution to the equation of motion of the cubic action. In their papers, however, whether the equation of motions allows any nontrivial curved space is not well discussed.

In this talk[11] we address ourselves to the question of what sorts of geometries would be allowed as solutions to the equation derived from the cubic action. We will find in the following a set of equations of motion for background fields, which agrees to those previously obtained in the context of conformal invariant $\sigma$-model[1,2]. In our approach, however, all of these conformal invariant solutions are within a space which is covered by a single theory and hence the cubic action theory is shown to be qualified as a nontrivial, self-contained and geometry independent field theory.

All arguments in this talk will be performed in the frame work of Kyoto-CERN[4,9] formalism, since it is the only avairable covariant theory of closed string. Although the formalism contains an unphysical width parameter and a problematical feature in loop amplitudes, the width parameter dependence can be factored out in tree amplitudes. Since we are interested in a neighborhood of the string vertex in this discussion, our results are not affected by the unphysical parameter being involved in the theory. According to recent developments by Neveu and West and others[12], however, the width parameter is shown to be washed out by the Parisi-Sourlas mechanism[13]. It is then possible to modify our approach so that no problem happens in loop amplitudes. We will however do not discuss these points except for a comment on loop corrections which will be given later.

Our discussion will be also developed in the framework of HLRS[10] formalism. However, as has been studied by Horowitz and Strominger[14], there are a number of delicate problems caused by that Witten's field theory[5] does not explicitly include the closed string field. These aspects we do not consider here.

Let us first review the relevant results of HIKKO[9]. The purely cubic action in the bosonic closed string theory is given by

$$S = \frac{2}{3g^2} \Psi \cdot ( \Psi * \Psi ) \tag{2}$$

which is invariant under the local gauge transformation $\delta \Psi = 2 \Psi * \Lambda$, where the $*$-product is defined by using the 3-string vertex operator $| V >$ as follows,

$$| \Psi_1 * \Psi_2 (3) > = \int < \Psi_1 (1) | < \Psi_2 (2) | | V (1,2,3) > d1\, d2 \tag{3}$$

As pointed out by Friedan[8], (2) is formally invariant under the general coordinate transformation

$$X^\mu (\sigma) \longrightarrow Y^\mu (\sigma) = Y^\mu (X (\sigma))$$

provided the string field $\Psi$ transforms as a half-density, where Y is an arbitrary function of $X^\mu$.

The variation of $\Psi$ in (2) leads us to the equation of motion

$$\Psi * \Psi = 0 \tag{4}$$

Provided $\Psi_0$ being a solution to (4), let us represent the full field by $\Psi = \Psi_0 + g \Phi$ and substitute it back into the action (2). The action then reduces to the familiar form (1), if there exists an operator Q that satisfies

$$\Psi_0 * \Phi = \frac{1}{2} Q \Phi \tag{5}$$

for an arbitrary field $\Phi$. Furthermore, to make the induced action (1) to be a string field theory Q must satisfy all properties that BRST charge does, i.e., the nilpotency

$$Q^2 = 0 \tag{6}$$

and the distribution law

$$Q ( \Phi * \Psi ) = Q \Phi * \Psi + (-)^\Phi \Phi * Q \Psi. \tag{7}$$

The action (1), which is now derived from the cubic action (2), then defines a string action for a certain background fields whose information is contained in $\Psi_0$, hence in Q.

HIKKO shows that the solution to (4) is given by

$$\Psi_0 = - \frac{1}{2} Q \Gamma \tag{8}$$

provided that a field $\Gamma$ obeys the equation

$$\Gamma * \Phi = \{ N_{FP} + 1 - \frac{\alpha}{|\alpha|} - \alpha \frac{d}{d\alpha} \} \Phi \qquad (9)$$

and

$$N_{FP} \, \Gamma = -2\Gamma \qquad (10)$$

for arbitrary $\Phi$, where $N_{FP}$ is the Faddeev-Popov ghost number operator and $\alpha$ the width parameter of $\Phi$. The explicit form of $\Gamma$ is known. It is easy to show that (8) satisfies both (4) and (5);

$$\Psi_0 * \Psi_0 = \frac{1}{2} Q(\Gamma * Q\Gamma) = \frac{1}{4} Q \{ N_{Fp} + 1 - \frac{\alpha}{|\alpha|} - \alpha \frac{d}{d\alpha} \} Q \Gamma$$

$$\propto Q^2 \Gamma = 0$$

where the distribution law (7) and the nilpotency (6) have been used, and

$$\Psi_0 * \Phi = - \frac{1}{2} \{ Q ( \Gamma * \Phi ) - \Gamma * Q\Gamma \}$$

$$= \frac{1}{2} [ N_{FP}, Q ] \Phi = \frac{1}{2} Q\Phi$$

where $[ N_{FP}, Q ] = Q$, and (7) have been used. The uniqueness of $\Psi_0$ for a given Q is also shown.

Summing up above arguments, one can conclude that, if any non-trivial operator satisfying both the nilpotency and the distribution law is found, $\Psi_0$ formed by (8) is a solution to the equation of motion (4) and defines a string field theory associated with the geometry specified by Q.

In this work, instead of solving (4), we look for the BRST operator that satisfies (6) and (7). This is equivalent to solving the equation of motion (4).

To do this we first prepare a test operator $Q [ G_{\mu\nu}, B_{\mu\nu}, \Phi ]$ as a function of unspecified metric field $G_{\mu\nu}$, antisymmetric tensor $B_{\mu\nu}$ and dilaton field $\Phi$ . Then we impose the nilpotency (6) and the distribution law (7), which is more explicitly represented as

$$\sum_{r=1}^{3} Q^{(r)} [ G, B, \Phi ] \; | V(1, 2, 3) > = 0 \qquad (11)$$

where the superscript r refers the string 1, 2 or 3. Since the vertex operator $| V >$ is given by the overlapping condition, (11) requires conditions over Q.

As for the nilpotency condition (6), Banks, Nemeschansky and Sen[2] studied its implication in the context of non-linear $\sigma$-model and obtained those equations for background fields that have been found in ref.1.

Here we address ourselves to the condition (11), which has a straightforward connection with the cubic theory. In choosing the test operator Q we take advantage of the non-linear σ-model as a supplementary tool,

$$S = \frac{1}{4\pi\alpha'} \int d^2\sigma \, [\, \sqrt{g}\, g^{ab}\, G_{\mu\nu}(X)\, {}_aX\, \partial_b X^\nu + i\varepsilon^{ab} B_{\mu\nu}(X)\, \partial_a X^\mu \partial_b X^\nu$$

$$+ \alpha' \sqrt{g}\, R^{(2)}\, \Phi(X)\, ] \qquad (12)$$

where $g_{ab}$ and $R^{(2)}$ represent the metric tensor and the curvature scalar of the two dimensional worldsheet. The test operator Q is then defined by

$$Q^{(r)}[G,B,\Phi] = \int_0^{\pi\alpha_r} d\sigma\, j(\sigma)$$

$$= \int_0^{\pi\alpha_r} d\sigma\, \{C(\sigma)(T^X_{++} + \tfrac{1}{2}\, T^{gh}_{++}) + C(\sigma)(T^X_{--} + \tfrac{1}{2}\, T^{gh}_{--})\} \qquad (13)$$

where $C(\sigma)$ represents the ghost field and $T_{ab}$ is the two dimensional stress tensor defined by

$$T_{ab}(\sigma) = \frac{2}{\sqrt{g}} \frac{\delta S}{\delta g^{ab}(\sigma)}$$

The superscripts X and gh of T in (13) stand for the string coordinate and ghost parts of T, respectively.

When one operates the test operator on $|\, V >$ , one should first express (13) in terms of coordinates $X^\mu$ and the canonical momentum $P_\mu = \delta S/\delta \dot{X}^\mu$ , which is no more equal to $\dot{X}^\mu/2\pi\alpha'$ due to the background fields. The straightforward operator calculation is not so simple because of this nonlinear effect. We have, however, a simple alternative way, which is called the Lagrangian method. We consider the vertex function, each string channel of which is stretched along the time-like direction [fig. 1]. The charge density $j(\rho=\tau+i\sigma)$ is now operated along contour $C_0$

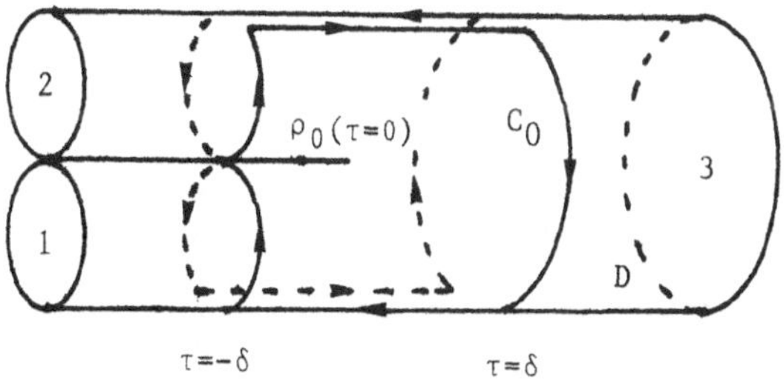


Figure 1. The three string vertex on which Q is operated and calculated by taking a limit $\delta \to 0$.

at $\tau = \pm \delta$ as shown in fig. 1, and then one takes the limit $\delta \to 0$ provided that the interaction of string takes place at $\rho = \rho_0(\tau = 0)$. The shift of $j(\sigma)$ from $\tau = 0$ to $\tau = \pm\delta$ should be made by use of the full action (12). Using path integral this can be expressed as

$$Q \mid V > = \lim_{\delta\to 0} \int D\,X\,D\,C\,D\,\bar{C} \oint_{C_0} d\rho j(\rho)\; e^{-S_D} \mid V > \tag{14}$$

where $S_D$ stands for the action (12) defined over the strip domain D in fig. 1. The equivalence of (14) to the operator Hamiltonian method is known as Matthew's theorem[15] in the local field theory. The validity of the Lagrangian method can be also partially confirmed by calculations below.

In what follows we assume the slope parameter $\alpha'$ is small and make the normal coordinate expansion of $X^\mu = x^\mu + \sqrt{2\pi\alpha'}\,\xi^\mu$ around a constant classical solution $x^\mu$(the center of mass). The evaluation of the right-hand-side of (14) is made up to $O(\alpha')$. The calculation is performed on the complex z-plane instead of $\rho$-plane after making the Mandelstam mapping

$$\rho(\tau,\sigma) = \sum_{r=1}^{3} \alpha_r \ln(z - z_r). \tag{15}$$

It is important to note that we are calculating the conformal anomaly term under BRST operation. Let us look at a matrix element of (14). The contour $C_0$ in $\rho$-plane is mapped onto $C_0'$ in z-plane [fig. 2].

$$\begin{aligned} &{}_{1,2,3}\!< 0 \mid \sum_{r=1}^{3} Q^{(r)} \mid V(1,2,3) > \\ &= -\frac{i\sqrt{\pi}}{2} \oint_{C_0'} dz \left[\frac{d\rho(z)}{z}\right]^{-1} {}_{1,2,3}\!< 0 \mid C(z)[-T^X_{++}(z) + 2\partial_z C(z)C(t)] \mid V > \\ &= \frac{\sqrt{\pi}}{64}\,[D - 26 - \frac{3}{2}\alpha'(R + 4\nabla^\mu\nabla_\mu\Phi - 4\nabla^\mu\nabla_\mu\Phi - \frac{1}{3} S_{\mu\nu\rho}S^{\mu\nu\rho})] \\ &\quad \cdot [\frac{1}{a} C''(z_0)C(z_0) - \frac{6}{a^2} C'(z\;)C(z_0)] \end{aligned} \tag{16}$$

where a and b are certain constants, and R and $S_{\mu\nu\rho}$ represent the curvature scalar of $G_{\mu\nu}$ and the field strength of $B_{\mu\nu}$, respectively. Although the detail of calculations[11] is not shown we give some comments here. (i) In the perturbation calculation the $\sigma$-model corrections occur only on the contour $C_0'$. There are potential sources of non-local corrections such as from the diagram shown in fig. 3. In the amplitude, however, some propagators which link $L_{int}(w)$ with $T^{(0)}_{++}(z)$ and the conformal factor $\phi(u) = \ln \mid \partial\rho/\partial z \mid^2$ collapse into delta functions due to

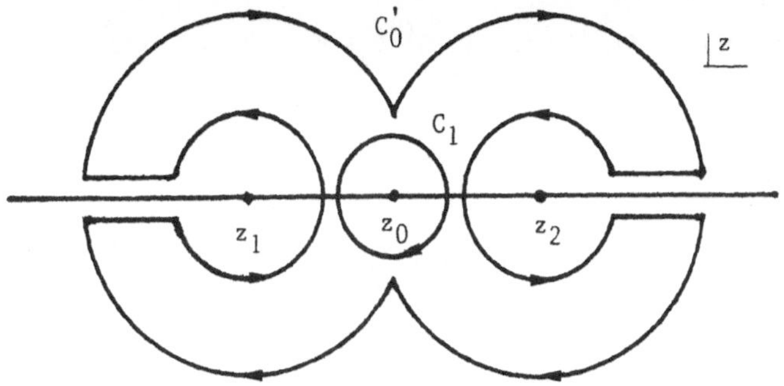


Figure 2. The three string vertex mapped on z-plane, where $z_0$ represents the image of interaction point $\rho_0$.

derivatives on the propagator, leaving a local correction to $T_{++}^{(0)}(z)$. This supports that the Lagrangian formalism takes account of the effect of full canonical momentum $P_\mu$ in accordance with Matthew's theorem.
(ii) Careful inspections of the integrand in the second line of (16) show that no singularities appear except at the image of interaction point $\rho_0$, so that the z-integration can be shrunken to a small circle $C_1$ around $z_0$[fig.2]. Singularities at $z_0$ appear from two sources, one from $(d\rho/dz)^{-1}$ in (16) which comes from the Jacobian and the transformation factor of j, and the other from the conformal factor $\phi = \ln|d\rho/dz|$ in z-plane.

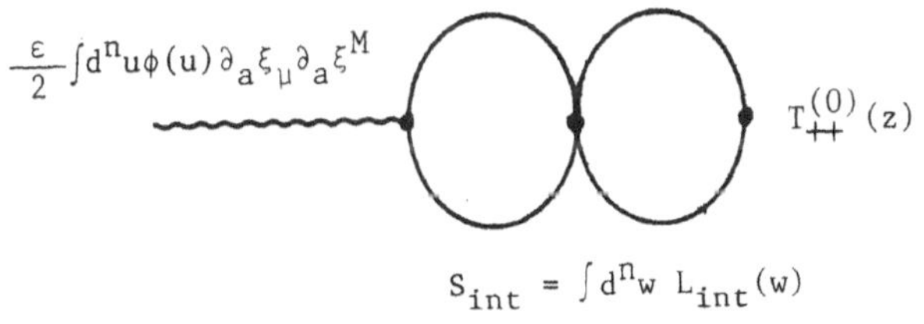


Figure 3. A diagram which gives a potential source on non-local correction.

(iii) Collecting these singular terms and performing the Cauchy integral we obtain the final expression in (16), which is nothing but the conformal anomaly.

If the BRST charge should eliminate the vertex function as is required by (11), the last expression of (16) should vanish, i.e.,

$$D - 26 - \frac{3}{2}\alpha' (R + 4\nabla^\mu\nabla_\mu\Phi - 4\nabla^\mu\Phi\nabla_\mu\Phi - \frac{1}{3} S_{\mu\nu\rho}S^{\mu\nu\rho}) = 0 \qquad (17)$$

From another matrix element of (11), namely

$$\underset{1,2,3}{<0} \mid \xi_\mu (1)\, \xi_\nu (2) \sum_{r=1}^{3} Q^{(r)} \mid V(1,2,3) > = 0 \qquad (18)$$

we have obtained

$$R_{\mu\nu} + 2\nabla_\mu \nabla_\nu \Phi - S_{\mu\rho\sigma} S_\nu{}^{\rho\sigma} = 0 \qquad (19)$$

$$\nabla_\rho S_{\mu\nu}{}^\rho - 2S_{\mu\nu\rho} \nabla^\rho \Phi = 0 \qquad (20)$$

All matrix elements other than (16) and (18) are shown to be trivially zero up to $O(\alpha')$. In the calculation of anti-symmetric tensor contributions one needs special care. The details will be published in ref. 16.

Now one can conclude that the background fields $G_{\mu\nu}$, $B_{\mu\nu}$ and $\Phi$ should obey equations (17), (19) and (20). Our results agree to those obtained in ref. 1 and ref. 2. It should be emphasized that the same equations have been obtained from a new requirement which is linear[17] in Q, and that these equations are sufficient to define a field theory of the string.

The method developed above is able to be extended to higher orders in $\alpha'$. Using the sigma model as a supplementary tool one constructs a test operator $Q[G, B, \Phi]$, then derives the equations for background fields to any order in $\alpha'$ from the conditions of nilpotency and the distribution law. Substituting a set of solutions to these background field equations one determines a BRST operator

$$Q = Q^{(0)} + \alpha' Q^{(1)} + (\alpha')^2 Q^{(2)} + \cdots \qquad (21)$$

The string action (1) associated with (21) then provides the field theory of quantum string with the background fields.

Finally some comments are in order.

(i) Although we have taken account of massless fields as background field in the above discussion, it is easy to extend the method so that massive fields are included[18]. To take account of the ground state tachyon, all one should do is to add an extra external source term

$$\frac{1}{4\pi\alpha'} \int d^2\sigma \sqrt{g}\, \Phi_T(X(\sigma))$$

to the action (12), where $\Phi_T(X)$ stands for the tachyon field. The BRST charge $Q[G, B, \Phi, \Phi_T]$ constructed from the new action then provides conditions over $G_{\mu\nu}$, $B_{\mu\nu}$, $\Phi$ and $\Phi_T$.

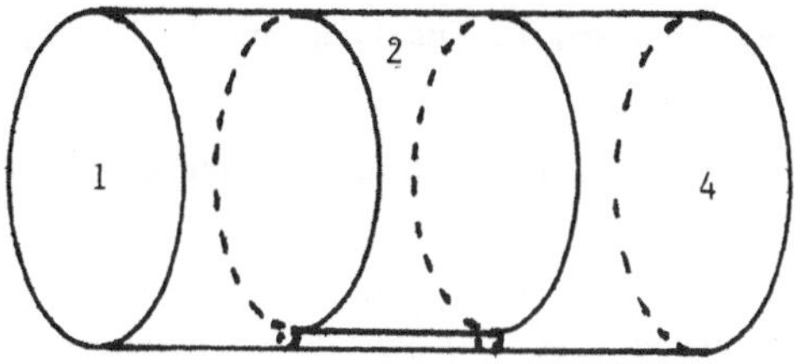


Figure 4. A tad-pole emission diagram which gives rise to the Fischer-Sisskind effect, wher the width parameter of the string 3 is approching to zero.

(ii) The string loop correction has to be taken account in higher orders. As was discussed by Fischler and Susskind[19] the divergences in amplitudes having non-trivial topology require new counter terms in (12), and the form of BRST charge has to be modified thereby [fig. 4]. In order to evaluate the loop effects one has to modify HIKKO's method as has been pointed out previously[12]. We will report on this subject elsewhere.

## References


1. E.S. Fradkin and A.A. Tseytlin, Phys. Lett. 158B (1985) 316; 160B (1985) 69; Nucl. Phys. B261 (1985) 1.
   C.G. Callan, D. Friedan, E.J. Martinec and M.J. Perry, Nucl. Phys. B262 (1985) 593.
   A. Sen, Phys. Rev. Lett. 55 (1985) 1846; Phys. Rev. D32 (1985) 2102.

2. T. Banks, D. Nemeschansky and A. Sen, Nucl. Phys. B277 (1986) 67.

3. M. Kaku and K. Kikkawa, Phys. Rev. D10 (1974) 1110 and 1823.
   E. Cremmer and J.L. Gervais, Nucl. Phys. B76 (1974) 209; B90 (1975) 410.

4. H. Hata, K. Itoh, T. Kugo, H. Kunitomo and K. Ogawa, Phys. Lett. 172B (1986) 186; Phys. Rev. D35 (1987) 1318 and 1356.
   A. Neveu and P. West, Phys. Lett. 168B (1986) 192; Nucl. Phys. B278 (1986) 601.

5. E. Witten, Nucl. Phys. B268 (1986) 253.

6. M. Kato and K. Ogawa, Nucl. Phys. B212 (1983) 443.

7. N. Sakai and Y. Tanii, Nucl. Phys. B287 (1987) 457.
   K.S. Narain, Phys. Lett. 169B (1986) 41.
   H. Kawai, D.C. Lewellen and S.H.H. Tye, Phys. Rev. Lett. 57 (1986) 1832.

8. D. Friedan, Nucl. Phys. B271 (1986) 540.
   T. Yoneya, talk at "Workshop on Unified Theories", RIFP Kyoto (1985), see ref. 5 also.

9. H. Hata, K. Itoh, T. Kugo, H. Kunitomo and K. Ogawa, Phys. Lett. B175 (1986) 138.

10. G.T. Horowitz, J. Lykken, R. Rohm and A. Strominger, Phys. Rev. Lett. 57 (1986) 238.

11. This talk is based on K. Kikkawa, M. Maeno and S. Sawada, "String Field Theory in Curved Space" Osaka Univ, preprint OU-HET-106 (1987), to be published in Phys. Lett..

12. A. Neveu and P. West, Nucl. Phys. B293 (1987) 266.
J.G. Taylor, Phys. Lett. B186 (1987) 57.
S. Uehara, Phys. Lett. B190 (1987) 76.
T. Kugo, a private communication (1987).

13. G. Parisi and N. Sourlas, Phys. Rev. Lett. 43 (1979) 744.

14. G.T. Horowitz and A. Strominger, Phys. Lett. B185 (1987) 45,
A. Strominger, Phys. Lett. B187 295; Phys. Rev. Lett. 58 (1987) 629.

15. A nice proof of the theorem is given by C. Bernard and A. Duncan, Phys. Rev. D11 (1975) 848.

16. M. Maeno and S. Sawada "String Field Theory in Curved Space: Non-Liniear σ-Model Calculation" Osaka Univ. preprint (1987)

17. The conformal covariance of the tachyon emission vertex was studied by S.R. Wadia, Proc. of XXIII International Conference on High Energy Physics (1986), p.369, Berkely and S. Jain, G. Mandel and S.R. Wadia, Phys. Rev. D35 (1987) 3116.

18. R. Brustein, D. Nemeschansky and S. Yankielowicz, "Beta Functions and S-matrix in String Theory", preprint USC-87/004.
H. Ooguri and N. Sakai, "String Loop Corrections from Fusion of Handles and Vertex Operators" TIT/HEP-117 and UT-512 (1987)

19. W. Fischler and L. Susskind, Phys. Lett B17 (1986) 383; B173 (1986) 262.

# THE GLUING THEOREM IN THE OPERATOR FORMULATION OF STRING FIELD THEORY

Christian R. Preitschopf*
*Department of Physics and Astronomy*
*University of Maryland, College Park, Maryland, 20742*

## Abstract

The gluing theorem, which describes the sewing of vertex operators in string field theory, is formulated and proven at tree level of string perturbation theory. Examples of vertices are given and some applications of the theorem are briefly described.

## 1. Introduction

One of the main unsolved problems in string theory is the vacuum problem, *i.e.* the calculation of the ground state of the string. The solution should enable us to decide which one of the many conformally invariant string theories that exist is preferred by theory and maybe by nature. The most conservative attack on this problem proceeds via the construction and, eventually, solution of string field theory. Unfortunately, our ability to calculate in string field theory, or for that matter in any off shell formalism, is rather limited, even though quite powerful methods have been developed for the calculation of on shell scattering amplitudes.

In the present paper I will give a short review of the work of A. LeClair, M. Peskin and myself[1] which addresses this deficiency. I will describe a formalism that links the operator formulation of string (field) theory and the path integral approach of Polyakov.[2] It is shown how to divide up the free conformal field theories that comprise the bosonic string path integral on some Riemann surface into pieces much like an atlas divides up the surface into overlapping charts. The pieces are vertex operators and the reconstruction process we called gluing. In this sense the formalism constitutes the inverse of factorization. We use strongly the results of Belavin, Polyakov and Zamolodchikov[3] and their application to string theory by Friedan[4] and Friedan, Martinec and Shenker.[5] The vertices may be pictured as Riemann surfaces with holes, and the gluing theorem that will be formulated and proved states that upon contracting two legs of different vertices with the BPZ inner product, another vertex is obtained which represents precisely the Riemann surface that arises from sewing the two surfaces along the boundaries of the holes. The

⋆ Work supported by the National Science Foundation, contract PHY-86-05207

geometrical images that are conjured up in these pictures are quite familiar when we use light cone vertices and propagators. One obtains world sheets in the form of strips with cuts. The gluing theorem, which has been proven so far only at the tree level of string theory, is then part of the answer to the problem Cremmer and Gervais[6] pose: how to extend their analysis of the four point function on the light cone to all the Koba-Nielsen amplitudes.[7] In the operator formalism this was never done. Of course, Mandelstam gave the answer using the path interal in the light cone gauge,[8] and for Witten's proposal of a covariant open string field theory[9] Giddings[10] calculated the four point function and in collaboration with Martinec[11] and Witten[12] outlined how the Polyakov path integral emerges in this picture of the string world sheet. The gluing theorem provides the key to establishing these results in an explicit and more general form in the operator formalism, paving the way for the calculation of off shell amplitudes. It can also be applied to the problem of associativity of Witten's vertex to yield the first complete proof of the gauge invariance of his string field action.[13]

The outline of this paper is as follows: in section 2, I will motivate and discuss our construction of vertices. The picture of a Riemann surface with holes will be explained and several examples will be given. In section 3 the gluing procedure will be introduced in a precise way. I will state the gluing theorem and in section 4 describe the main ingredients in its proof. The actual work will be left to ref. [1]. However, the subtle points are all at least mentioned. Section 5 concludes with an outlook on further work and applications.

For a self-contained presentation of the notation I use, see ref. [1]. Whenever possible, I will restrict myself to the matter sector of the string, *i.e.* to the conformal fields $\partial_z x(z)$ and $e^{ipX(z)}$. For a complete collection of formulae for the bosonic string, and of references, see ref. [1].

## 2. Vertices

Actions in string field theory are schematically written as

$$S = \int A \bullet \mathcal{K} A + \int A * A * A + \int A \circ A \circ A \circ A + \cdots , \qquad (2.1)$$

where $\mathcal{K}$ denotes the (gauge invariant or gauge fixed) kinetic operator, and in general higher point interactions may be necessary for gauge or BRST invariance. Various products of string fields are indicated by the symbols $\int$ and $\bullet, *$, and $\circ$. They are usually defined in terms of pictures of particular two-dimensional world sheets. For example, the expression $\int A \bullet B$ may be thought of as a path integral defined on a strip of infinite length with boundary conditions that correspond to the states $A, B$. In Fig. 1 this world sheet is mapped to the complex plane.

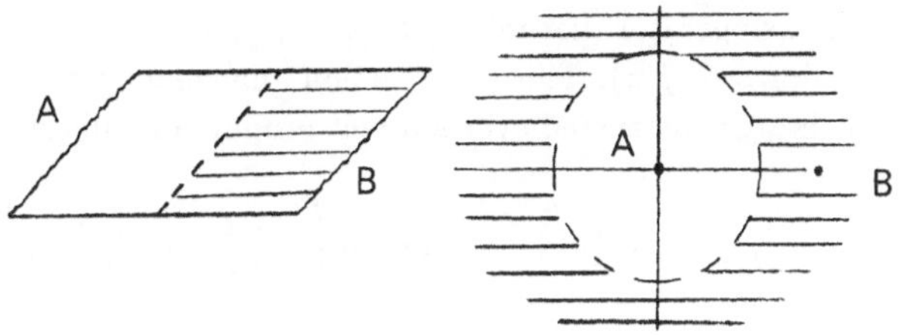


Fig. 1 The world sheet defining the two point function, represented as a strip and as the complex plane

The boundary conditions now appear as operator insertions at 0 and $\infty$. For example, the state $a^{\mu}_{-n}|p\rangle$ is represented by the operator expression

$$a^{\mu}_{-n}e^{ipX(0)}|0\rangle = \oint \frac{dz}{2\pi i} z^{-n}\partial_z x(z)e^{ipX(0)}|0\rangle \ . \tag{2.2}$$

defined on a neighbourhood of 0. In order to put similar boundary conditions at other points of the complex plane one has to map them with vertex operators. Take a function $f(z)$, analytic around some $z \neq \infty$, but not necessarily globally analytic on the complex plane. The conformal transformation $z \to f(z)$ is implemented on any operator as $\mathcal{O} \to U_f \mathcal{O} U_f^{-1} \equiv f[\mathcal{O}]$, where

$$U_f = \exp\left(\sum_{n\geq -1} v_{-n}L_n\right) \quad \text{with} \quad v(z)\partial_z f(z) = v\left(f(z)\right) \ . \tag{2.3}$$

$v(z)$ is the vector field that infinitesimally generates the conformal transformation. Let us define $\langle f| \equiv \langle 0|U_f$. Since $U_f^{-1}|0\rangle = |0\rangle$, the mapped form of the correlation function $\langle \mathcal{O}_1 \cdots \mathcal{O}_n\rangle$ can be written as

$$\langle f[\mathcal{O}_1]\cdots f[\mathcal{O}_n]\rangle = \langle f|\,\mathcal{O}_1\cdots\mathcal{O}_n|0\rangle \ . \tag{2.4}$$

The normal ordered version of the state $\langle f|$ is obviously

$$\langle f| = \langle 0|\exp\left(-\frac{1}{2}\sum_{n\geq 0} a_n N^{ff}_{nm} a_m\right) \ , \tag{2.5}$$

with

$$\begin{aligned} N^{ff}_{00} &= \ln f'(0), \\ N^{ff}_{0m} &= \frac{1}{m}\oint \frac{dw}{2\pi i} w^{-m}\left(f'(w)\right)\frac{1}{\left(f(w)-f(0)\right)}, \\ N^{ff}_{nm} &= \frac{1}{n}\oint \frac{dz}{2\pi i} z^{-n}\left(f'(z)\right)\frac{1}{m}\oint \frac{dw}{2\pi i} w^{-m}\left(f'(w)\right)\frac{1}{\left(f(z)-f(w)\right)^2}, \end{aligned} \tag{2.6}$$

since this expression reproduces the correlation functions. The coefficients $N^{ff}_{nm}$ are just the Fourier modes of the Green function $\langle\partial_z f[x(z)]\partial_w f[x(w)]\rangle$. We see that the state $\langle f|$ constitutes a vertex operator that takes states on a coordinate patch around 0 and performs the mapping $z \to f(z)$. This construction immediately generalizes to multipoint vertices, which are constructed over a direct product of Hilbert spaces:

$$\langle V_{1\cdots k}| = \prod_{I=1}^{k} \left[ \int d^d p_I \langle -p_I|_I \right] (2\pi)^d \delta^d(p_1+\cdots+p_k) \exp\left( -\frac{1}{2} \sum_{I,J} a_n^I N_{nm}^{IJ} a_m^J \right) , \quad (2.7)$$

where, for example,

$$N_{00}^{IJ} = \ln (f_I(0) - f_J(0)) \quad (2.8)$$

and the other Neumann function coefficients are obvious generalizations of (2.6). Then

$$\langle V_{1\cdots k}| [\mathcal{O}_1 |0\rangle_1 \otimes \cdots \otimes \mathcal{O}_k |0\rangle_k] = \langle f_1[\mathcal{O}_1] \cdots f_k[\mathcal{O}_k] \rangle \quad (2.9)$$

and hence $\langle V_{1\ldots k}|$ maps k states into k different regions, as indicated in Fig. 2.

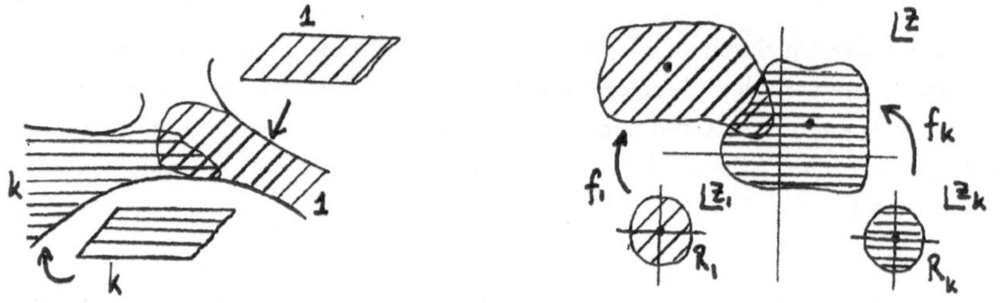


Fig. 2 The picture associated with $\langle V_{1\cdots k}|$. The maximal radii of convergence of the functions $f_I$ are denoted by $R_I$

The correlation function (2.9) is a priori completely general and will depend on the moduli of the underlying Riemann surface. The questions of the correct integration measure and its emergence from natural operator manipulations comprise string perturabtion theory, which is treated in detail in paper III of ref. [1]. Here I will restrict attention to the situation where all the vertices are constructed over the Riemann sphere for the closed string and the half sphere for the open string. Some examples of vertices are the BPZ two point function (Fig. 1), defined by

$$f_1(z) = z; \quad f_2(z) = -\frac{1}{z} , \quad (2.10)$$

the Caneschi-Schwimmer-Veneziano (CSV) vertex[14] (Fig. 3), with

$$f_1(z) = z; \quad f_2(z) = \frac{1}{1-z}; \quad f_3(z) = \frac{z}{z-1} , \quad (2.11)$$

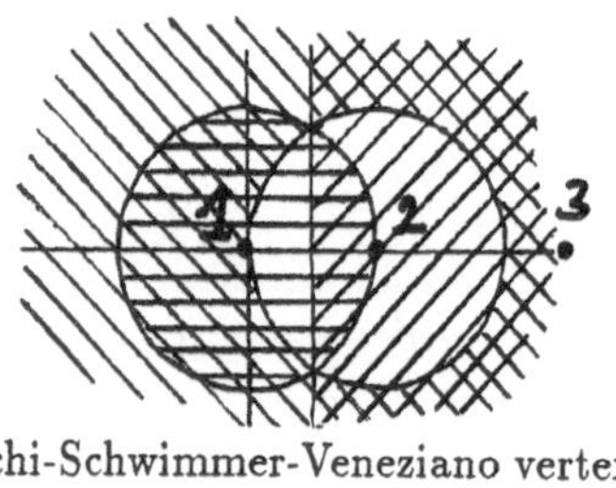


Fig. 3 Caneschi-Schwimmer-Veneziano vertex

and a generalized light cone vertex (Fig. 4) specified by

$$f_1(z) = \rho^{-1}(\alpha_1 \ln z_1 + Re\rho(-\beta));$$

$$f_2(z) = \rho^{-1}(\alpha_2 \ln z_1 + \rho(-\beta));\qquad (2.12)$$

$$f_3(z) = \rho^{-1}(-\alpha_3 \ln z_1 + Re\rho(-\beta) + i\pi\alpha_3)\ ,$$

where $\alpha_1$, $\alpha_2$ and $\alpha_3$ are the (positive) string lengths which are related to two complex parameters $\beta$ and $\gamma$ by $\alpha_1/\alpha_2 = \sqrt{\beta\gamma}$; $\alpha_3/\alpha_2 = \sqrt{(1+\beta)(1+\gamma)}$. The mapping $\rho(z)$ is then given by

$$\rho(z) = -\alpha_2 \Big\{ \ln\left[2\mathcal{R}(z) + 2z + \beta + \gamma\right] + \sqrt{(1+\beta)(1+\gamma)}\ln(z-1) - \sqrt{\beta\gamma}\ln z$$
$$-\sqrt{(1+\beta)(1+\gamma)}\ln\left[2\sqrt{(1+\beta)(1+\gamma)}\mathcal{R}(z) + (2+\beta+\gamma)z + 2\beta\gamma + \beta + \gamma\right]$$
$$+\sqrt{\beta\gamma}\ln\left[2\sqrt{\beta\gamma}\mathcal{R}(z) + (\beta+\gamma)z + 2\beta\gamma\right]\Big\}$$

(2.13)

with $\mathcal{R}(z) = \sqrt{z^2 + (\beta+\gamma)z + \beta\gamma}$. For $\alpha_1 + \alpha_2 = \alpha_3$ this is the usual light cone vertex, while for $\alpha_1 = \alpha_2 = \alpha_3 = 1$ Witten's vertex emerges (Fig. 4).

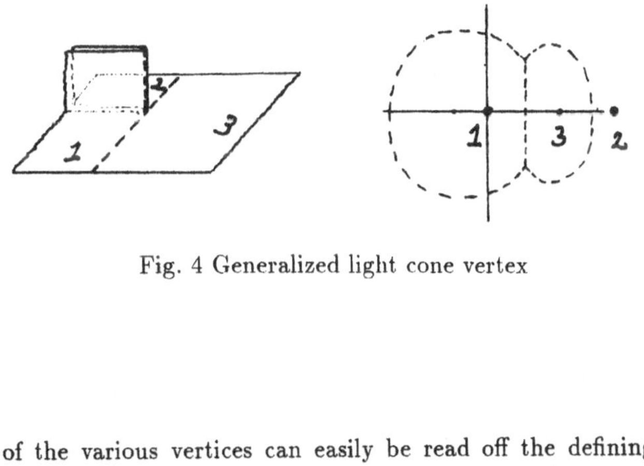


Fig. 4 Generalized light cone vertex

Symmetries of the various vertices can easily be read off the defining coordinate charts. (2.9) together with $SL(2)$ invariance of correlation functions implies the same symmetry for the vertex. Then one may derive for example the cyclicity of the CSV vertex by the observation $f_i(z) = T^{\circ(i-1)}(z)$, where $T(z) \in SL(2)$ and $T^{\circ 3}(z) \equiv T \circ T \circ T(z) = z$. Two-dimensional BRST invariance of the vertex also is a simple consequence of (2.9) and the fact that $Q_{BRST}$ is conformally invariant, *i.e.* $f[Q_{BRST}] = Q_{BRST}$.

## 3. Gluing

Any calculation that starts with a string field action like (2.1) will generically involve products of string fields like $(A * A) \bullet (A * \Phi)$. Often the meaning of such products is emodied in pictures like Fig. 5, but not every vertex has such a convenient representation.

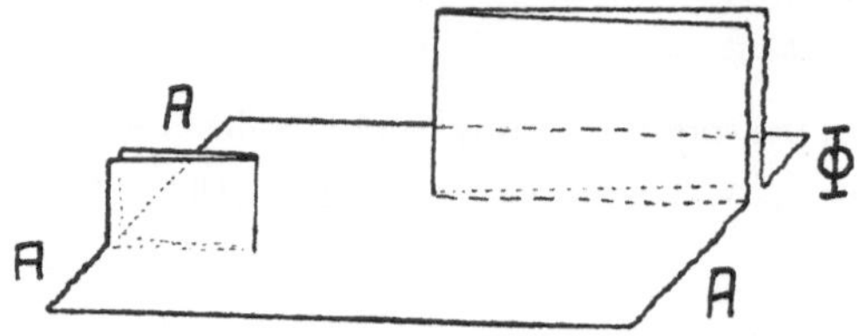


Fig. 5 Product of strings in the strip representation

A more abstract approach to the problem of joining strings starts from the BPZ inner product $|\mathcal{I}_{AB}\rangle$, which is defined by

$$\langle \mathcal{I}_{AB} | \mathcal{I}_{BC} \rangle = \delta_{AC} \; . \tag{3.1}$$

Given two vertices $\left\langle V_{\{A_1 \cdots A_k\}C} \right|$ and $\left\langle V_{\{B_1 \cdots B_l\}D} \right|$ specified by mappings $h_{A_i}(z)$, $h_C(z)$ and $h_{B_j}(z)$, $h_D(z)$, respectively, we define the glued vertex

$$\left\langle V_{\{A_1 \cdots A_k\}\{B_1 \cdots B_l\}} \right| \equiv \left\langle V_{\{A_i\}C} \right| \left\langle V_{\{B_j\}D} | \mathcal{I}_{CD} \right\rangle \; . \tag{3.2}$$

The geometry underlying this vertex is decribed by the following sequence of pictures. The inner product involves taking the matrix element of operators in the Hilbert space $C$, and therefore we perform the mapping $h_C^{-1}$ on Fig. 2:

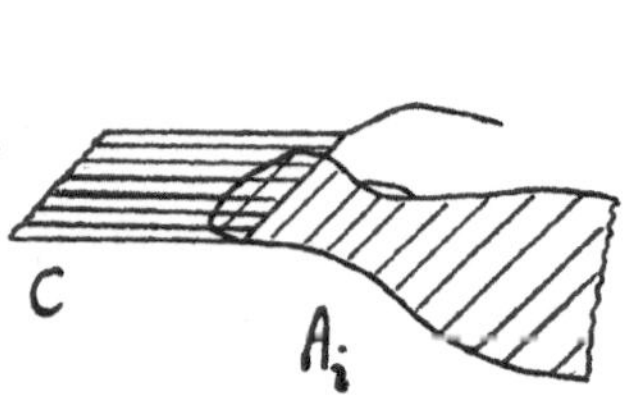


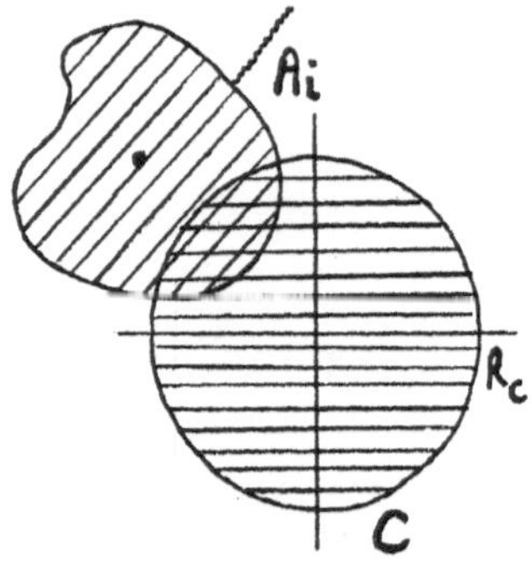


Fig. 6 Mapping Fig. 2 for $\left\langle V_{\{A_1 \cdots A_k\}C} \right|$ with $h_C^{-1}$

Of course, $h_C^{-1}$ is not single valued over the whole sphere, and hence cuts appear on the complex plane, outside the radius of convergence $R_C$ of $h_C$ and $h_C^{-1}$. Operators in Hilbert space D are mapped by $I(z) = -\frac{1}{z}$, and hence Fig. 2 is transformed by $I \circ h_D^{-1}$:

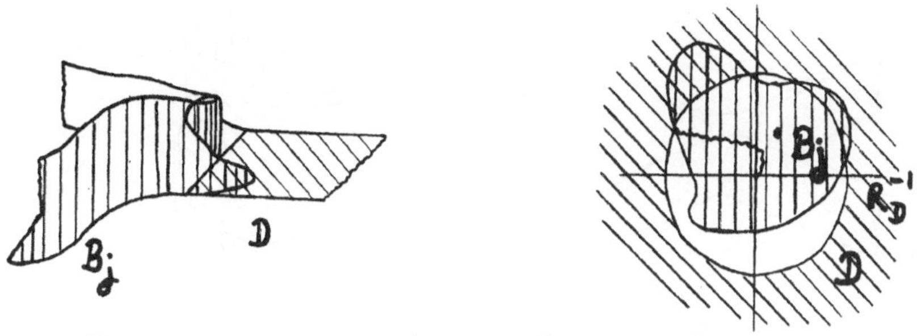

Fig. 7 Mapping Fig. 2 for $\langle V_{\{B_1 \cdots B_t\}D} |$ with $I \circ h_D^{-1}$

Due to the inversion the cuts now reside inside $R_D^{-1}$. Evaluating the matrix element between the Hilbert spaces C and D erases the corresponding regions and superimposes the rest,

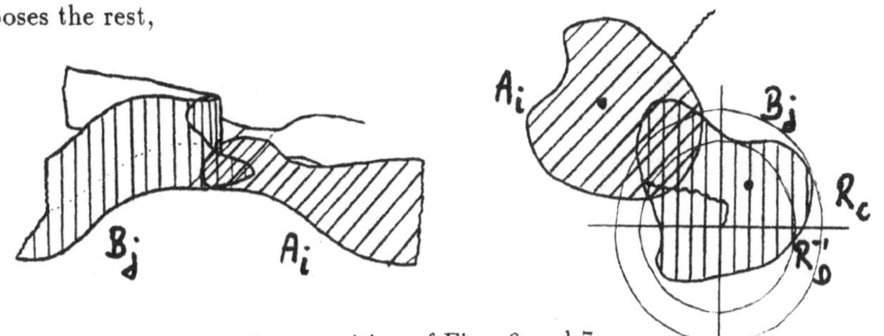

Fig. 8 Superposition of Figs. 6 and 7

leaving a Riemann surface with cuts that still has the topology of a sphere. Hence there exists a map $g(z)$ that smoothes out the cuts:

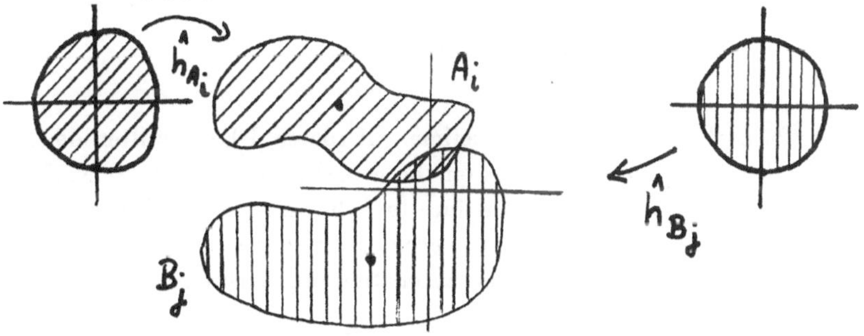

Fig. 9 Worldsheet after smoothing out the cuts in Fig. 8 with the map $g(z)$

The gluing theorem states that the glued vertex (3.2) is given in terms of the functions

$$\hat{h}_{A_i} = g \circ h_C^{-1} \circ h_{A_i} \; ; \quad \hat{h}_{B_j} = g \circ I \circ h_D^{-1} \circ h_{B_j} \tag{3.3}$$

by formula (2.7). The application to the contraction of two of Witten's three-point functions results in a picture like Fig.10.

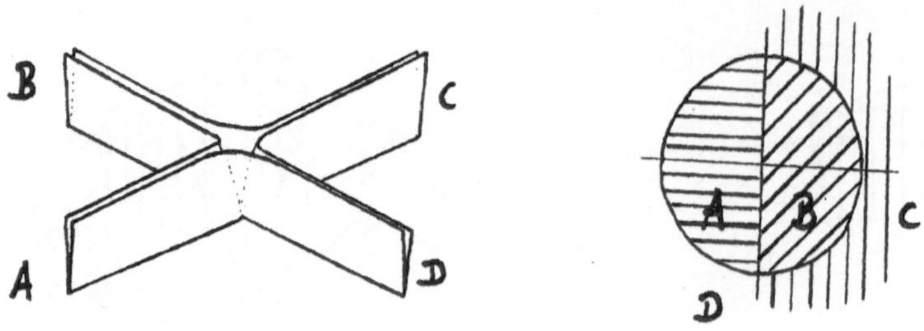


Fig. 10 Associativity of Witten's $*$ product

The geometrically obvious four-fold symmetry is realized on the complex plane in terms of $SL(2)$ transformations in a way we already encountered in the CSV vertex. If one performs the contractions in (3.2), it becomes apparent that the gluing theorem has two parts. First, it is claimed that

$$
\begin{aligned}
N^{\hat{A}_i\hat{A}_j}_{n\ m} &= \left[N^{A_iA_j} + N^{A_iC}EN^{DD}E(1-N^{CC}EN^{DD}E)^{-1}N^{CA_j}\right]_{nm} \\
N^{\hat{B}_i\hat{B}_j}_{n\ m} &= \left[N^{B_iB_j} + N^{B_iD}EN^{CC}E(1-N^{DD}EN^{CC}E)^{-1}N^{DB_j}\right]_{nm} \qquad (3.4)\\
N^{\hat{A}_i\hat{B}_j}_{n\ m} &= \left[N^{A_iC}E(1-N^{DD}EN^{CC}E)^{-1}N^{DB_j}\right]_{nm} .
\end{aligned}
$$

This equation implies that the exponents of the glued and the hatted vertices are identical. For the second part of the theorem, we note that in (2.7) there is no extra c-number factor, whereas the contractions yield one:

$$
\left[\det(1-N^{CC}EN^{DD}E)\right]^{-\frac{d}{2}} , \qquad (3.5)
$$

where $d = 26$ is the dimension of spacetime. Of course, I have written down only the part of the vertex which depends on the nonzero modes of $X$. Therefore the above determinant does not have to reduce to 1, but has to cancel its counterpart from the ghost sector. The matrices one encouters there are quite different from the ones given in (2.6), and the miraculous cancellation that occurs is the content of the second part of the gluing theorem.

## 4. Proof of the Gluing Theorem

The identities (3.4) are shown by studying the analytic properties of the underlying Green function. After dropping the integrals that select the $n$-th and $m$-th Fourier modes to obtain functions with arguments $u, v$ and the subsequent change of variables

$$
h_C^{-1} \circ h_{A_i}(u) = z; \quad I \circ h_D^{-1} \circ h_{B_j}(v) = w \qquad (4.1)
$$

one has to show

$$
\begin{aligned}
&\partial_z\partial_w \ln(g(z)-g(w))\\
&=\partial_z\partial_w \ln(h_C(z)-h_C(w))\\
&\quad+\frac{1}{n}\oint\frac{du}{2\pi i}u^{-n}\partial_z\partial_u \ln(h_C(z)-h_C(u))\left[EN^{DD}E(1-N^{CC}EN^{DD}E)^{-1}\right]_{nm}\\
&\quad\cdot\frac{1}{m}\oint\frac{dv}{2\pi i}v^{-m}\partial_v\partial_w \ln(h_C(v)-h_C(w)) \quad \text{for } |z|,|w|<R_C
\end{aligned}
\tag{4.2}
$$

$$
\begin{aligned}
&=\partial_z\partial_w \ln(h_D\circ I(z)-h_D\circ I(w))\\
&\quad+\frac{1}{n}\oint\frac{du}{2\pi i}u^{-n}\partial_z\partial_u \ln(h_D\circ I(z)-h_D(u))\left[EN^{CC}E(1-N^{DD}EN^{CC}E)^{-1}\right]_{nm}\\
&\quad\cdot\frac{1}{m}\oint\frac{dv}{2\pi i}v^{-m}\partial_v\partial_w \ln(h_D(v)-h_D\circ I(w)) \quad \text{for } |z|,|w|>R_D^{-1}
\end{aligned}
\tag{4.3}
$$

$$
\begin{aligned}
&=-\frac{1}{n}\oint\frac{du}{2\pi i}u^{-n}\partial_z\partial_u \ln(h_C(z)-h_C(u))\left[E(1-N^{DD}EN^{CC}E)^{-1}\right]_{nm}\\
&\quad\cdot\frac{1}{m}\oint\frac{dv}{2\pi i}v^{-m}\partial_v\partial_w \ln(h_D(v)-h_D\circ I(w)) \quad \text{for } |z|<R_C,\ |w|>R_D^{-1}.
\end{aligned}
\tag{4.4}
$$

We found two methods to prove the last two equalities. The first one consists of taking Fourier integrals in the annulus $R_D^{-1}<|z|,|w|<R_C$, which applies with equal ease to the fermionic ghost system, while the second one leads to a more convenient treatment of the bosonized ghosts. There one rewrites factors like

$$
\frac{1}{n}\oint\frac{du}{2\pi i}u^{-n}\partial_z\partial_u \ln(h_C(z)-h_C(u)) \tag{4.5}
$$

which possess a jump discontinuity as $z$ crosses the integration contour in terms of expressions like

$$
\frac{1}{n}\oint\frac{du}{2\pi i}u^{-n}\partial_z\partial_u \ln\left(\frac{h_C(z)-h_C(u)}{z-u}\right), \tag{4.6}
$$

which don't. The result is the same for (4.2), (4.3) and (4.4). Therefore these three expressions are but different representations of the same function, which is analytic on the sphere and has a double pole at $z=w$. That proves the first equality. For the $X$-system, the proof of the first part of the gluing theorem is complete. In the ghost sectors, one in addition has to check the proper normalization of certain functions. For example,

$$
\begin{aligned}
\langle 0|\,U_{h_C}c(z)c(w)c(u)U^{-1}_{I\circ h_D\circ I}\,|0\rangle &=(z-w)(w-u)(z-u)\\
&\quad+O\left(|z-w|^2,|w-u|^2,|z-u|^2\right)
\end{aligned}
\tag{4.7}
$$

must hold. Note that

$$|I \circ h_D \circ I\rangle \equiv U^{-1}_{I \circ h_D \circ I} |0\rangle = \langle h_D | \mathcal{I}_{D.} \rangle \tag{4.8}$$

and that the Green functions (4.2) to (4.4) arise in matrix elements like

$$\langle h_C | \, \partial_z x(z) \partial_w x(w) \, | I \circ h_D \circ I\rangle \; . \tag{4.9}$$

Therefore the appearance of expressions like (4.7) should not be too surprising. The proof goes as follows: without restricting generality we may assume that $U_{h_C}$ is an exponential of $L_n$ with $n \geq 2$, whereas $U^{-1}_{I \circ h_D \circ I}$ contains only Virasoro operators with index less than 1. These sets span the whole Virasoro algebra as a direct sum, and hence[15] there exist exponentials $U_f, U_g$ containing only $L_n$ with $n \geq 2$ and $n \leq 1$ respectively such that $U_{h_C} U^{-1}_{I \circ h_D \circ I} = U_g U_f$. Therefore the left hand side of (4.7) can be written as

$$\langle 0 | \, h_C[c(z)] h_C[c(w)] h_C[c(u)] U_g \, |0\rangle = \langle \tilde{g}[c(z)] \tilde{g}[c(w)] \tilde{g}[c(u)] \rangle \tag{4.10}$$

for some function $\tilde{g}(z)$ which has a convergent Laurent expansion sufficiently far from either 0 or $\infty$, and now (4.7) is easily derived. Of course, this result is based on the fact that the conformal anomly vanishes, *i.e.* that we are in the critical dimension $d = 26$ (otherwise there would be an additional c-number factor). But then we also observe that

$$\langle h_C | I \circ h_D \circ I\rangle = \left\langle U_{h_C} U^{-1}_{I \circ h_D \circ I} \right\rangle = 1 \tag{4.11}$$

by virtue of $\langle L_n \rangle = 0$ for all $n$. This equation, if written out, contains the cancellation of determinants, completing the proof of the gluing theorem.

## 5. Outlook

From Fig.8 it should be clear how to insert propagators: the annulus $R_D^{-1} < |z| < R_C$ can be enlarged by the insertion of a factor $\exp(-\tau L_0)$. However, at this point Giddings and Martinec[11] instruct us to add a factor $b_0$ in order to generate a density on moduli space in the proper way. The details of these calculations are given in ref.[1,III]. It is also fairly clear how to treat the superstring, at least in two-dimensional component language: the gluing of the general bosonized ghost system is performed in detail in ref. [1,II]. How fermions and their associated spin operators are dealt with is contained implicitly in the existing literature[16] on the operator approach to superstring field theory. An explicit treatment should render the cut structure of the Green functions more transparent for general vertices. It is worth noting that the methods developed above apply (so far) to any free conformal field theory without anomaly. These include gauge choices that are not conformal (e.g. harmonic type gauges) as long as one can achieve an effective conformal field theory by appropriate field redefinitions.[17] We hope to exploit the general formalism developed here for the construction of new string field theories, in particular one for the closed bosonic string. Get some UHU, keep on gluing!

## ACKNOWLEDGEMENTS

I wish to thank first and foremost my collaborators, for their time and effort that went into the work I reviewed. Both SLAC and the University of Maryland provided a pleasant and inspiring background for my research, and this workshop an exciting opportunity to present it.

## REFERENCES

1. A. LeClair, M.E. Peskin, C.R. Preitschopf, String Field Theory on the Conformal Plane I,II,III; SLAC-PUB 4306, 4307, 4308
2. A.M. Polyakov, Physics Letters 103B (1981) 207, 211
3. A.A. Belavin, A.M. Polyakov and A.B. Zamolodchikov, Nuclear Physics B241 (1983) 333
4. D. Friedan, Introduction to Polyakov's string theory, in *Recent Advances in Field Theory and Statistical Mechanics*, eds. J.B. Zuber and R. Stora. Proceedings of 1982 Les Houches Summer School (Elsevier, 1984), pp. 839;
   D. Friedan, Notes on string theory and two dimensional conformal field theory, in *Workshop on Unified String Theories, 29 July - 16 August 1985*, eds. M. Green and D. Gross (World Scientific, Singapore, 1986) pp. 162
5. D. Friedan, E. Martinec, S. Shenker, Nuclear Physics B271 (1986) 93
6. E. Cremmer and J.-L. Gervais, Nuclear Physics B90 (1975) 410
7. Z. Koba and H.B. Nielsen, Nuclear Physics B10 (1969) 633
8. S. Mandelstam, The interacting string picture and functional integration, in *Workshop on Unified String Theories, 29 July - 16 August 1985*, eds. M. Green and D. Gross (World Scientific, Singapore, 1986) pp. 46, 57
9. E. Witten, Nuclear Physics B268 (1986) 253
10. S.B. Giddings, Nuclear Physics B278 (1986) 242
11. S.B. Giddings and E. Martinec, Nuclear Physics B278 (1986) 91
12. S.B. Giddings, E. Martinec and E. Witten Physics Letters 176B (1986) 362
13. the proofs of associativity of Witten's * product given by D. Gross and A. Jevicki, Nuclear Physics B283 (1987) 1, Nuclear Physics B287 (1987) 225 and Nuclear Physics B293 (1987) 29, do not address certain c-number factors that arise in the product of string states. For the bosonic string these factors are, according to our proof, in fact irrelevant for associativity. For the superstring they diverge and destroy the gauge invariance of the string field action. This has been shown recently by C. Wendt, SLAC-PUB 4442 .
14. L. Sciuto, Lett. Nuo. Cim. 2 (1969) 411;
    L. Caneschi, A. Schwimmer and G. Veneziano, Physics Letters 30B (1969) 351
15. S. Helgason, Differential Geometry and Symmetric Spaces, Academic Press, New York and London, 1962, Lemma 2.4
16. for example, in
    D.J. Gross and A. Jevicki, Nuclear Physics B293 (1987) 29;

H. Hata, K. Itoh, T. Kugo, H. Kunitomo and K Ogawa, Kyoto University preprint KUNS 854 HE(TH) 87/01;
K. Suehiro, Kyoto University preprint KUNS 857 HE(TH) 87/04

17. L. Baulieu, W. Siegel and B. Zwiebach, Nuclear Physics B287 (1987) 93;
W. Siegel and B. Zwiebach, Nuclear Physics B288 (1987) 332

# CONFORMAL FIELD THEORY METHODS AND CONSTRUCTION OF WITTEN'S SUPERSTRING FIELD THEORY

Antal Jevicki*

Department of Physics
Brown University
Providence, RI 02912

ABSTRACT

This is a review of the operator construction of open superstring field theory. Witten's approach to interactions is followed and the construction is given using conformal field theory methods. Using the mappings from the upper half plane one obtains the superstring correlation functions in the scattering plane. These give the vertex operators of the theory.

The outline of Witten's open superstring field theory goes as follows [1]. In the Neveu-Schwarz-Ramond framework one has, in addition to the coordinates $x_\mu(z)$, also the fermionic variable $\psi_\mu(z)$ and, in addition to the ghost variables $c, b$ the bosonic superghost variables $\gamma(z), \beta(z)$. In the Neveu-Schwarz sector the variables are half-integer moded, while for Ramond, one has integer moding; the superghost zero modes $\gamma_o, \beta_o$ in general play an important role. The physical states of the Neveu-Schwarz sector $| A_{NS} >$ have $N_{gh} = -\frac{1}{2}$ coming from the ghosts while in the Ramond sector $| A_R >$ is taken to have ghost number $N_{gh} = O$ (here the $-\frac{1}{2}$ ghost number of the ghosts is cancelled by a $+\frac{1}{2}$ contribution assigned to superghosts). The string field $A = (A_{NS}, A_R)$ then has ghost number $(-\frac{1}{2}, O)$ and the free action reads:

$$S_o = < A_{NS} \mid Q \mid A_{NS} > + < A_R \mid Y(\pi/2)\, Q \mid A_R > \tag{1}$$

with $Y$ being the inverse picture changing operator.

Witten has generalized his $\int$ and $*$ operations to superstrings in the following way. For the integration one defines

$$\oint (A_{NS}, A_R) = \int Y(\pi/2)\, A_{NS} \tag{2}$$

where on the right hand side one still has a folding of a string about $\pi/2$. The

* Work supported by the Department of Energy, Contract DE-AC02-76ER03130.A022 - Task A

generalized star product reads

$$(A_{NS}, A_R) \star (A'_{NS}, A'_R) = \left[X(\pi/2)\left(A_{NS} * A'_{NS}\right) + A_R * A'_R \atop X(\pi/2)\left(A_{NS} * A'_R + A_R * A'_{NS}\right)\right. \tag{3}$$

Here $X$ is the picture changing operator. In terms of the generalized $\star$ and $\oint$ operations the gauge invariant action for open superstrings takes the same form as for the bosonic string

$$S = \oint \left(A \star A + \frac{2}{3} A \star A \star A\right). \tag{4}$$

This lagrangian is again gauge invariant under $\delta A = Q\Lambda + A \star \Lambda - \Lambda \star A$ where the gauge parameter $\Lambda = (\Lambda_{NS}, \Lambda_R)$ carries ghost number $\left(-\frac{3}{2}, -1\right)$. The theory is also invariant under supersymmetry transformations

$$\begin{aligned} \delta A_{NS} &= Q_\alpha A_R \\ \delta A_R &= X\left(\frac{\pi}{2}\right) Q_\alpha A_{NS} \end{aligned} \tag{5}$$

The supersymmetry charges are to be given by the fermion emission vertex

$$Q_\alpha = \int do S_\alpha(\sigma)\, e^{-\frac{\phi}{2}} \tag{6}$$

where $S_\alpha(\sigma)$ is the spin field and $\phi$ is the bosonized superghost.

The operator construction which follows makes the above general picture more explicit. The Lagrangian is written as

$$\begin{aligned} S = &< A_{NS} \mid Q \mid A_{NS} > + < A_R \mid YQ \mid A_R > \\ &+ \frac{2}{3} < V_3^{NSNSNS} \mid\mid A_{NS} >\mid A_{NS} \mid A_{NS} > \\ &+ 2 < V_3^{NSRR} \mid\mid A_{NS} >\mid A_R >\mid A_R > \end{aligned} \tag{7}$$

with the explicit forms for the $NS-NS-NS$ and also the $NS-R-R$ vertex operators. In this representation on shell three point amplitudes are directly computable and the BRST symmetry can be directly verified.

One also has the two-vertex operators corresponding to $NS-NS, R-R$ and the $NS-R$ string overlaps. The Neveu-Schwarz and the Ramond two-vertex states $|V_2^{NSNS}>$ and $|V_2^{RR}\rangle$ represent scalar products and define the hermiticity properties of the corresponding string fields, which read

$$\begin{aligned} < A_{NS}| &=< V_2^{NSNS}|A_{NS} > \\ < A_R| &=< V_{2\,o}^{RR}|A_R > \end{aligned} \tag{8}$$

In terms of two-string overlaps the free action can be written as

$$S_o =< V_2^{NSNS}||A_{NS} > |QA_{NS} > - < V_2^{RR}|A_R > |QA_R > \tag{9}$$

with $|V_2^{RR} >= Y(\pi/2)|V_{20}^{RR} >$. These scalar product operators have a simple form. The mixed $NS-R$ two-string overlap vertex $|V_2^{NSR}\rangle$ also exists and is nontrivial. It is relevant for giving an alternative form for supersymmetry transformations.

Finally, we have the integration state $|I>$. In superstring field theory one only defines an integration for Neveu-Schwarz strings $|I_{NS}>$:, there is no need for a Ramond integration due to the form of Wittens generalized integration $\oint$ which has no Ramond contribution.

The constructions reported here summarize Ref. [2] and also Ref. [3]. Some equivalent results were derived in [4]. There is, furthermore, the work of [5] which is different; relying on bosonization of all fermionic degrees of freedom the superstring problem can be transcribed into a bosonic one and for this no additional constructions are necessary.

The procedure followed in general to obtain the vertices was to perform a conformal transformation that mapped the diagram to the upper half-plane. On the upper half-plane the scattering amplitude of free strings coupled by the vertex could easily be constructed using the methods of conformal field theory. Compared with the bosonic string where the variables $x_\mu, c, b$ had integer conformal weights $J = 0, -1$ and 2, the present case we have half-integer conformal weights $J_\psi = \frac{1}{2}$, $J\gamma = -\frac{1}{2}, J_\beta = \frac{3}{2}$. A conformal field transforms in general as

$$\psi_J'(z') = \left(\frac{\partial z}{\partial z'}\right)^J \psi_J(z) \tag{10}$$

The overlap condition for a general string variable can then be deduced by equating its value at $z$ with its value at $z' = -\frac{1}{z}$, (this being the conformal transformation that corresponds to $\sigma \to \pi - \sigma$ with $\rho = \ln z$). Thus the overlap condition for the identity for a variable of conformal weight $J$ reads

$$\psi_J(\sigma) = (-)^J \psi_J(\pi - \sigma). \tag{11}$$

For the Neveu-Schwarz variables then

$$\begin{aligned} \left(\psi_+(\sigma) - i\psi_+(\pi - \sigma)\right) \mid I\rangle = 0, \qquad (0 \leq \sigma \leq \frac{\pi}{2}), \\ \left(\psi_-(\sigma) + i\psi_-(\pi - \sigma)\right) \mid I\rangle = 0, \qquad (0 \leq \sigma \leq \frac{\pi}{2}). \end{aligned} \tag{12}$$

The opposite signs in the $\psi_+$ and $\psi_-$ equations are dictated by the boundary conditions.

In the case of the bosonic string the overlap conditions implied that the Virasoro generators obeyed $L_\pm(\sigma) = L_\pm(\pi - \sigma)$. The Fourier components of this equation, $K_n \equiv L_n - (-1)^n L_{-n}$, were derivations of the integration defined by $|I\rangle$. The supersymmetric string has a supersymmetric generalization of the Virasoro algebra, which contains the fermionic charges $G_r$. The fermionic charge density is given by $G_\pm(\sigma) = P_\pm(\sigma) \cdot \psi_\pm(\sigma)$ and the overlap conditions for $\psi$ and $P$ imply that

$$G_\pm(\sigma) \mp iG_\pm(\pi - \sigma) = 0 \tag{13}$$

in addition to the $L(\sigma)$ equation. Correspondingly, we have a subalgebra of the full superconformal algebra that annihilates $|I\rangle$.

The Neveu-Schwarz part of the integration state $|I>$ is expressed as

$$|I^{\psi}\rangle = \exp[\frac{1}{2}\sum_{s,r\geq\frac{1}{2}} \psi_{-r} I_{rs} \psi_{-s}]|o> \tag{14}$$

with a standard Fock vacuum $|0>$. The overlap equations for the antisymmetric matrix $I_{rs}$ can again easily be worked out from Eq. (12), the solution is obtained in terms of the corresponding Neumann function.

On the circle the Neumann function is

$$K(\omega,\omega') = \langle\psi(\omega)\psi(\omega')\rangle = \frac{1}{(\omega-\omega')}. \tag{15}$$

We transform back to the strip using the transformation properties of $\Psi$ and

$$K(\rho,\rho') = \left(\frac{\partial\omega}{\partial\rho}\right)^{\frac{1}{2}} \frac{1}{\omega-\omega'} \left(\frac{\partial\omega'}{\partial\rho'}\right)^{\frac{1}{2}}, \tag{16}$$

Explicitly using the mapping relevant to the integration $\omega = z^2 = -\left(\frac{1+ie^{\rho}}{1-ie^{\rho}}\right)^2$ we have

$$K(\rho,\rho') = \left(\frac{z^2+1}{i}\right)^{\frac{1}{2}} \frac{\sqrt{z}\sqrt{z'}}{z^2-z'^2} \left(\frac{z^2+1}{i}\right)^{\frac{1}{2}}. \tag{17}$$

The expansion of this Neumann function in Fourier Series reads

$$K(\rho,\rho') = \sum_{r=\frac{1}{2}}^{\infty} e^{-ir(\sigma-\sigma')} + \sum_{r,s=\frac{1}{2}}^{\infty} e^{ir\sigma} e^{is\sigma'} I_{rs}. \tag{18}$$

The first term corresponds to the propagination of free strings, while the second scattering coefficients of plane waves, $e^{ir\sigma}e^{is\sigma'}$, give the integration vertex coefficents $I_{rs}$. The overlap equations follow from the fact that the Neumann function itself satisfies the appropriate overlaps, namely when $\rho = i\sigma$

$$K(\sigma,\sigma') = \begin{cases} iK(\pi-\sigma,\sigma'), & o \leq \sigma \leq \pi/2 \\ -iK(\pi-\sigma,\sigma'), & \pi/2 \leq \sigma \leq \pi \end{cases} \tag{19}$$

On the integration state one has the representation

$$\psi_+(\sigma)|I^{\psi}\rangle = \int \frac{d\sigma'}{2\pi} K(\sigma,\sigma')\psi_+^{cr}(\sigma')|I^{\psi}\rangle , \tag{20}$$

The operator overlap equations then follow from the matching conditions obeyed by the Neumann function $K(\sigma,\sigma')$.

Let us now discuss the scalar product operation $|V_2>$. For example for two Neveu-Schwarz strings one has the scalar product

$$\int A^1 * A^2 = < A^1 | < A^2 | V_2^{NSNS} > \tag{21}$$

with the NS-NS vertex obeying $[\psi^1_\pm(\sigma) \pm i\psi^2_\pm(\pi - \sigma)] |V^2_{NS,NS}\rangle = 0$. In this case there is no singularity at the midpoint, so the above overlap equation holds for all $0 \leq \sigma \leq \pi$. The solution is simple, namely

$$|V^2_{NS,NS}\rangle = \exp\left[ i \sum_{r \geq \frac{1}{2}} \psi^{1\mu}_{-r} (-)^r \psi^{2\mu}_{-r} \right] |0\rangle_{12} \,. \tag{22}$$

This quadratic form also trivially follows from the Neumann function

$$K^{ab}(\sigma, \sigma') = \left( \frac{\partial z}{\partial \varsigma_a} \right)^{\frac{1}{2}} \frac{1}{z - z'} \left( \frac{\partial z'}{\partial \varsigma'_b} \right)^{\frac{1}{2}} \tag{23}$$

with the two-string mapping $\rho = ln \frac{z-i}{z+i} - i\frac{\pi}{2}$ and the parametrization $\rho = \varsigma_1$ on string 1 and $\rho = -\varsigma_2 + i\pi$ on string two; for each string $\varsigma = \tau + i\sigma$. An identical form follows for the R-R scalar product and also the superghost degrees of freedom.

The three-string vertex is written in the form

$$|V_3^\psi >= \exp\left\{ \frac{1}{2} \sum_{r,s \geq 1/2} \psi_r^{\mu(a)} K^{ab}_{rs} \psi_s^{\mu(b)} \right\} |0>_{123} \tag{24}$$

and it corresponds to the overlap equations

$$\psi^a_\pm(\sigma) = \pm i \psi^{a-1}(\pi - \sigma) \qquad a = 1, 2, 3 \ldots \tag{25}$$

The quadratic form is given by

$$\Delta^\psi = \int \frac{d\sigma}{2\pi} \int \frac{d\sigma'}{2\pi} \psi^{(a)}_{an}(\sigma) K^{ab}(\sigma, \sigma') \psi^{(b)}_{an}(\sigma'). \tag{26}$$

where $K^{ab}(\sigma, \sigma')$ represents the correlation function of a fermion on the $a^{th}$ and a fermion on the $b^{th}$ string. The Fourier coefficients $K^{ab}_{rs}$ are obtained from the explicit form of $K(\rho, \rho\prime)$ using the three string mapping $\rho = \ln \frac{\omega^{\frac{3}{2}} - i}{\omega^{\frac{3}{2}} + i} - i\frac{\pi}{2}$. The basic function entering in these expansions is $g(\sigma) = \omega_a^{1/4} \left( \frac{1 + ie^{i\sigma}}{1 - ie^{i\sigma}} \right)^{1/6}$. This is recognized as the mapping appropriate for a 12-string rearrangement and we have the twelve string analogues of our earlier six string and three string coefficients $a_n, b_n$. The coefficients $K^{ab}_{rs}$ are explicitly given in Ref.(2). From (eq. 26) we have the integral formula

$$K^{ab}_{rs} = \int_{\omega_a} \frac{d\omega}{2\pi i} \int_{\omega_b} \frac{d\omega'}{2\pi i} e^{-n\rho} e^{-m\rho'} \left( \frac{d\omega}{d\rho} \right)^{-1/2} \left( \frac{d\omega'}{d\rho'} \right)^{-1/2} \frac{1}{\omega - \omega'} \tag{27}$$

which can be used for various general considerations.

For the Ramond sector one has a similar construction. Of main interest is the interaction operator for $R-R-NS$ strings. This and the $NS$ identity are coefficient to give all the Ramond terms in the action. Namely, the Ramond kinetic term is constructed through $\langle I_{NS}|V^3_{RRNS}\rangle = |V^2_{RR}\rangle$, and the vertex operator itself gives the coupling of Ramond and Neveu-Schwarz fields.

It is given by the following overlap equations

$$\begin{aligned} &\left(\chi^1_{\mp}(\sigma) \mp i\psi^3_{+}(\pi-\sigma)\right)|V^3_{RRNS}\rangle = 0 \\ &\left(\chi^2_{\pm}(\sigma) - i\chi^1_{\pm}(\pi-\sigma)\right)|V^3_{RRNS}\rangle = 0 \qquad (0 \le \sigma \le \frac{\pi}{2}) \\ &\left(\psi^3_{\pm}(\sigma) - i\chi^2_{\pm}(\pi-\sigma)\right)|V^3_{RRNS}\rangle = 0. \end{aligned} \tag{28}$$

The signs are consistent with the boundary conditions on Ramond strings $\chi^1, \chi^2$ and the Neveu-Schwarz string $\psi^3$. The quadratic form for the vertex is again given by the Neumann function which is now constructed as

$$\begin{aligned} R(\rho,\rho') = \frac{1}{2}\left(\frac{\partial\omega}{\partial\rho}\right)^{1/2} \frac{1}{\omega-\omega'}\left(\frac{\partial\omega'}{\partial\rho'}\right)^{1/2} \Bigg[\left(\frac{(\omega-\omega_1)(\omega'-\omega_2)}{(\omega'-\omega_1)(\omega-\omega_2)}\right)^{1/2} &+ \\ \left(\frac{(\omega'-\omega_1)(\omega-\omega_2)}{(\omega-\omega_1)(\omega'-\omega_2)}\right)^{1/2}\Bigg]&. \end{aligned} \tag{29}$$

The additional factors are there to insure the integer moded Fourier expansions appropriate for Ramond strings.

All the foregoing constructions apply also to superghosts. First one has the bosonized form where the commuting superconformal ghost $\gamma(\sigma), \beta(\sigma)$ operators are replaced by two scalar fields $\phi(\sigma)$ and $\overline{\phi}(\sigma)$. In the bosonized approach no additional constructions are needed. The vertex operators for the new degrees of freedom are the same as the bosonic coordinate operators. The only additional effects that have to be taken into account are the insertions at the midpoint. This bosonized approach is quite elegant since it explicitly exhibits the conformal and superconformal anomalies of the Riemann surfaces in question and was given in detail in [2]. However, in order to expand string functional fields in terms of local fields as well as for other reasons one would like to have a formulation of the theory in terms of the basic ghost coordinates $b, c$ and $\beta, \gamma$.

It is well known that the (super)ghost system $\gamma, \beta$ allows an infinite set of different vacuum states. These "picture changed" vacua labelled by $|0;n\rangle$ are characterized by different ghost numbers $N_g = n$ and different sets of annihilation (creation) operators. We gave a formulation of the theory in this more general representation. This formulation is useful for various applications; one being manifest BRST symmetry. In general it is in the Neveu-Schwarz sector that one needs the picture changing. In the Ramond sector things are actually simpler, for instance, there is no Ramond integration and no picture changing insertion $X(\frac{\pi}{2})$ in the NS-R-R vertex. Of special interest is the construction based on the $SL(2,R)$ invariant vacuum state $|\Omega\rangle = |0;1\rangle$.

Its creation operators are given by

$$\left.\begin{array}{ll}\beta_{-r} & r\geq \frac{3}{2} \\ \gamma_{-s} & s\geq -\frac{1}{2}\end{array}\right\} \quad \Omega. \tag{30}$$

and it gives the simplest form for the superghost integration state $\mid I^{\gamma\beta}\rangle$. This is achieved by the correlation function

$$K^{(+1)}(\rho,\rho') = \left(\frac{\partial\omega}{\partial\rho}\right)^{-1/2} \frac{1}{\omega-\omega'}\left(\frac{\partial\omega'}{\partial\rho'}\right)^{3/2}\left(\frac{\omega+1}{\omega'+1}\right)^{2} \tag{31}$$

which corresponds to the canonical $<\gamma\beta>$ correlation function with an additional factor. The complete superghost integration state was given then by

$$\mid I^{\gamma\beta}\rangle = Y\left(\frac{\pi}{2}\right)\exp\Big(\sum_{\substack{r\geq 3/2\\ s\geq -1/2}} \beta_{-r}K^{(+)}_{rs}\gamma_{-s}\Big)\mid\Omega\rangle \tag{32}$$

There are several other representations for the superghost integration state that we have constructed. One can perform a construction choosing any of the $\mid 0;n\rangle$ states as the vacuum; in general the quadratic form will be different but the insertions at the midpoint will also change. Of particular additional interest are the constructions using the states $\mid 0;0\rangle$ and the $SL(2,R)$ conjugate state $\mid\overline{\Omega}\rangle = \mid 0;-1\rangle$. Either of the forms for the integration state is appropriate for demonstrating the BRST invariance. The discussion is simplest in the $SL(2,R)$ formulation (see Ref. 2).

Let us now turn to a consideration of the superghost interaction vertex. The simplest quadratic form is found for the vacuum $\mid 0;0\rangle_1 \mid 0;0\rangle_2 \mid 0;0\rangle_3 \equiv \mid 0\rangle_{123}$ where one has the Neumann function:

$$\tilde{K}(\rho,\rho') = \frac{1}{2}\left(\frac{z}{z'}+\frac{z'}{z}\right)\left(\frac{\partial\omega}{\partial\rho}\right)^{1/2}\frac{1}{\omega-\omega'}\left(\frac{\partial\omega'}{\partial\rho'}\right)^{1/2} \tag{33}$$

It corresponds to the overlap equations

$$\left(\beta^{a}_{+}(\sigma)+i\beta^{a-1}_{+}(\pi-\sigma)\right)=0$$

$$\left(\gamma^{a}_{+}(\sigma)+i\gamma^{a-1}_{+}(\pi-\sigma)\right)=0 \tag{34}$$

The change in sign in comparison with the orbital case is achieved by the additional factor $\frac{1}{2}\left(\frac{z}{z'}+\frac{z'}{z}\right)$. The superghost $\left(\tilde{K}^{ab}_{rs}\right)$ Neumann coefficients are shown to be essentially equal to the orbital coefficients $K^{ab}_{rs}$ but with some changes in signs between the direct and reflected scattering contributions. In this basis the superghost vertex reads

$$\left[X\left(\frac{\pi}{2}\right)e^{-\phi}\right]\exp\Big\{\sum_{\substack{r\geq 1/2\\ s\geq 1/2}}\beta^{a}_{-r}\tilde{K}^{ab}_{rs}\gamma^{b}_{-s}\Big\}\prod_{a=1}^{3}\mid 0:0\rangle_a \tag{35}$$

with the appropriate insertions at the midpoint. A representation which has simpler insertion can be constructed utilizing the vacuum $|\,\overline{\Omega}\rangle$. This corresponds to transferring $e^{-\phi}$ to the ground state.

For the Ramond sector of the theory one constructs the analogous ghost vertex operator as follows. Consider as in the orbital case a R-R-NS configuration. The three vacuua taken are of ghost number zero. The vertex then reads

$$|\,V_{3,gh}^{NSRR}\rangle = \exp\{\sum_{a,b}\sum_{n,m}\beta^{a}_{-n}\tilde{R}^{ab}_{nm}\gamma^{b}_{-m}\}\,|\,0\rangle_{NSRR} \tag{36}$$

The summation for the third string (NS) goes over all positive half-integers while for the Ramond strings (1 and 2) we have $n = 0, 1, \ldots \infty$ and $m = 1, \ldots \infty$. The following Neumann function

$$\tilde{R}_{\gamma\beta} = \left(\frac{\partial\omega}{\partial\rho}\right)^{1/2}\frac{1}{\omega-\omega'}\left(\frac{\partial\omega'}{\partial\rho'}\right)^{1/2}\sqrt{\frac{\omega}{\omega'}}\sqrt{\frac{(\omega'-\omega_1)(\omega'-\omega_2)}{(\omega-\omega_1)(\omega-\omega_2)}} \tag{37}$$

can be seen to obey all the needed requirements.

We now summarize the construction of supersymmetry generators in Witten's theory using NS-R overlaps [3]. These give a natural form for the generators in which the equations necessary for supersymmetry are easily established. As such this offers an alternative to the bosonized spin field formulas. In addition, one has a precise derivation of the fermion emission vertex from Neveu-Schwarz and Ramond string overlaps.

One has a one parameter set of overlap equations which correspond to the parameter of the emission vertex. Consider the fermionic variables defined on the whole range $-\pi \leq \sigma \leq \pi$ by

$$\psi(\sigma) = \begin{cases} \psi_{-}(\sigma) & 0 \leq \sigma \leq \pi \\ \psi_{+}(-\sigma) & -\pi \leq \sigma \leq 0 \end{cases} \tag{38}$$

and similarly for $\chi$. We define the general NS-R overlap $|\,V_2^{NSR}(\theta)\rangle$ to obey the following equations

$$\psi^1(\sigma) + i\chi^2(\pi-\sigma) = 0, \quad -\pi \leq \sigma \leq \theta$$

$$\psi^1(\sigma) - i\chi^2(\pi-\sigma) = 0, \quad \theta \leq \sigma \leq \pi \tag{39}$$

Again due to a change of sign at $\sigma = \theta$ one has assured that the NS and R boundary conditions are obeyed, the above set of equations is consistent and has a unique solution in terms of a quadratic form. After a trivial twisting transformation these are seen to be related to the coefficients defining the fermion emission vertex [6]. The overlap equations are especially useful for discussing supersymmetry in the operator form of Witten's superstring field theory. One can write the transformations in the form

$$\delta \mid A_{NS}\rangle_1 = \bar{\epsilon}\, \hat{S}_{12}\rangle_1 \mid A^R\rangle_2$$

$$\delta \mid A_R\rangle_1 = \bar{\epsilon}\, \hat{S}^+_{12} \mid A^{NS}\rangle_2 \tag{40}$$

with the supersymmetry generators given by

$$\hat{S} \equiv {}_2\langle S_{RNS}\rangle_1 = {}_{32}\langle V_2^{NSR} \mid V_2^{NSNS}\rangle_{31} \tag{41}$$

and the conjugate

$$\hat{S}^+ = {}_2\langle S_{NSR}\rangle_1 = {}_{32}\langle V_2^{RNS} \mid X \mid V_2^{RR}\rangle_{31} \tag{42}$$

The invariance of the action under the supersymmetry transformations defined above goes as follows. First for the free part of the action one finds that the variation equals

$$\delta S_0 = 2\bar{\epsilon}\langle V_2^{NSR} \mid (Q^1_{NS} + Q^2_R) \mid A^{NS}\rangle \mid A^R\rangle \tag{43}$$

which is zero because the two string NS-R overlap vertex $\mid V_2^{NSR}\rangle$ is constructed to be BRST invariant. For the invariance of the interaction term one obtains necessary relationships between the NS-NS-NS and R-R-NS vertex operators. Name the condition is

$${}_{423}\langle V_3^{NSNSNS} \mid \hat{S}_{14} + {}_{143}\langle V^{RRNS} \mid \hat{S}^+_{24} = 0 \tag{44}$$

Here in the first NS-NS-NS vertex the $\hat{S}$ operator converts the configuration into a R-NS-NS overlap while in the second R-R-NS term one changes the second Ramond string (into NS) obtaining again a R-NS-NS overlap. Consequently by adjusting a sign one has the required cancellation. A more detailed analysis is given in [3].

## REFERENCES


1. E. Witten, *Nucl. Phys.* **B276** (1986) 291.
2. D. J. Gross and A. Jevicki, *"Operator Formulation of Interacting String Field Theory III: Superstrings,"* to appear in Nucl. Phys. B.
3. A. Jevicki and B. Sazdović, *"Supersymmetry in the Operator Formulation of Witten's Field Theory,"* Brown preprint HET-638 (1987).
4. K. Suehiro, Kyoto preprint KUNS 857 HE(TH) 87104; KUNS 875 July 1987.
5. S. Samuel, CCNY-HEP-8712.
6. Y. Kazama, CERN-TH-4518/86 (1986).

# GAUGE INVARIANCE, BRST INVARIANCE, AND WARD IDENTITIES, IN STRING FIELD THEORY*

S. P. de Alwis

Theory Group
Dept. of Physics
University of Texas
Austin, Texas 78712

Abstract

The derivation of Ward identities from BRST invariant gauge fixed versions of string field theory is discussed for both open and closed strings. However, in the latter case some problems are encountered in attempting to justify the gauge fixed action by the Batalin Vilkovisky method.

I will first briefly discuss the deriviation of Ward identities in Witten's open string field theory[1]. Then I will discuss some attempts at understanding closed string field theory.

In string theory the Ward identity is in effect the statement that spurious states (represented by commutators of the first quantized BRST charge with some vertex operator) decouple. In the first quantized version, where the perturbation expansion is given as a sum over Riemann surfaces, the Ward identity may be proved by conformal field theory techniques[2]. Now of course the demonstration by Giddings Martinec and Witten,[2] of the perturbative equivalence of quantum string field theory and the first quantized theory, would then imply that the Ward identities are valid in the former as well. However, a direct demonstration of this fact[3] is I believe instructive for two reasons. Firstly, it is at least formally non-perturbative and secondly, it gives some insight into the structure of the theory.

The gauge fixed action for Witten's open string field theory is given by[4]

$$I[\phi, B] = \frac{1}{2}\int \phi * Q\phi + \frac{1}{3}\int \phi * \phi * \phi + \int B_0 * b_0\phi. \tag{1}$$

In the above $\phi = \sum_{n=-\infty}^{+\infty} A_{n-\frac{1}{2}}$ and $A_{n-\frac{1}{2}} = \Phi_{n-\frac{1}{2}}^{\{\cdots\}} \left|\{\cdots\}n - \tfrac{1}{2}\right\rangle$ with $n-\frac{1}{2}$ being the ghost number (in the first quantized theory) and the ket vector is a state in the Fock space

*Invited talk given at the NATO advanced research workshop on superstrings, University of Colorado (Boulder) July 27 - August 1, 1987.

of the first quantized theory. $Q$ is the BRST charge of the first quantized theory. It is important to note that the Grassman parity of $\phi_{n-\frac{1}{2}}$ is fixed such that $\phi$ is Grassman odd. $I$ is invariant under the following BRST transformations

$$s\phi = Q\phi + \phi * \phi - b_0 B, \qquad sB = 0. \tag{2}$$

These transformations are nill-potent only on-shell.

The generating functional for Green's functions is

$$Z[J,L,K,] = e^{iW} = \int [d\phi][dB] \exp i\{I[\phi,B] + \int J * \phi + \int L * B + \int K * s\phi\} \tag{3}$$

where the measure $[d\phi] = \Pi \left[ d\Phi^{\{\cdots\}}_{n-\frac{1}{2}} \right]$ can be shown to be invariant[3,4] under (2). From the latter and the invariance of $I$ under (2), it is easy to see that,

$$\Omega Z[J,L,K] = 0 \tag{4}$$

where

$$\Omega \equiv \int \left( J * \frac{\delta}{\delta K} + K * [Q - i\frac{\delta}{\delta J}, F] \right)$$

and $F \equiv \delta I/\delta\phi$ with $\phi \to -i\delta/\delta J$, $B \to -i\delta/\delta L$. (4) is the Slavnov-Taylor identity of the theory. Note that since $s^2 = 0$ only on-shell $\Omega^2$ has, in addition to the usual $0(J^2)$ terms $0[(J,K) \times F]$ terms as well.

Now let $\phi_c$ be a classical string field and $\phi_0$ be a classical solution of the free field equation

$$L\phi_0 \equiv (L_0 - 1)\phi_0 = 0 \tag{5}$$

Where $L_0$ is the total hamiltonian for the first quantized theory for coordinates and ghosts. The S-matrix generating functional is then given by

$$S[\phi_0] = e^{\Sigma_{\phi_c}} Z[J,0,0,] |^{\phi_c \to \phi_0}_{J=0} \tag{6}$$

where $\Sigma_{\phi_c} = \int \phi_c * c_0 L \frac{\delta}{\delta J}$, and we also impose the gauge conditions $b_0\phi_0 = 0$, $\hat{Q}\phi_0 = 0$. In the above $(b_o)c_0$, is the zero mode of the first quantized (anti-) ghost, and $\hat{Q}$ is the piece independent of these zero modes in the BRST charge. $(Q = c_0 L + b_0 M + \hat{Q})$. Now we have

$$[\Sigma, \Omega] = \int \phi * c_0 L \frac{\delta}{\delta K,} \tag{7}$$

so that

$$[\Sigma, \Omega]\, Z[J,L,K] = i \int [d\phi][dB] \int \phi_c * c_0 L (Q\phi + \phi * \phi - b_0 B) e^{iI + \cdots}$$

In the limit $\phi_c \to \phi_0$ only the one-string poles contribute since $\int \phi_0 * cL(\cdots)$ vanishes away from these poles. If we ignore wave function renormalization, only the linear term contributes so that (using $[L,Q] = 0$, and $b_0\phi = 0$ inside the functional integral)

$$[\Sigma_{\phi_c}, \Omega]\, Z[J,K,L,] \Big|^{\phi_c \to \phi_0}_{J,K,L=0} = \Sigma_{s_0\phi_c} Z[J,K,L] \Big|^{\phi_c \to \phi_0}_{0} \tag{8}$$

where $s_0\phi_c = \hat{Q}\phi_c$.

The Ward identity that we seek is in effect the statement that $S[\phi_0]$ is invariant under $\phi_0 \to \phi_0 + s_0\psi$. Now using (8) with $\phi_0 \to \psi$ (with $L\psi = 0$) we find that

$$\begin{aligned} S[\phi_0 + s_0\psi] &= e^{\Sigma_{s_0\psi}} e^{\Sigma_{\phi_0}} Z[J,L,K]\,|_0 \\ &= e^{[\Sigma_\psi,\Omega]} e^{\Sigma_{\phi_0}} Z[J,L,K]\,|_0 \\ &= S[\phi_0] + \sum_{n=1}^{\infty} \frac{1}{n!} [\Sigma_\psi, \Omega]^n \, e^{\Sigma_{\phi_0}} Z\,|_0 \end{aligned}$$

In the sum above terms of the type $\cdots \Omega e^{\Sigma_{\phi_0}} Z|_0 = 0$, from (8) with $s_0\phi_0 = \hat{Q}\phi = 0$ and the Slavnov-Taylor identity (4). Terms of the type $\Omega \cdots e^{\Sigma_{\phi_0}} Z|_0 = 0$ because $\Omega = 0$ when $J = 0$. All other terms have at least one factor of $\Omega^2$ and vanish when $J = K = L = 0$ and the equation of motion is used inside the functional integral. If we ignore questions of regularization of divergences, it is possible to extend this argument beyond tree level but I will refer the reader to reference (3) for that.

Let us now discuss a covariant gauge invariant formalism for closed strings. The discussion presented here is applicable to both the Witten type of closed strings interaction[1,7] as well as to the Kyoto type[8]. Both have their problems as is evident from the comparison with the first quantized theory. The purpose here is however to try and elucidate the problems within the framework of the Batalin Vilkovisky[9] method of quantizing the theory.

A fully gauge invariant closed string field theory (with no constraints on fields or gauge parameters) may be formulated on the space $F_+ \otimes F_-$, where $F_\pm$ are the Fock spaces for right and left movers. The creation and annihilation operators are $a_n^\pm, b_n^\pm$ for the coordinates, ghosts and anti ghosts. The ghost zero mode vacuum is defined by

$$b_0{}^+ \,|\downarrow\downarrow\rangle = b_0{}^- \,|\downarrow\downarrow\rangle = 0$$

$$c_0{}^+ \,|\downarrow\downarrow\rangle = |\uparrow\downarrow\rangle\,, \text{ etc,}$$

and

$$\langle\uparrow\uparrow \,|\, \downarrow\downarrow\rangle = 1$$

The (first quantized) BRST charge is

$$\begin{aligned} Q &= c_0{}^+ L_+ + c_0{}^- L_- + b_0{}^+ M_+ + b_0{}^- M_- + \hat{Q}_+ + \hat{Q}_- \\ &= c_0 H + \frac{1}{2}\tilde{c}_0 K + \frac{1}{2} b_0 M + \tilde{b}_0 \tilde{M} + \hat{Q} \end{aligned} \tag{9}$$

where

$$\begin{array}{ll} H = L_+ + L_- & K = L_+ - L_- \\ 2c_0 = c_0{}^+ + c_0{}^- & \tilde{c}_0 = c_0{}^+ - c_0{}^- \\ b_0 = b_0^+ + b_0{}^- & \tilde{b}_0 = \frac{1}{2}(b_0{}^+ - b_0{}^-). \end{array}$$

It is useful now to introduce an oscillator basis and 2-string and 3-string vertex operators. The two vertex is

$$\langle V_{12}| \equiv_1 \langle\downarrow\downarrow|_2 \langle\downarrow\downarrow|_{12} \langle o|\, V_{12}$$

$$V_{12} = V_{12}{}^+ V_{12}{}^- \delta(p_1 + p_2)$$

$$V_{12}{}^\pm = (c_0{}^{\pm(1)} + c_0{}^{\pm(2)}) \exp \sum_{n>0} \left( a_n{}^{\pm(1)} a_n{}^{\pm(2)} + b_n{}^{\pm(1)} c_n{}^{\pm(2)} + b_n{}^{\pm(2)} c_n{}^{\pm(1)} \right)$$

Note that $\langle V_{21}| = -\langle V_{12}|$ and $\langle V_{12}|(Q_1+Q_2) = 0$. We introduce the bilinear product,

$$(A|B) \equiv \langle V_{12}|A\rangle_1|B\rangle_2$$

and the star product

$$|A*B\rangle_3 = \langle V_{123'}|V_{3'3}\rangle A\rangle_1|B\rangle_2$$

where the three string vertex is defined by[10]

$$\langle V_{123}| = \prod_{i=1}^{3} {}_i\langle\uparrow\uparrow|_i\langle o|V^+_{123}V^-_{123}\delta(p_1+p_2+p_3)$$

$V^{\pm}_{123}$ being 3-open string vertex operators for right and left movers. By the BRST invariance of these it follows that $< V_{123}|(Q_1+Q_2+Q_3) = 0$.

It should be noted that we take $|\downarrow\downarrow\rangle$, to be fermionic and the same for $|\uparrow\uparrow\rangle$, $\langle\downarrow\downarrow|$ and $\langle\uparrow\uparrow|$, but $|\uparrow\downarrow\rangle$ etc. are bosonic. Writing $g_A$ for the Grassman parity of a state $A$ we then have the following axioms for a closed string field theory.

$$\begin{aligned}
(A|B) &= (-1)^{g_A g_B}(B|A) \\
(QA|B) &= -(-1)^{g_A}(A|QB) \\
(b_nA|B) &= (-1)^{g_A}(A|b_{-n}B) \\
(c_nA|B) &= -(-1)^{g_A}(A|c_{-n}B \\
g_{A*B} &= g_A+g_B+1 \\
-Q(A*B) &= QA*B+(-1)^{g_A}A*QB \\
(A*B|C) &= (-1)^{g_A(g_B+g_C)}(B*C|A) \\
&= (-1)^{g_C(g_A+g_B)}(C*A|B)
\end{aligned} \tag{10}$$

Since we have not committed ourselves to a particular form of the star product, it is not necessarily commutative or associative. If for instance we had used the Witten open string 3-vertex for $V^{\pm}_{123}$ then our star product will be associative, but we will not need this in the following discussion.

Let us first take a look at the free closed string theory[10,11],

$$I_0 = (\bar{A}_0|QA_{-1}). \tag{11}$$

This has the following gauge invariances

a) $\delta_a\bar{A}_0 = Q\Lambda_{-1}$ , $\delta_a A_{-1} = 0$
b) $\delta_b\bar{A}_0 = 0$ , $\delta_b A_{-1} = Q\Lambda_{-2}$

Because of the large gauge invariance it can be shown that this has the correct closed string spectrum in spite of the apparent doubling of states. In fact the above action can be partially gauge fixed, without generating dynamical Faddeev-Popov ghosts[11], to the Neveu-West action[12]

$$I_0' = (\bar{A}_0|Q\tilde{b}_0A_0),$$

which has a gauge invariance $\delta\bar{A}_0 = Q\Lambda_{-1}$ only if the constraint $K \equiv L_+ - L_- = 0$ is imposed.

We will now consider the natural, unique, generalization of (11) to an interacting theory[13]

$$I = (\bar{A}_0|QA_{-1}) + \frac{1}{3}(A_{-1} * A_{-1}|A_{-1}) \tag{12}$$

Using our axioms it can be shown that $I$ has the following gauge invariances.

$$\begin{array}{ll} \text{a)} & \delta_a \bar{A}_0 = Q\Lambda_{-1} \quad , \quad \delta_a A_{-1} = 0 \\ \text{b)} & \delta_b \bar{A}_0 = A_{-1} * \Lambda_{-2} - \Lambda_{-2} * A_{-1} \quad , \quad \delta_b A_{-1} = Q\Lambda_{-2}. \end{array} \tag{13}$$

The gauge algebra closes as follows

$$\left[\delta_b^{\Lambda}, \delta_b^{\Lambda^1}\right] = -\delta_a{}^{[\Lambda,\Lambda^1]}$$

$$[\delta_a^{\Lambda}, \delta_b^{\Lambda^1}] = 0 \quad , \quad [\delta_a^{\Lambda}, \delta_b^{\Lambda^1}] = 0. \tag{14}$$

As with the open string field theory the Faddeev-Popov method cannot be used to gauge fix the theory, but one may ask, in analogy with the gauge fixed open string action [1] whether the natural generalization $\bar{A}_0 \to \bar{\phi} = \sum_{-\infty}^{+\infty} \bar{A}_n, \quad A_{-1} \to \phi = \sum_{-\infty}^{+\infty} A_{+n}$ with the action

$$\begin{aligned} I &= (\bar{\phi}|Q\phi) + \frac{1}{3}(\phi * \phi|\phi) \\ &+ (B|b_0\phi) + (C|\phi - \tilde{b}_0\bar{\phi}) \end{aligned} \tag{15}$$

gives a sensible quantum theory[14]. Indeed using the axioms (10) we may show that (15) has a BRST invariance under the following transformations.

$$s\phi = -Q\phi + \tilde{b}_0 c \quad , \quad s\bar{\phi} = Q\bar{\phi} - \phi * \phi - b_0 B - \tilde{b}_0 C$$

$$sB = sC = 0. \tag{16}$$

As in the open string case these transformations are nill-potent only on-shell, and we may still derive Ward identities by following the method discussed earlier for the open string case, with some obvious (and trivial) modifications. However it seems that (unlike in the open string case) this gauge fixed action cannot be justified by the Batalin Vilkovisky ($BV$) method[9].

In the $BV$ procedure one first enlarges the set of classical fields $A \to \{\phi_i\} \supset A$ and adds anti-fields $\{\phi_i{}^*\}$, which have opposite Grassman parity $g_{\phi_i}* = g_{\phi i} + 1$. Gauge fixing corresponds to choosing the so-called gauge fermion $\psi = \psi(\phi_i)$ and then putting $\phi_i{}^* = \partial\psi/\partial\phi_i$. The gauge quantum theory is then defined by the partition function

$$Z = \int [d\phi_i] e^{\frac{i}{\hbar} W[\phi_i, \phi_i^*]} |_{\phi_i^* = \partial\psi/\partial\phi_i} \tag{17}$$

provided that $W$ is a solution of

$$\sum_i \frac{\partial^\ell W}{\partial\phi_i} \frac{\partial^r W}{\partial\phi_i *} = i\hbar \sum_i \frac{\partial^2 W}{\partial\phi_i \partial\phi_i *} \tag{18}$$

with the boundary condition

$$W[\phi_i, \phi_i{}^* = 0] = I_{cl}[A] \tag{19}$$

where $I_{c\ell}[A]$ is the classical action. Actually in many cases the left and right hand sides of (18) are separately zero the LHS bing the statement of BRST invariance of the action and the RHS the statement of the BRST invariance of the measure.

In our closed string theory we may introduce the following fields

$$\phi_- = A_{-1} + A_{-2} + A_{-3} + \cdots \quad , \quad \bar{\phi}_- = \bar{A}_0 + \bar{A}_{-1} + \bar{A}_{-2} + \cdots \tag{20}$$

and anti-fields

$$\phi_-{}^* \equiv \bar{\phi}_+ = \bar{A}_{+1} + \bar{A}_{+2} + \cdots \quad , \quad \bar{\phi}_-^* \equiv \phi_+ = A_0 + A_{+1} + A_{+2} + \cdots \tag{21}$$

so that $\phi = \phi_- + \phi_+ = \phi_- + \bar{\phi}_-{}^*$ and $\bar{\phi} = \bar{\phi}_- + \bar{\phi}_+ = \bar{\phi}_- + \phi_-{}^*$. Our initial ansatz for $W$ is simply (15) without the last two (gauge fixing) terms.

$$\begin{aligned} W &= (\bar{\phi}|Q\phi) + \frac{1}{3}(\phi * \phi|\phi) \\ &= (\phi_-{}^*|Q\phi_-) + (\bar{A}_0|QA_{-1}) + (\bar{\phi}_-|Q\bar{\phi}_-{}^*) \\ &\quad + \frac{1}{3}\left((\phi_- + \bar{\phi}_-{}^*) * (\phi_- + \bar{\phi}_-{}^*)|\phi_- + \bar{\phi}_-{}^*\right) \\ &= W(\phi_-, \bar{\phi}_-; \phi_-{}^*, \bar{\phi}_-{}^*) \end{aligned} \tag{22}$$

Using ghost number conservation it is easily seen that this satisfies the boundary condition

$$W(\phi_-, \bar{\phi}_-; 0, 0) = I_{c\ell}(A_-, \bar{A}_0)$$

Also using our axioms (12) it is possible to see that

$$\begin{aligned} \sum_i \frac{\delta^\ell W}{\delta\phi_i} \frac{\delta^r W}{\delta\phi_i^*} &= \sum_{n<0} \left( \frac{\delta^\ell W}{\delta A_{-n}} | \frac{\delta^r W}{\delta A_{-n}^*} \right) \\ &\quad + \sum_{n\leq 0} \left( \frac{\delta^\ell W}{\delta \bar{A}_{-n}} | \frac{\delta^r W}{\delta \bar{A}_{-n}^*} \right) \\ &= (Q\phi|\phi * \phi) = 0. \end{aligned} \tag{23}$$

It is also easily seen that the measure term is zero. Thus we have a solution to the $BV$ equation (18). It should be emphasized here that if we had started with the partially gauge fixed theory (with $\phi = \tilde{b}_0\bar{\phi}$) then we would not have had a solution of the $BV$ equation.

Let us now discuss the gauge fixing terms. Introduce new fields $\hat{\phi}_+$ , $\tilde{\phi}_+$ (with $\hat{\phi}_+(\tilde{\phi}_+)$ Grassman odd (even)) and their anti-fields $\hat{\phi}_+{}^*$ and $\tilde{\phi}_+{}^*$ and add the terms $(B_+|\hat{\phi}_+{}^*) + (C_+|\tilde{\phi}_+{}^*)$ to the action. The $BV$ equation (18) is still satisfied. Introducing the "gauge fermion" $\psi = -(\hat{\phi}_+|b_0\phi_-) - (\tilde{\phi}_+|\phi_- - \tilde{b}_0\bar{\phi}_-)$, the gauge fixing conditions become,

$$\begin{aligned} \hat{\phi}_+{}^* &= \frac{\partial^r\psi}{\partial\hat{\phi}_+} = b_0\phi_- \\ \tilde{\phi}_+{}^* &= \frac{\partial^r\psi}{\partial\tilde{\phi}_+} = \phi_- - \tilde{b}_0\bar{\phi}_- \\ \phi_-{}^* &= \frac{\partial^r\psi}{\partial\phi_-} = b_0\hat{\phi}_+ - \tilde{\phi}_+ \\ \bar{\phi}_-{}^* &= \frac{\partial^r\psi}{\partial\bar{\phi}_-} = \tilde{b}_0\tilde{\phi}_+ \end{aligned} \tag{24}$$

Writing $|\phi\rangle = \gamma\,|\downarrow\downarrow\rangle + \cdots$ and $\gamma = \gamma_+ + \gamma_-$ where $\gamma_+$ contains the components with non-negative ghost number, the gauge fixed action of the $BV$ formalism becomes (after imposing (24) and integrating out the auxiliary fields $B_+$ and $C_+$)

$$\begin{aligned} I_{G.F.} &= (\gamma\,|\downarrow\downarrow\rangle\,|c_0{}^+c_0{}^- H\gamma\,|\downarrow\downarrow\rangle) + (\tilde{\phi}_+|K\gamma_-\,|\downarrow\downarrow\rangle) \\ &+ ((\gamma_-\,|\downarrow\downarrow\rangle + \tilde{b}_0\tilde{\phi}_+) * (\gamma_-\,|\downarrow\downarrow\rangle + \tilde{b}_0\tilde{\phi}_+)|\gamma_-\,|\downarrow\downarrow\rangle + \tilde{b}_0\tilde{\phi}_+) \end{aligned} \qquad (25)$$

The gauge fixed kinetic term in the above agrees with that given by the analysis of the free theory in reference (9) and thus yields the right propagator. However the interaction term is not of the expected form (15). In particular $\tilde{\phi}_+$ is a non-propagating field and it is not clear how the perturbation expansion of the first quantized theory can be reproduced.

This problem is not just an artifact of our formulation. The latter is a general framework which encompasses different versions of the star product. The advantage of working with our doubled representation is that it enables us to find a solution to the $BV$ equation, and indeed it is difficult to see how the gauge fixing of a closed string field theory can be performed without using this formulation.

It is of course quite likely that the problem encountered in quantizing closed string field theory pointed out here, is related to the well-known difficulties of the (present version) of the theory related to its failure to give a cover of moduli space. However at present I do not know how to clarify this connection.

## Acknowledgments

This work was done partly in collaboration with L. Mezincescu and M. Grisaru (reference 3) and partly in collaboration with N. Ohta (references 10,11,13,14). I am grateful to the organizers of the workshop, P.G.O. Freund and K.T. Mahantappa, for the opportunity to present these results. This work was supported in part by the Robert A. Welch Foundation and NSF grant No. PHY-86-05978.

## References

1. E. Witten, Nucl. Phys. B268, (1986) 253.

2. D. Friedan, E. Martinec and S. Shenker, Nucl. Phys. B271, (1986) 93.

3. S.P. de Alwis, M.T. Grisaru, L. Mezincescu, "Ward identities in open string field theory" preprint, BRX-TH-237, UMTG-141, UTTG-20-87, to be published in Phys. Lett. B.

4. M. Bochichio, Phys. Lett. 188B (1987) 332, 192B (1987).

5. C.B. Thorn, Nucl. Phys. B287 (1987) 61.

6. A. Bogojevic, preprint BROWN-HET-615 (1987).

7. J. Lykken and S. Raby, Nucl. Phys. B278 (1986) 258.

8. H. Hata, K. Itoh, T. Kugo, H. Kunimoto and K. Ogawa, Phys. Lett. 172B, 195 (1986) and Kyoto preprint RIFP-673 (1986).

9. I.A. Batalin and G. Vilkovisky, Phys. Rev. D28 (1983) 2567.

10. S.P. de Alwis and N. Ohta, Phys. Lett. 173B (1986) 388

11. S. P. de Alwis and N. Ohta, preprint CERN-TH 4798/87, UTTG-19-87.

12. A. Neveu and P.C. West, Nucl. Phys. B278 (1986) 601.

13. S.P. de Alwis and N. Ohta, Phys. Lett. 188B 425.

14. S.P. de Alwis and N. Ohta, work in progress.

# CLOSED STRINGS AND WITTEN'S STRING FIELD THEORY

Joel A. Shapiro

Department of Physics and Astronomy
Rutgers, the State University of New Jersey
P. O. Box 849, Piscataway, New Jersey, 08855-0849

This talk is an introduction to work done in the last year[1–3], concerning closed strings as they enter an open string theory. That open string theory contained closed string singularities has been known since 1970[4]. In 1971, Lovelace pointed out[5] that these could be a healthy property of the theory if only we were willing to increase our dimensionality to 26. The detailed form of the closed string contributions was given, in terms of an open string – closed string transition[6] operator $\check{\Upsilon}_0$, by early 1973. The transition operator $\check{\Upsilon}_0$ was found by reexpressing the non-planar one-loop diagram for open strings in terms of a tree graph containing the open-closed transition twice and a modified closed string propagator. It did not have a straight-forward functional integral or overlap interpretation. All of this work was done in the "old covariant" formalism before the string paper of GGRT[7], which was crucial in converting people to the light-cone.

In the old covariant approach, it is essential that operators which describe string transitions reflect and transmit the gauges $L_n$, which act as Ward identities eliminating spurious states. The necessity of worrying about Ward identities comes from chopping the world sheet into pieces, vertices and propagators, and expressing amplitudes as matrix elements of products of these operators. Modern two-dimensional approaches have avoided these complications by concentrating on the on-shell amplitudes for physical states. These are represented holistically by functional integrals of physical state vertex operators, in a manner which is independent of conformal transformations. When we consider field theory, however, it is not enough to have only transitions between on-shell physical states; the factorization of amplitudes implicit in a covariant field theoretic formulation introduces unphysical states. Then the expressions we find for transition amplitudes are not unchanged under conformal transformations. We need to make choices of the conformal structure of the regions upon which our transition amplitudes and propagators are defined, which is really a choice of gauge. All of our considerations will be done in fixed gauge, generally with a world sheet metric taken to be $\delta_{ij}$.

With the coming of the light-cone path integral formulation of strings, it was possible to consider the transition amplitude as an overlap integral between states of closed and open strings, giving the operator $\Upsilon_\perp$, an operator acting on states described in terms of only transverse modes. In light-cone gauge there are no unphysical states, but one pays the price of lost manifest Lorentz invariance. $\Upsilon_\perp$ is rather different in form from $\check{\Upsilon}_0$ as well. The fact that two very different operators are found in these two approaches to describe the same transition operator is not surprising — they are only supposed to give the same physical amplitudes, and so can differ by conformal transformations, acting as gauges, on each of the external states in the covariant picture. The old $\check{\Upsilon}$ conformal frame is very convenient for describing the closed string states when the open strings are described *à la* Witten.

We first turned[1] to the extention of this light-cone treatment to the modern, covariant treatment using ghost fields. From the light-cone we take the notion that the transition is represented simply by the ends of the string joining, and that otherwise the transition is simply an overlap of the string state described in open string oscillators with one described in terms of closed string oscillators. Note that questions of string widths do not complicate the two point function here. We evaluated the overlap in the ordinary coordinates using the methods of Mandelstam[8]. That is, if $\langle 0|\,\Upsilon\,|open\rangle\,|closed\rangle$ represents the overlap amplitude for a state $|open\rangle$ in the open string Hilbert space to be the same configuration as $|closed\rangle$, then the operator

$$\langle 0|\,\Upsilon e^{-H_o T_o} \otimes e^{-H_c T_c}\,|open\rangle\,|closed\rangle$$

is a functional integral $\int \mathcal{D}\, X_\mu(\rho)$ over

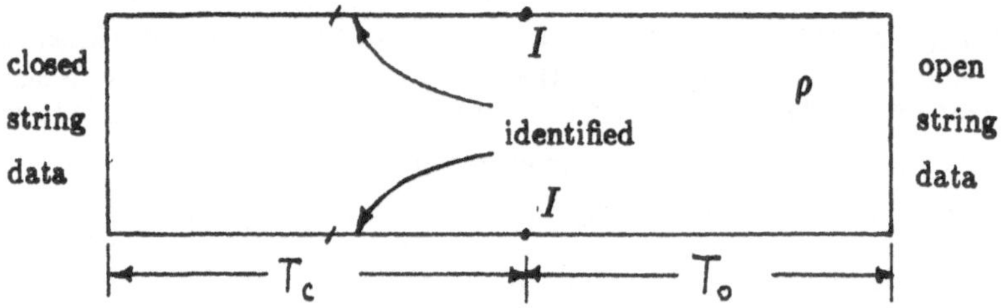


where the open and closed string data are specified by boundary conditions on the right and left boundaries, respectively. The horizontal boundaries to the left of the interaction points $I$ are identified; there is no constraint to the right. The evaluation for ordinary coordinates $\Upsilon_0$ is the same as for the transverse modes in the evaluation of $\Upsilon_\perp$. In order to include ghost modes, we pursued two approaches. The first uses bosonized ghosts, $\phi$, for which the calculation is the same as for ordinary oscillators, except that the ghost field is quantized on a compact space. This means that the zero mode momentum is quantized, and the ghost field can have winding number about the closed string. Thus $\phi$ is not single-valued at the interaction point, where there is a curvature $\delta$ function. Despite these complications, evaluating the functional integral is not difficult, and leads to a full transition operator $\Upsilon$ in terms of bosonized ghosts.

The second approach uses fermionic ghosts. The operator $\Upsilon_0$ is an exponential of a quadratic in the coordinate oscillators $a$, $A$, $\tilde{A}$, where $A$ and $\tilde{A}$ are the left and right moving modes for the closed string. Assuming the fermionic ghosts enter the exponential as a bilinear in $(b, B, \tilde{B}) \otimes (c, C, \tilde{C})$, and requiring BRST invariance

$$\langle 0|\,\Upsilon(Q_{open} + Q_{closed}) = 0$$

uniquely defines the full operator $\Upsilon$, which includes the ghosts. To find the expression

$$\Upsilon = \Upsilon_0 \exp\left\{c^j_n b^k_m M^{jk}_{nm}\right\},$$

(where $j, k$ take the values open, left, and right), we needed to find the Ward identities of $\Upsilon_0$, which give the reflection of the Virasoro gauges

$$\langle 0| \Upsilon_0(L^j_{-n} + \sum M^{jk}_{nm} L^k_m + k^j_n) = 0.$$

These are required in order that a spurious state in one sector is reflected and transmitted into gauge transformations, which vanish provided we have physical states in the other sectors. Then the reflection (and transmission) coefficients $M$ are just the required coefficients for the $cb$ term in the exponential in $\Upsilon$.

In a light-cone formulation there are only physical states, so the question of Ward identities of $\Upsilon$ had, apparently, not been addressed in the 13 years since $\Upsilon_\perp$ was derived. For the older $\check{\Upsilon}$ they were given[9] within a year. One can use these (with sign mistake corrected) to find a BRST invariant $\check{\Upsilon}$ from $\check{\Upsilon}_0$ with fermionic ghosts, proceding exactly as for $\Upsilon$.

Another thing we could not find in the literature is an explicit check that $\Upsilon_0$ gives the same transition amplitude for one closed string to $N$ open strings as $\check{\Upsilon}_0$, when all the external particles are on-shell and physical. As $\check{\Upsilon}_0$ was derived from factorization to give the correct answer for this process, this is a check on $\Upsilon_0$. We have performed this check when all the open strings are tachyons and the closed string is either a tachyon or a graviton. We have also checked that if one evaluates the non-planar loop using two $\Upsilon$ transitions, the additional ghost modes bounce back and forth over the ordinary closed string propagator to give the extra two powers of the partition function which had to be supplied by hand in the modified closed string propagator of Ref. 6.

Getting back to our evaluation of the two forms of $\Upsilon$ using bosonic and fermionic ghosts, we may ask if the two forms agree. They do, as far as the identification of bosonized and fermionic ghost is uniquely defined, which is only up to a constant. We were somewhat surprised, however, to find that the normalizing constant $N$ in $b \sim N : e^{\phi} :$ must be chosen differently in the closed and open string operator expansions. There are also some subtleties involving the Klein factors between the three sets of fermionic operators.

There is one more thing I would like to mention before refocussing on the Witten frame. The existence of a closed string — open string transition means that in general there will be mass shifts in each sector due to the intermediate states in the other. These will be strong effects if the particles are degenerate. In current, nonorientable open string models, the open string states degenerate with closed string states, being at odd excitation levels, have no singlet component, so there can be no mixing between degenerate states. For the $U(N)$ model prevalent when this mixing was first considered[10] there is a singlet photon, which mixes with the antisymmetric tensor $B_{\mu\nu}$. In the treatment by Cremmer and Scherk, who used $\check{\Upsilon}_0$, the open string self energy operator has the form

$$\langle 0| a_1^{\mu} a_1^{\prime\nu} \left(g_{\mu\nu} - \frac{p_\mu p_\nu}{p^2}\right) \Pi(p^2),$$

just as in ordinary field theory. In contrast to the operator $\check{\Upsilon}_0$, $\Upsilon$ does not involve the zero modes of the ordinary momentum. This has an interesting effect on the way the photon and the antisymmetric tensor massless states mix to give masses. For $\Upsilon$, the $p \cdot a$ terms cannot arise, and the way gauge invariance is maintained is with an admixture of ghost modes

$$\langle 0| a_1 \cdot a_1' + c_1 b_1' + c_1' b_1 - 2c_1 c_1' b_0 b_0'/p^2 .$$

I now want to change the frame for describing open strings, considering the corresponding operator in the conformal frame of Witten's string field theory. In Witten's theory, the fundamental interaction is the star product in which the right half of one string coincides and annihilates the left half of another, attaching the two remaining halves to form an outgoing string. This leads to the Witten vertex shown in Fig. 1. Two of these can be joined to form the non-planar one-loop correction to the open string propagator, which is shown in Fig. 2a. We call this region the turret. The propagation times $T_1$ and $T_2$ of the intermediate open string states are integrated independently over $[0, \infty]$. When both times are small, there is a singular limit from which the closed string intermediate state singularities arise. As the times shrink to zero (Fig. 2b), the effective interaction between the two external propagators is confined to a point — for each half the effect is to impose identification on the opposite halves of the strings, and to have some interaction operator act at the midpoint. If we cut the identified ends but remember that these points are really identified, we get the parameterization of the top half of the turret which we call the $\hat{\rho}$-plane, Fig. 3.

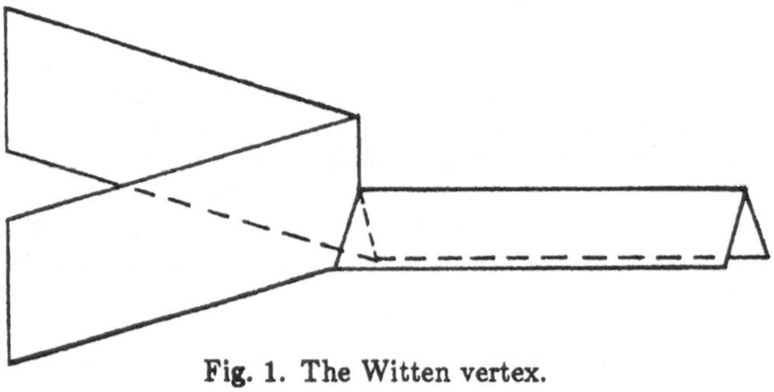

Fig. 1. The Witten vertex.

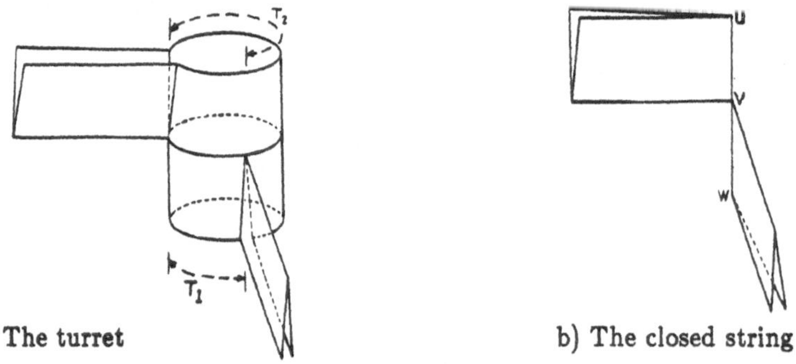


a) The turret

b) The closed string limit

Fig. 2. The non-planar one-loop propagator correction in Witten's open string theory, exhibiting the closed string singularity.

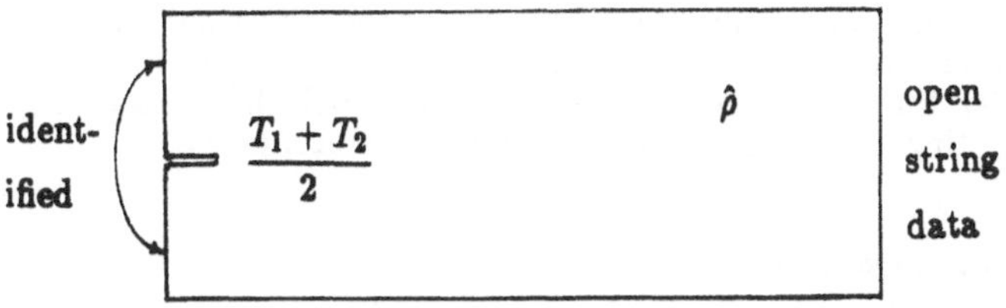


Fig. 3. Half of the turret, coordinatized by $\hat{\rho}$

In the limit $T_1, T_2 \to 0$, we have a half strip with initial open string data out near $\infty$ on the right, open string edges, and on the left the two halves are identified, which is what is called the identity state of the open string. At its midpoint, some interaction vertex representing the other half of the turret acts.

This is a very different picture than what happens in the light cone picture or in the old Koba-Nielsen picture. In old language, the graph is an annulus with one external particle on each boundary. The propagation times are

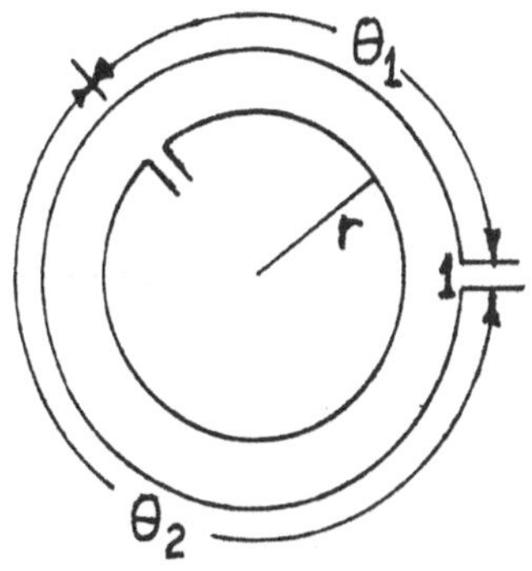


$$T_i = \frac{-\pi}{\ln r}\theta_i,$$

which gives the two open string cuts as $r \to 1$, but clearly shows the radial propagation of a closed string as $r \to 0$.

In the light cone picture, the nonplanar loop graph with two open string intermediate states is shown in Fig. 4a. As the propagator times approach 0, we do not get a singular surface. Instead, we find that we have only gone part way towards the closed string, to the $T_c = 0$ limit of another graph (Fig 4b), the diagram with one intermediate closed-string state propagator. The closed string poles come from the $T_c \to \infty$ limit of this graph. The closed string singularity is again manifestly the propagation of a closed string. These two separate light cone graphs are different regions of moduli space for the single Riemann surface which is unified in the Witten turret diagram. For Witten's theory, we see that the closed string limit does not appear so obviously to factor into transition–propagator–transition, although it does fall into two pieces connected by closed string data.

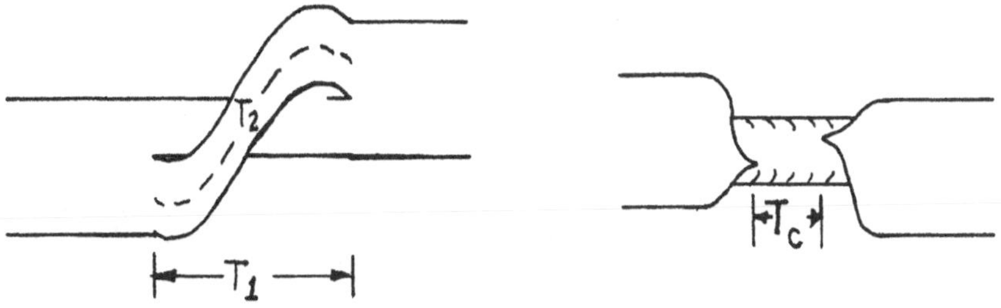


a) two open string intermediate state

b) one closed string intermediate state

Fig. 4. The corresponding diagrams in the light-cone formulation.

We first examined the question of evaluating the functional integral on the surface which is the limit as $T_1, T_2 \to 0$, with closed string data specified at the string midpoint. First, consider the tachyon closed string state, on-shell. It is clearly a source of momentum acting at the midpoint of the identity state, hence

$$\langle\langle 0|\hat{\Upsilon}\,|tachyon; p\rangle =$$

$$\langle\uparrow|\exp\left\{-\frac{1}{2}\sum_{n=1}^{\infty}(-)^n\frac{\alpha_n^2}{n} - \sum_{n=1}^{\infty}(-)^n c_n b_n\right\}\exp\left\{\sqrt{2}p\cdot\sum(-)^n\frac{\alpha_{2n}}{n}\right\},$$

where the second exponential is a midpoint vertex insertion acting on the preceding identity state. $\hat{\Upsilon}$ is the transition operator described in the $\hat{\rho}$ picture. This state is exactly what one gets after applying $Q$ to the closed string ground state described by Strominger[11]. To find the general form of $\hat{\Upsilon}$, note that it is conformally related to the previous light-cone picture in the $\rho$ plane. We may define an operator $\langle 0|\,\Upsilon\hat{C}$, where $\hat{C}$ generates a conformal transformation on the open string data. The conformal transformation is

$e^{\rho} = \cosh\hat{\rho}$,

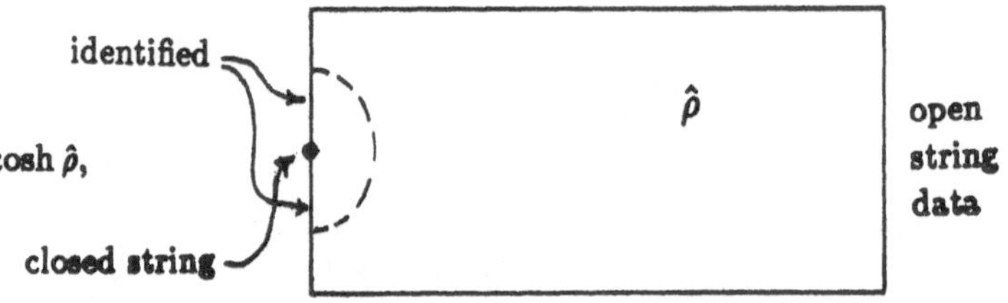


and leaves $\rho \to \infty$ (essentially) unchanged. Thus $\hat{C}$ is generated by $L_n, n \geq 0$, which has no effect on on-shell physical states. In fact, we have not described how we are to specify the closed string data: for this it is useful to use a different parameterization $\breve{\rho}$, with

$e^{-\rho} = \cosh\breve{\rho}$,

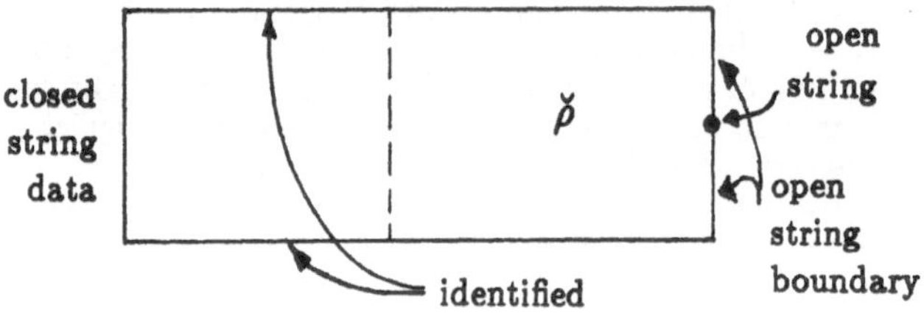


which leaves the closed string data in the same form as used in evaluating $\Upsilon$. Thus we define an operator

$$\langle 0|\,\hat{\Upsilon} = \langle 0|\,\Upsilon\breve{C}\hat{C}$$

where $\breve{C}$ is a conformal transformation on the closed string data implementing $\rho \to \breve{\rho}$. This operator corresponds to doing the functional integral on the right half of the $\hat{\rho}$ diagram, and the left half of the $\breve{\rho}$ diagram, with values matched along the dashed line.

The full expression for $\hat{\Upsilon}$ is derived in Ref. 2. Including ghosts, we find

$$\begin{aligned}\langle 0|\,\hat{\Upsilon} = 2^{2p^2-5}\,\langle 0|\exp\Big\{-\sum_{m=1}^{\infty}\Big[&(-)^m\frac{1}{m}\big(\frac{1}{2}a_m^2-\sqrt{2}p\cdot a_{2m}\big)+(-)^m c_m b_m\\ &+\frac{1}{m}A_m\cdot\tilde{A}_m+C_m\tilde{B}_m+\tilde{C}_m B_m+(-)^m\big(\tilde{C}_m-C_m\big)\big(\tilde{B}_0-B_0\big)\Big]\\ -\sum_{m=1}^{\infty}\sum_{n=0}^{\infty}\Big[&\frac{(-)^m}{m}\eta_n^{2m}\,a_n\cdot\big(A_m\,i^n+\tilde{A}_m\,i^{-n}\big)\\ &+2(-)^n\big(\eta_{2n+1}^m-\eta_{2n-1}^m\big)c_m\big(i^m\tilde{B}_n+(-i)^m B_n\big)\\ &+\frac{1}{4}(-)^m\big(\eta_{n+1}^{2m}-\eta_{n-1}^{2m}\big)\big(i^n\tilde{C}_m+(-i)^n C_m\big)b_n\Big]\Big\}.\end{aligned}$$

where $\eta_n^k$ is defined by

$$\left(\frac{1+x}{1-x}\right)^k=\sum_{n=0}^{\infty}\eta_n^k x^n.$$

This operator does not completely solve the factorization of the one-loop diagram, however, because we have not written the diagram as $\Upsilon^\dagger\Delta_{cl}\Upsilon$. In fact, we have done a functional integral over the region which is the limit $T_1, T_2 \to \infty$, ignoring corrections of order $T_1, T_2$ coming from the difference between our limiting region and the turret. These will give subleading behavior and even unphysical poles.

In Ref. 3, we look at the full one-loop graph. In order to evaluate the amplitude, we look for a conformal map from a topologically identical cylinder to the turret. We choose to represent the cylinder as a rectangle $x \in [-\frac{1}{2}, \frac{1}{2}], y \in [-\tau/4, \tau/4]$ in the $z = x + iy$ plane, with $x = -\frac{1}{2}$ identified with $x = +\frac{1}{2}$. The external particles $A$ and $B$ are mapped from points on a smooth boundary. A small contour around $A$ gives a change of $\pi$ in $\operatorname{Im}\rho$, which is the width of the string. Thus $d\rho/dz = \pm 1/z$ at $A$ and $B$. Around the points which are mapped into Witten midpoints, a small circle gets mapped into the sum of three semicircles, which requires $\rho - \rho_0 \sim (z - z_0)^{3/2}$. Thus $d\rho/dz$ has square-root branch points at $V$ and $W$.

The open string boundary conditions on the top and bottom edges lead us to consider the double of the turret and the rectangle. Imposing symmetry of reflection about the lower boundary and periodicity with period $\tau$ gives the required conditions on the open string boundaries, and we see that $d\rho/dz$ is a doubly periodic function with two square-root branch cuts. Thus its square on the doubled region gives a quadratic differential; we have a doubly periodic function with 2 double poles and 4 simple zeros. The only thing which we cannot evaluate analytically is where the two zeros $W$ and $V$ must be to insure that the midpoints are really at the midpoint of each string. To examine the closed string poles, we expand about the limit as $T_1, T_2 \to 0$, $(\tau \to \infty)$. The conformal map shows that $T_1, T_2 \sim q^{1/4}$, where $q = e^{i\pi\tau}$. There is, of course, a factor $q^{p^2/2}$ from the momentum term in the Neumann function. Thus the integration over modulus $q \to 0$ can give us the closed string poles. In general, the ratio of $\theta$ functions appearing in the Neumann functions are power series in $q$, but the mapping $z \to q$ is an expansion in $q^{1/4}$. This gives us a surprising effect: for off-shell

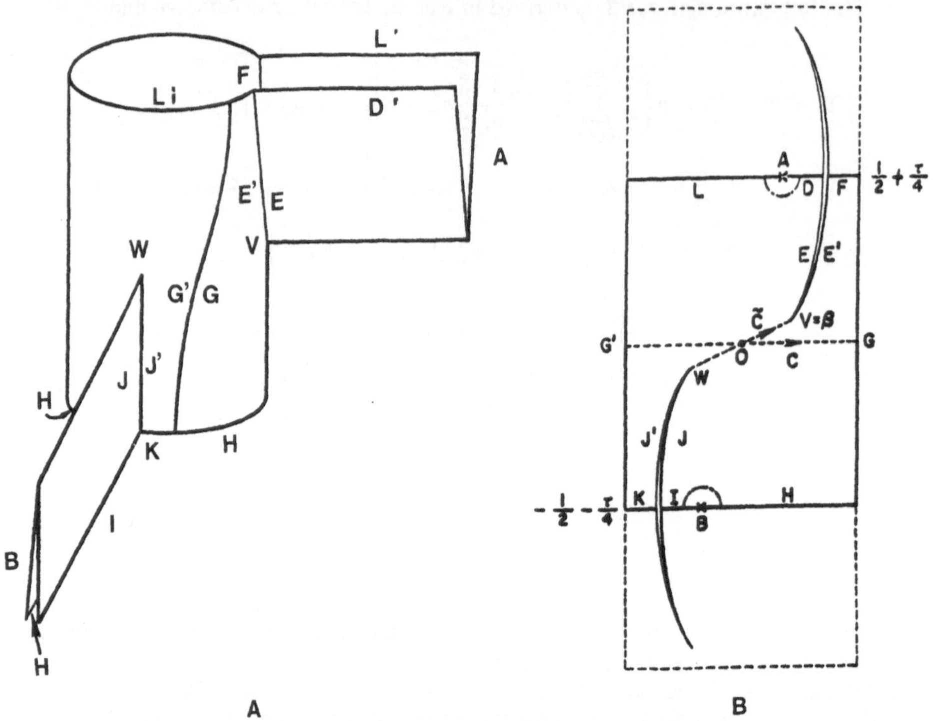


**Fig. 5. The turret is topologically a cylinder, represented in $z$ uy the solid rectangle, with left and right edges identified. The map $z \to \rho$ has a derivative with simple poles at $A$ and $B$, and square-root branch points at $V$ and $W$. We consider the doubled region (dashed lines) on which $(d\rho/dz)^2$ is doubly periodic.**

or unphysical open string external states, there are poles in addition to the expected ones. We have known for 15 years that the physical poles are at

$$\alpha' M^2_{phys} = 4n - 4, \qquad n = 0, 1, 2, \cdots,$$

but we find far more. For off-shell unphysical open string external states, the 1PI open string self energy has poles 8 times as dense

$$\alpha' M^2 = \frac{n}{2} - 4, \qquad n = 0, 1, 2, \cdots.$$

The additional poles are due to two effects. If we take the diagram and factor it along two closed string cuts, we see that the singularity as $T_1, T_2 \to 0$ is given by an hourglass-shaped propagator for the closed string, as shown in Fig. 6. This region can be conformally mapped to a cylinder, so for physical states, the detailed shape does not matter and we have an ordinary closed string propagator. But unphysical states can scatter off the central region after having propagated for only half the usual time. Thus we get

$$\langle 1 | \, e^{-tH/2} V e^{-tH/2} \, | 2 \rangle \sim e^{-t(E_1+E_2)/2}.$$

As scattering off V does not need to be diagonal or respect $L_0 = \tilde{L}_0$, we get powers $(e^{-t})^{n/2}$ instead of the usual $(e^{-t})^{2n}$, giving the integer poles.

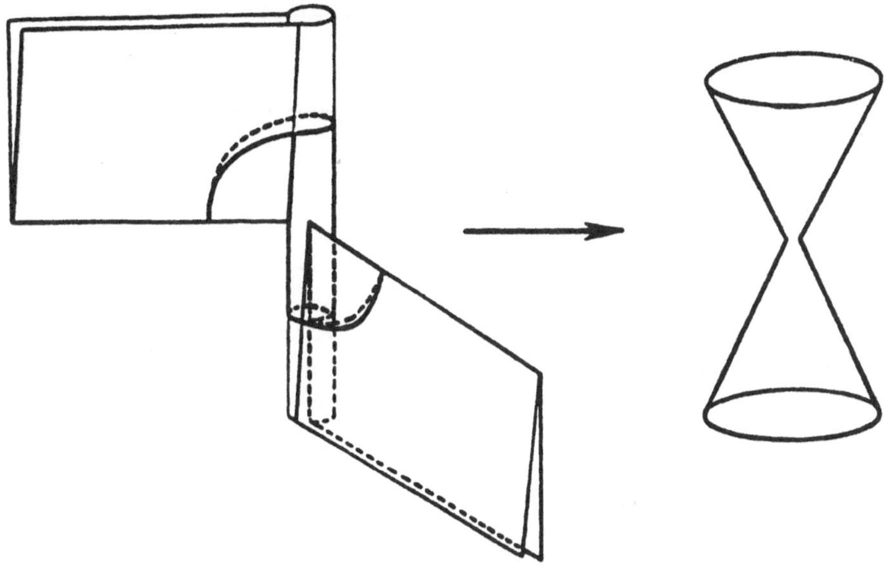

Fig. 6. The singular structure in the closed string limit can be seen in the middle piece when the tank diagram is factored into three parts by cutting as shown along two loops. The middle piece is a region which becomes two cones interacting at their apexes. It is conformally equivalent, but not metrically equivalent, to the standard closed string propagator, a right circular cylinder.

The half integer poles have a different cause — they arise from the precise definition of truncating the external legs of the diagram at $T = 0$. The powers of $q^{1/4}$ come from expanding $\rho(z)$, where $\mathrm{Re}\,\rho = T_A$ is held fixed. Things are considerably improved if we take the origin of $\hat{\rho}$ at the furthest point of the cylinder,

$$\hat{\rho} = \rho + \frac{T_1 + T_2}{2},$$

and hold $\mathrm{Re}\,\hat{\rho} = \hat{T}_A$ fixed. Then the expansion becomes a series in $q^{1/2}$, and we have only the integer poles. The change of $\rho$ would make no difference for the on-shell behavior of the open strings, even for unphysical states, because they depend only on the $T \to \infty$ residue, which is unaffected by the small $(T_1 + T_2)/2$ shift. However, in evaluating the one-open-string irreducible diagram, we are instructed to amputate the external propagators, fixing $T = 0$ rather than $\hat{T} = 0$, and so this diagram does have the half-integer poles.

In fact, we discovered a much simpler context in which this effect arises — the four-point tree amplitude in a light-cone type calculation (Fig. 7). If $\alpha_2 \neq \alpha_3$, $s$-channel and $t$-channel poles come from $\rho_- - \rho_+ \to \pm\infty$. Then the amplitude has the usual poles even for off-shell particles. But when $\alpha_2 = \alpha_3$, the $t$-channel poles arise from $\rho_- - \rho_+ \to 0$. The amplitude with external propagators cut off becomes (with all $\alpha$'s equal)

$$\int_0^1 dx\, x^{-t-2}(1-x)^{-s-2}(1+\sqrt{x})^{-2\sum_{i=1}^4 (p_i^2 - 1)},$$

and we have poles at $t = (n/2) - 1$, when $\sum_1^4 (p_i^2 - 1) \neq 0$.

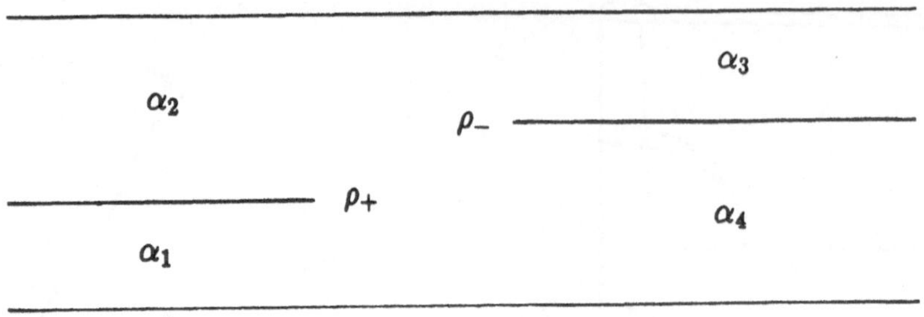


Fig. 7. The tree four-point function in the light-cone. The $s$ channel poles always come from the region $\rho_- - \rho_+ \to \infty$. When $\alpha_2 \neq \alpha_3$, the $t$ channel poles come from $\rho_- - \rho_+ \to \infty$, but when $\alpha_2 = \alpha_3$, they come from $\rho_- - \rho_+ \to 0$. When the external legs are truncated, one finds only the usual poles in the former case, but half-integer poles in the latter.

## REFERENCES


1. Joel A. Shapiro and Charles B. Thorn, "BRST Invariant Transitions between Closed and Open Strings", Phys. Rev. **D36** (1987) 432.
2. Joel A. Shapiro and Charles B. Thorn, "Closed String – Open String Transitions and Witten's String Field Theory", Phys. Lett. **B194**, (1987) 43.
3. Daniel Z. Freedman, Steven B. Giddings, Joel A. Shapiro, and Charles B. Thorn, "The Nonplanar One-Loop Amplitude in Witten's String Field Theory", to be published, Nucl. Phys. **B**.
4. Frye and Susskind, Phys. Lett. **31B** (1970) 589; Gross, Neveu, Scherk and Schwarz, Phys. Rev. **D2** (1970) 697; Hsue, Sakita and Virasoro, Phys. Rev. **D2** (1970) 2857.
5. C. Lovelace, Phys. Lett. **34B** (1971) 500.
6. E. Cremmer and J. Scherk, Nucl. Phys. **B50** (1972) 222; L. Clavelli and J. A. Shapiro, Nucl. Phys. **B53** (1973) 490.
7. P. Goddard, J. Goldstone, C. Rebbi and C. B. Thorn, Nucl. Phys. **B56**, (1973) 109.
8. S. Mandelstam, Nucl. Phys. **B64**, (1973) 205.
9. L. Brink, D. Olive, and J. Scherk, Nucl. Phys. **B61** (1973) 173.
10. E. Cremmer and J. Scherk, Nucl. Phys. **B72** (1974) 117.
11. A. Strominger, private communication.

# GEOMETRIC STRING FIELD THEORY:

# DERIVING STRING THEORY FROM FIRST PRINCIPLES

Michio Kaku

Physics Dept., City College of CUNY
New York, N.Y. 10031

## I. INTRODUCTION

String field theory [1] gives us the greatest promise of a non-perturbative approach to string theory, but it is still plauged by a frustratingly large number of arbitrary conventions, folklore, and rules-of-thumb. Originally, string field theory was introduced to give us a unifying, comprehensive formulation of string theory. Ironically, it has spawned not one, but two covariant theories. This is because both the light cone and the BRST theories [2-4] are gauge fixed theories. Our goal, by contrast, should be to follow the example of general relativity and Yang-Mills theory, which can be uniquely derived from two simple geometrical principles!

If we were creating a geometric string field theory from scratch, what features would we want? In anology to general relativity and Yang-Mills theory, we might desire the following:

1) the entire theory should be derived from a *single local gauge group.* This gauge group should have a simple physical, intuitive interpretation, which allows us to derive the theory from first principles.

2) this gauge group should be defined in *loop space*, i.e. the space of physical, space-time strings, without any reference to any prefered parametization length, and without any reference to a background metric for space-time.

3) Following gauge theory, the derivation of the action should follow the sequence:

$$\textit{Gauge Group} \rightarrow \textit{Connection} \rightarrow \textit{Curvatures} \rightarrow \textit{Action}$$

4) the connections should transform as irreps of the gauge group.

5) the unique action, built out of curvatures, should yield both BRST theories when gauge-fixed.

6) the theory must be totally free of artifacts arising from gauge-fixing and fixing a parametrization, i.e. Faddeev-Popov ghosts, ghost number, midpoints, string lengths.

7) the theory must explain the rather baffling presence of the four-string interaction, which occurs in one BRST theory [3,4] but not the other [2].

Geometric string field theory [5-8] was created to satisfy all these principles.

## II. THE UNIFIED STRING GROUP

We begin by defining {C} to be the space of oriented physical, space-time strings. The strings {C} are not parametrized at all. Three strings form a *triplet* if they can be arranged as in fig. 1.* Any two strings C and $\bar{C}$ are *conjugate* to each other if they appear in the same triplet. (A string C has an infinite number of conjugates.) Let us associate an operator $L_C$ for each string C. Let $\tilde{C}$ represent the string C with reversed orientation. An anti-triplet differs from a triplet because all the string orientations are reversed. Let us now define the *string group* with the following algebra:

$$SG: \quad [L_{C_1}, L_{C_2}] = f^{C_3}_{C_1, C_2} L_{C_3}$$

We sum over repeated indices. Let us define:

$$f^{C_1 C_2 C_3} = \begin{cases} +1 \text{ if } C_i \textit{ form a triplet} \\ -1 \text{ if } \textit{they form an anti-triplet} \\ 0 \textit{ otherwise} \end{cases}$$

The geometric vertex function is the simplest of all possible vertex functions: ±1, 0:

$$f^{C_3}_{C_1 C_2} = f^{C_1 C_2 \tilde{C}_3}$$

It is anti-symmetric in 1 and 2. *The Jacobi identities are satisfied even if the joining point for several strings do not match.* For fig. 2, the identities yield 1 - 1 + 0 = 0:

$$f^{C_4}_{[C_1, C_2} f^{C_5}_{C_3], C_4} = 0$$

Notice that the number of elements in this algebra is the set of all possible loop space strings. Thus, it is much "larger" than the Kac-Moody algebras!

Let us now construct representations and invariants for the string group. We begin with a contravariant field $\phi^C$ which transforms as the adjoint representations of the group. Then a specific representation of the algebra is:

$$L_C = f^{C_2}_{C C_1} \phi^{C_1} \frac{\delta}{\delta \phi^{C_2}}$$

Covariant and contravariant tensors transform as:

$$\delta\phi^C = f^{C}_{C_1 C_2} \phi^{C_1} \Lambda^{C_2}$$

$$\delta\phi_C = f^{C_2}_{C_1 C} \Lambda^{C_1} \phi_{C_2}$$

(Reversing the orientation of a string is the raising and lowering operator.)

Let an element of this group be represented as:

$$O = e^{\Lambda^C L_C}$$

Let it act on a set of basis states $| e^C \rangle$ and its dual $\langle e_C |$:

* Figures appear at the end of this article.

$$<e_{C_1}|e^{C_2}> = \delta_{C_1}^{C_2}; \quad <e_{C_1}| O |e^{C_2}> = O_{C_1}^{C_2}$$

We define the action of the operator $L_C$ on the basis states as:

$$L_{C_1} |e^{C_2}> = f_{C_3 C_1}^{C_2} |e^{C_3}>$$

where: $1 = \sum_C |e_C><e^C|$. O forms the adjoint representation which preserves the metric $\delta_{C_1}^{C_2}$. In this particular representation, the generators have the following representation:

$$L_C = |e^{C_2}> f_{C_2 C}^{C_3} <e_{C_3}|$$

Let $|\phi> = \phi_C |e^C>$. Then the invariants under the SG are $\phi_C \phi^C = <\phi|\phi>$ and $\phi^{C_1}\phi^{C_2}\phi^{C_3} f_{C_1 C_2 C_3}$.

It is crucial to note that the invariants $\phi^2$ and $\phi^3$ can be constructed without any mention of any parametrization or background metric.

Let us now introduce a parametrization for our strings. We want a theory which is based on the physical, invariant length of a string:

$$L = \int_0^{\pi\alpha} \sqrt{X'^{\mu}(\sigma)^2}\, d\sigma$$

and not its parametrization length, which is a fiction:

$$\pi\alpha = \int_0^{\pi\alpha} d\sigma$$

The fictitious parametrization length can be changed at will at any time.

Let X represent a parametrization of the the string C. Thus, X and Y belong to the same equivalence class (i.e. parametrizae the same loop space string) if they satisfy:

$$X \sim Y \text{ if } X(\sigma) = Y(\sigma) + \int d\sigma\, \epsilon(\sigma) Y'(\sigma)$$

Lastly, we now replace $L_C$ with an infinite number of parametrized generators:

$$L_C \rightarrow L_X = L_{X_\mu(\sigma_1) X_\mu(\sigma_2) \cdots X_\mu(\sigma_N)}$$

where we have divided up the parametrization evenly into N points, $N \rightarrow \infty$.

Now we can write down the universal string group, which is the semi-direct product of the Virasoro group and the string group:

$$USG: \begin{cases} [L_\sigma, L_\rho] = f_{\sigma\rho}^{\omega} L_\omega \\ [\epsilon(\sigma) L_\sigma, L_X] = \epsilon(\sigma) X'_{\mu\sigma} \dfrac{\delta}{\delta X_{\mu\sigma}} L_X \\ [L_X, L_Y] = f_{XY}^{Z} L_Z \end{cases}$$

$$f_{\sigma\rho}^{\omega} = \delta(\rho-\sigma)\delta'(\rho-\omega) + 2\delta(\omega-\sigma)\delta'(\sigma-\rho)$$

Quite simply, the USG is the group which maps the points along a string C into itself and into all its conjugates:

$$USG = \begin{cases} C \to C \\ C \to \bar{C} \end{cases}$$

When it is combined with space-time supersymmetry, it becomes the unified string group. Our fundamental claim is that the action for geometric string field theory is the unique solution to the following physical principles:

**Global Symmetry: the theory should propagate ghost-free irreps of Diff(S), where S = [0, $\pi$ ] or $S_1$.**

**Local Symmetry: it should be invariant under local USG transformations.**

The tensor calculus arising from this group differs considerably from that of the usual Lorentz group, which explains why string field theory appears on the surface to be so different from the usual point-particle field theory. Surprisingly, the tensor calculus for the USG is so stringent and restrictive that only one unique action is possible!

## III. THE ACTION

Unfortunately, the representations of the unified string group USG are largely unknown. We have only been able to identify three:

(1) the adjoint representation of the SG, given by field functionals $\phi^X = \phi(X)$

(2) Verma modules V: $|e^\alpha\rangle = L_{-\alpha_1}^{\mu_1} L_{-\alpha_2}^{\mu_2} \cdots L_{-\alpha_n}^{\mu_n} |0\rangle$ labelled by Greek letters $\alpha$, $\beta$, $\gamma$. (We will use $\mu$, $\nu$, $\lambda$ for Lorentz indices.)

(3) the string representation S, represented by $\phi^\sigma$, where Greek letters near the end of the alphabet $\sigma$, $\rho$, $\omega$ range from 0 to the string parametrization length $\pi\alpha$.

The variation of a field $\phi^{A,X}$ where A can represent either V or S, is given by:

$$U \ \phi^{A,X} \ U^{-1} = \phi^{A,X+\delta X} + f^{\ A,X}_{B,Y;C,Z} \ \Phi^{B,Y} \ \Lambda^{C,Z}$$

where $\delta X^{\mu\sigma} = \epsilon^\sigma(X) X'^{\mu\sigma}$ and all f's are the structure constants of the unified string group USG for different representations. They are all given in [5-6]. For example, the symmetrized $f^{X_1, X_2, X_3}$ is given by:

$$f^{X_1, X_2, X_3} = \prod_{r=1}^{3} \prod_{0 \leqslant \bar{\sigma}_r \leqslant l_r} \delta(X_r(\bar{\sigma}_r) - X_{r-1}(\pi |\alpha_{r-1}| - \bar{\sigma}_{r-1})$$

where $\bar{\sigma} = \sigma + \zeta^\sigma(X)$. Notice that functionally integrating over $\zeta^\sigma(X)$ sums over all possible equivalence classes of the same physical triplet configuration. *It is essential to note that the structure constant includes, as special cases, both the vertex of [2] and [3-4].* Also $f^{\alpha\beta\gamma} = \langle e^\alpha | \langle e^\beta | \langle e^\gamma | V\rangle$ where $\langle e^\alpha |$ is defined only in the ghost sector of our geometric vertex $|V\rangle$.

*The "ghost sector" has an elegant interpretation in geometric string field theory: it simply corresponds to the tangent space of the geometric theory,* i.e. the orbital indices A of our fields $\phi^{X,A}$.

$$Ghost\ Sector \ \rightarrow \ Tangent\ Space$$

The important point to note is that the "ghost vertex," which must be postulated in the

BRST formalism, is a direct consequence of group theory, arising as the unique Clebsch-Gordon coefficient of the tensor product of three Verma modules.

Because the gauge parameter $\Lambda^{A,X}$ is a functional of the string X, derivatives like $\partial_{\mu\sigma} = \frac{\delta}{\delta X_\mu(\sigma)}$ do not transform properly under the unified string group USG. This forces us to introduce two distinct connection fields, one for the SG and one for Diff(S). This is remarkably similar to the connection fields $A_\mu^a$ for SU(N) Yang-Mills and $\omega_\mu^{ab}$ for gravity:

$$SU(N): \ A_\mu^a \quad \rightarrow \quad SG: \ A_\sigma^\alpha$$
$$Diff(M_4): \ \omega_\mu^{ab} \quad \rightarrow \quad Diff(S): \ \omega_{\mu\sigma,\alpha}^\beta$$

where the usual point particle fields are on the left, string fields are on right.

As in Yang-Mills and gravity, we now have two covariant derivatives:

$$D_\mu = \partial_\mu + A_\mu^a \rightarrow D_\sigma = \nabla_\sigma + A_\sigma^\alpha$$
$$\nabla_\mu = \partial_\mu + \omega_\mu^{ab} \rightarrow \nabla_{\mu\sigma} = \partial_{\mu\sigma} + \omega_{\mu\sigma,\alpha}^\beta$$

where $\nabla_\sigma \equiv \partial_\sigma + \frac{1}{2} L_\sigma$; $\partial_\sigma \equiv (-i\,\partial_{\mu\sigma} + X'_{\mu\sigma})^2$ and where $\partial_\sigma$ acts on the radial index X and $L_\sigma$ acts on the orbital index A.

The connection $A_\sigma^\alpha$ transforms under the string group as:

$$\delta A_\sigma^\alpha = (\nabla_\sigma \Lambda)^\alpha + (A_\sigma \times \Lambda)^\alpha - (\Lambda \times A_\sigma)^\alpha$$

The curvatures associated with these derivatives are:

$$F_{\sigma\rho} = [D_\sigma, D_\rho] = D_\sigma A_\rho + A_\sigma \times A_\rho - (\sigma \leftrightarrow \rho)$$
$$R_{\mu\sigma,\nu\rho,\alpha}^\beta = [\nabla_{\mu\sigma}, \nabla_{\nu\rho}]_\alpha^\beta$$

(where we will often delete V indices for clarity, and we use $\times$ to represent the structure constant of the SG.)

To multiply tensors under local Diff(S), we need a tensor calculus with a string vierbein. We need to generalize the vierbein $e_\mu^a$ of general relativity:

$$e_\mu^a \quad \rightarrow \quad e_{\mu\sigma}^{\nu\rho}$$

where the lower indices $\mu$ ( or $\mu\sigma$ ) transform nonlinearly under Diff($M_4$) ( or Diff(S) ) while the upper indices a ( or $\nu\rho$ ) transform linearly in the *tangent space* of O(3,1) (or Diff(S) ). The measure of integration now becomes a string density:

$$\det |e_\mu^a| dx \quad \rightarrow \det |e_{\mu\sigma}^{\nu\rho}| \prod_{\mu,\sigma} dX^{\mu\sigma}$$

Covariant and contravariant tensors must now transform as:

$$dX^{\mu\sigma} = \frac{\partial X^{\mu\sigma}}{\partial \bar{X}^{\nu\rho}} d\bar{X}^{\nu\rho}; \qquad \frac{\delta}{\delta X^{\mu\sigma}} = \frac{\partial \bar{X}^{\nu\rho}}{\partial X^{\mu\sigma}} \frac{\delta}{\delta \bar{X}^{\nu\rho}}$$

Actually, the curvature tensor and string density are invariant under general co-ordinate transformations $\delta X^{\mu\sigma} = \Lambda(X)^{\mu\sigma}$, which is much larger than the subgroup of local Diff(S): $\delta X^{\mu\sigma} = \epsilon^\sigma(X) X'^{\mu\sigma}$. To restrict ourselves to Diff(S), we set:

$$\nabla_{\mu\sigma} e_{\nu\rho}^{\lambda\omega} = 0; \quad R_{\mu\sigma,\nu\rho,\alpha}^\beta = 0$$

This determines the connection field $\omega_{\mu\sigma,\alpha}^{\beta}$ in terms of the vierbein, and the vierbein in terms of a vector field, i.e. $e_{\mu\sigma}^{\nu\rho}(\zeta^{\omega}(X))$.

Writing down the final action is complicated by the peculiar Clebsch-Gordon coefficients of the USG. For the Lorentz group, for example, constant tensors are $\delta_{\mu\nu}$, $\epsilon_{\mu\nu\lambda\kappa}$, and the Dirac matrix $(\gamma^{\mu})_{\alpha\beta}$. This means that $\Box$, $\not\partial$, and $F_{\mu\nu}^{2}$ are all invariant. Surprisingly, $\partial_{\sigma}^{2}$ and $F_{\sigma\rho}^{2}$ are *not* invariant under the USG, because $\delta^{\sigma\rho}$ is not a tensor. A true constant tensor under the USG is a "Dirac-like" matrix $\gamma^{\sigma}$ found in the tensor decomposition of $\mathbf{S}\otimes\mathbf{V}\rightarrow\mathbf{V}$.

$$(\gamma^{\sigma})^{\alpha\beta} = \langle e^{\alpha} | \gamma^{\sigma} | e^{\beta} \rangle$$

Our next choice for an action is: $F\ F^{*} = F_{\sigma\rho}\,\epsilon^{\sigma\rho\omega\theta}\,F_{\omega\theta}$, where $\epsilon^{\sigma\rho\omega\theta} = \gamma^{\sigma}\gamma^{\rho}\gamma^{\omega}\gamma^{\theta}$, but this is a total derivative. Thus, the *unique* choice for the action is the Chern-Simons form associated with $F\ F^{*}$:

$$L = A_{\sigma}\times(\nabla_{\rho}A_{\omega}) + \frac{2}{3}A_{\sigma}\times A_{\rho}\times A_{\omega}$$

(where we have again suppressed $\mathbf{V}$ indices.) The action functional now becomes:

$$Z = \int DA_{\sigma}^{\alpha}(X)\,D\zeta^{\sigma}(X)\,e^{i\int \det|e|\,L\,DX}$$

## IV. THE INTERPOLATING VERTEX

We now explicitly calculate the vertex function for the theory by calculating it in a specific gauge, i.e. when the strings have arbitrary parametrization length $\pi\alpha_r$:

$$\prod_{r=1}^{3}\ \prod_{0\leqslant\sigma_r\leqslant l_r}\left\{X_{\mu}^{r}(\sigma_r) - X_{\mu}^{r-1}(\pi|\alpha_{r-1}| - \sigma_r)\right\}|V\rangle = 0$$

where $\sum_{r=1}^{3}\alpha_r = 2l_2$; $\alpha_3\leqslant 0$; $\alpha_{1,2}\geqslant 0$; $0\leqslant\sigma_r\leqslant\pi|\alpha_r| = \pi(l_r + l_{r+1})$.

To calculate the Neumann representation of this vertex, we need an explicit conformal map which reproduces the correct topology. Let us now construct an explicit representation of the interpolating vertex by writing down the conformal transformation which takes us from the upper half plane to the multi-sheeted $\rho$ plane [7] (fig. 3):

$$\rho(z) = \alpha_1\log(z-1) + \alpha_2\log z + \sum_{r=1}^{3}\beta_r\log\left|(az^2 + bz + c)^{1/2} + a_r z + b_r\right|$$

$$a_1 = \frac{\alpha_1^2-\alpha_2^2+\alpha_3^2}{2\alpha_1};\quad a_2 = \frac{-\alpha_1^2+\alpha_2^2+\alpha_3^2}{2\alpha_2};\quad a_3 = -\alpha_3$$

$$b_1 = \frac{\alpha_1^2+\alpha_2^2-\alpha_3^2}{2\alpha_1};\quad b_2 = -\alpha_2;\quad b_3 = \frac{\alpha_1^2-\alpha_2^2-\alpha_3^2}{-2\alpha_3}$$

$$a = \alpha_3^2;\quad b = \alpha_1^2-\alpha_2^2-\alpha_3^2;\quad c = \alpha_2^2;\quad \beta_r = -\alpha_r$$

We take the cut from $\rho(z_0)$ to $\rho(\bar{z}_0)$ to be vertical. We take the square root to be + (-) for $z\rightarrow\infty\,(-\infty)$. It is tedious but straightforward to show:

$$\frac{d\rho}{dz} = -\alpha_3\frac{(z-z_0)^{1/2}(z-\bar{z}_0)^{1/2}}{z(z-1)}$$

Fortunately, it is possible to explicitly invert this map by introducing an infinite set of numbers $a_n^{(r,j)}$. We find [7]:

$$N_{mn,rs} = \sum_{j=1}^{2} \frac{nm\ \alpha_j}{n\ \alpha_s + m\ \alpha_r}\ a_n^{(r,j)}\ a_m^{(s,j)}$$

$$N_{m\,0,rs} = \sum_{j=1}^{2} \frac{\alpha_j}{\alpha_s}\ a_m^{(r,j)}\ c_s^{(j)}$$

We can check that in the light cone limit $l_2 \to 0$ we recover the vertex of [9]. In the symmetric limit where $\alpha_r \to \pm 1$, we obtain a new power series for the Neumann functions, which should be equivalent to [10]. (There are minor corrections to the vertex if we incorporate ghost states $c$ , $\bar{c}$ and if we shift the origin of our axis from $\rho(z_0)$ with $L_0$ factors.)

To show the vertex satisfies BRST invariance, we must show:

$$\sum_{r=1}^{3} \frac{Q_r}{\alpha_r} | V > = 0$$

Using contour integrals, we use the techniques of [9,11-13] to show:

$$\left\{ \left[ \frac{5D}{48} - \frac{29}{24} - \frac{3}{2} \right] \frac{dc\,(z_0)}{dz} + \frac{\rho_1}{\rho_0} \left[ \frac{5D}{72} - \frac{65}{36} \right] c\,(z_0) \right\} | V > = 0$$

which is only satisfied if D=26.

## V. THE FOUR STRING INTERACTION

Let us now fix the gauge. In addition to invariance under the unified string group, the action is invariant under:

$$\delta A_\sigma^\alpha = \Sigma_\sigma^\alpha ; \qquad (\gamma^\sigma)_{\alpha\beta}\, \Sigma_\sigma^\beta = 0$$

This allows us to reduce out the $\sigma$ index and replace $A_\sigma^\alpha$ with $\phi^\alpha$. Next, we must break local Diff(S), which can be done by using the important identity:

$$SG = \frac{USG}{Diff\ (S)}$$

We can fix local Diff(S) invariance by fixing the vierbein $e_{\mu\sigma}^{\nu\rho}$ in two ways: (1) we can leave the parametrization length $\pi\alpha$ arbitrary but fix the internal parametrization of $\sigma$ between 0 and $\pi\alpha$. This particular choice for the Verma module allows us to derive an extra Osp(D,2/2) symmetry [14-16], or (2) we can fix $\pi\alpha$ as well as the internal parametrization. We adopt the second choice here.

The action in the interpolating gauge becomes:

$$L = <\phi|_\alpha Q\ |\phi>_\alpha + \frac{2}{3} <\phi|_{\alpha_1} | <\phi_{\alpha_2} | <\phi_{\alpha_3} |\ | V_{\alpha_1, \alpha_2, \alpha_3} >$$

where $Q = \gamma^\sigma \nabla_\sigma$. We can always change the parametrization length by the following:

$$U_{\alpha_1,\alpha_2} |\phi>_{\alpha_2} = |\phi>_{\alpha_1} ; \quad U_{\alpha_1, \alpha_2} = e^{\sum_{n=1}^{\infty} \zeta_n (L_n - L_{-n})}$$

This action can always be brought back to the midpoint gauge by the following:

$$<\phi|_{\alpha_1}<\phi_{\alpha_2}|<\phi|_{\alpha_3}|V_{\alpha_1\alpha_2\alpha_3}> = <\phi|_\alpha<\phi|_\alpha<\phi|_\alpha \prod_{i=1}^{3} U_{\alpha,\alpha_i} \; |V_{\alpha_1\alpha_2\alpha_3}>$$

$$= <\phi|_\alpha<\phi|_\alpha<\phi|_\alpha|V_{\alpha\alpha\alpha}>$$

The final step in the derivation of the midpoint gauge is the proof that the irreducible Verma module can always be reprented by the "ghost number minus one-half projection" of the BRST field $\Phi(X, \theta, \bar{\theta})$ [ see 5-6].

By constrast, we note that our endpoint gauge, although it uses the same vertex function as [3,4], differs significantly from the latter formalism because our endpoint gauge *does not need an extra integration over* $\alpha$ *because of the presence of the vierbein.* Notice that our Green's functions differ crucially from [3,4]:

$$< \phi(X_{\alpha_1}), \phi(Y_{\alpha_2}) > = U_{\alpha_1\alpha_2} < \phi(X_{\alpha_2}), \phi(Y_{\alpha_2}) >$$

In [3,4], the Green's functions replaces $U_{\alpha_1\alpha_2}$ with $\delta_{\alpha_1\alpha_2}$. When calculating N-point functions, the Feynman rules for the endpoint gauge, although they use the same vertex function, differ crucially because of the presence of vierbein $U(\zeta^\sigma)$ factors. The Feynman rules for the interpolating gauge are:

$$Feynman\ Rules: \begin{cases} Vertex \rightarrow & |V_{\alpha_1\alpha_2\alpha_3}> \\ Propagator \rightarrow & U_{\alpha_i\alpha} D_\alpha U^{-1}_{\alpha_j\alpha} \end{cases}$$

for arbitrary $\alpha_i$. When explicitly cancelling these U factors, we find that we can trivially go back to the midpoint gauge. However, when we remove these U factors and go the endpoint gauge, a remarkable series of identities for the Neumann functions which allows us to *derive the four-string interaction.* We find that the four-string interaction does not have to be included in the action if we use the vierbein; it is a gauge artifact, the counterpart of the four-fermion instantaneous Coulomb term found in QED. The details of the proof are rather involved, and are presented in [8], where we show that the t-u graphs in the midpoint, interpolating, and endpoint gauge are related by a unified string group transformation as follows (see fig. 4):

$$A_{tu} = <V_{\alpha\alpha\alpha}|D_{5,\alpha}(t)|V_{\alpha\alpha\alpha}> + <V_{\alpha\alpha\alpha}|D_{7,\alpha}(u)|V_{\alpha\alpha\alpha}>$$

$$= <V_{\alpha_1\alpha_4\alpha_5}|D_{5,\alpha_5}(t)|V_{\alpha_2\alpha_3\alpha_6}> + <V_{\alpha_2\alpha_4\alpha_7}|D_{7,\alpha_7}(u)|V_{\alpha_1\alpha_3\alpha_8}>$$

$$+ \int_{\alpha_1-|\alpha_4|+2\delta}^{\alpha_1+|\alpha_4|-2\delta} <V_{\alpha_1\alpha_4\alpha_X}|V_{\alpha_2\alpha_3\alpha_X}> \mu \, d\alpha_X$$

where $|\alpha_3| \geqslant \alpha_1, \alpha_2 \geqslant |\alpha_4|$ and $-\alpha_5 = \alpha_6 = \alpha_1 - |\alpha_4| + 2\delta$ $-\alpha_7 = \alpha_8 = \alpha_2 - |\alpha_4| + 2\delta$, for the t-scattering $1 + 4 \rightarrow 5 \rightarrow 2 + 3$ and u-scattering $2 + 4 \rightarrow 7 \rightarrow 1 + 3$. When $\delta = 0$, the last term becomes the usual light cone four-string interaction [1]. When $\delta \neq 0$, then we have the four-string interaction for the interpolating gauge. When $|\alpha_i| = \alpha$ and $\delta = \alpha$, then we have the vanishing of the four-string interaction in the midpoint gauge:

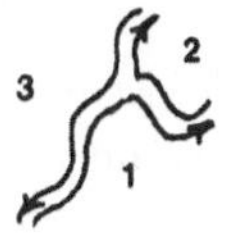


Fig.1

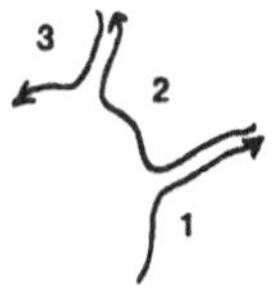


Fig.2

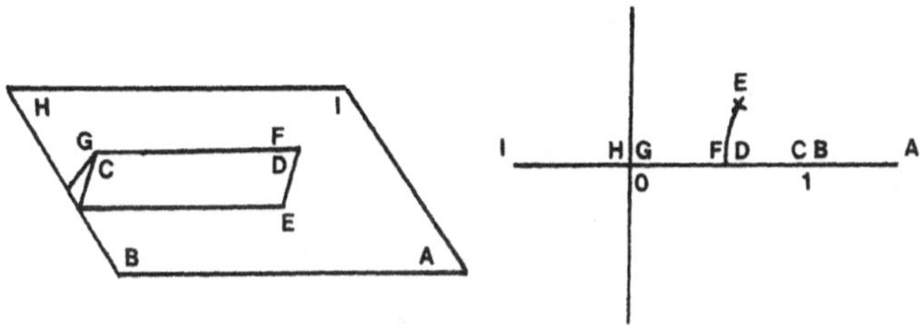


Fig. 3

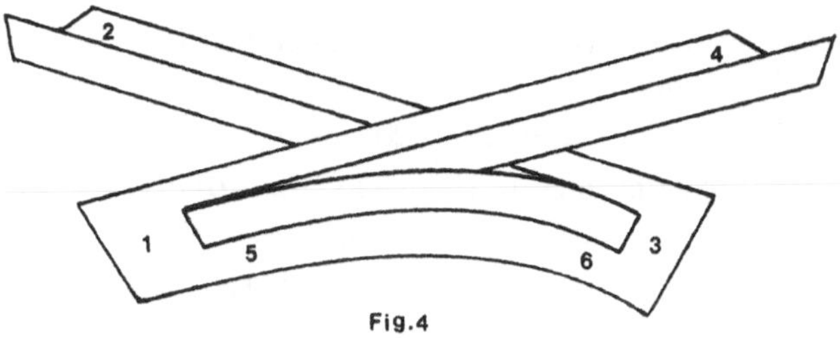


Fig.4

$$\begin{cases} \delta = 0 & \textit{endpoint gauge} \\ \delta \neq 0 & \textit{interpolating gauge} \\ \delta = \alpha \ \ (|\alpha_i| = \alpha) & \textit{midpoint gauge} \end{cases}$$

The four-string interaction, similar to the case of the four-fermion Coulomb term, arises because the propagator D is not invariant under unified string group gauge transformations. Precisely as in QED, we can always transform away certain propagators by a gauge transformation by eliminating its "time" dependence.

In summary, we have shown how to derive the geometric theory from two physical principles and how to fix the gauge to derive both BRST theories [2-4]. This is not surprising, because BRST by itself is not a physical principle. As an added bonus, the four-string interaction emerges as a gauge-artifact of the geometric theory. The four-string interaction in the interpolating gauge smoothly interpolates from the midpoint to the endpoint gauge.


### Acknowledgments

We are happy to acknowledge fruitful conversations with B. Sakita and S. Samuel. This work was supported in part by NSF-PHY-82-15364 and CUNY-FRAP-RF-13873.


### References


1. M. Kaku and K. Kikkawa, Phys. Rev. D10 (1974) 1110, 1823. See M. Kaku, Int. Jour. of Mod. Phys. A2, (1987), 1 for a review of string field theory. See also M. Kaku, *Introduction to Superstrings*, Springer-Verlag, N.Y., 1988 for a review of superstrings.
2. E. Witten, Nucl. Phys. B276 (1986) 291. See also: S.B. Giddings, Nucl. Phys. B278 (1986) 242 and D.J. Gross and A. Jevicki, Nucl. Phys. B283 (1987) 1.
3. A. Neveu and P.C. West, Phys. Lett. 168B (1986) 192; Nucl. Phys. B278 (1986) 601.
4. H. Hata, K. Itoh, T. Kugo, H. Kunitomo, K. Ogawa, Phys. Rev. D34 (1987) 2360.
5. M. Kaku, "Geometric Derivation of String Field Theory from First Principles I: Curvature Tensors and the Tensor Calculus," 10/86 and "II: Superstrings without Constraints," 2/87.
6. M. Kaku, "III: Closed Strings and the Midpoint Gauge," and "IV: Space-time Supersymmetry," in preparation.
7. M. Kaku, "Why Are There Two BRST String Field Theories?" to appear in Phys. Lett B.
8. M. Kaku, "Derivation Of The Four String Interaction From Geometric String Field Theory," CCNY preprint.
9. S. Mandelstam, Nucl. Phys. B64 (1973) 205.
10. E. Cremmer, A. Schwimmer, C.B. Thorn, Phys. Lett. B179 (1986) 57.
11. K. Kato and K. Ogawa, Nucl. Phys. B212 (1983) 443.
12. K. Itoh, K. Ogawa, and K. Suehiro, KUNS 846 HE(TH) 86/06.
13. K. Suehiro, KUNS 857 HE(TH) 87/04.
14. A. Neveu and P. West, Phys. Lett. B182 (1986) 343 and CERN-TH 4564/86.
15. W. Siegel and B. Zweibach, Nucl. Phys. B282 (1987) 125.
16. M. Kaku, "Deriving Osp(D,2/2) BRST from Geometric String Field Theory," CCNY preprint.

ON HIGHER ORDER INTERACTIONS FOR SUPERSTRINGS

J.G. Taylor

Department of Mathematics
King's College
London, U.K.

## INTRODUCTION

String field theory has recently become of considerable interest due to the need to distinguish between the plethora of different string theories[1] constructed in the last year or so. It is necessary to use both experimental data and theoretical ingenuity to distinguish between this host of possible models of quantum gravity. Part of this ingenuity may be by way of non-perturbative analyses (tunneling effects through instantons, etc.). In order to see if such non-perturbative effects are helpful a non-perturbative approach to strings is essential. Various avenues for this have been put forward, in particular the elegant formulation in terms of projective vector bundles over universal moduli space[2] and its related approach through Grassmannian[3] or loop space[4] techniques.

Yet the above developments have not lead to any clear formulation of a covariant field theory with interacting strings for the closed string case; even the beautiful approach of Witten[5] for open strings has proved difficult to extend to the closed string case[6]. In order to make progress in this quest for a covariant closed string field theory it may be helpful to return to the light cone (L.C.) gauge approach to the problem. There is already a complete L.C. field theory of closed bosonic strings[7], and that was used as a guide in the construction of a covariant string field theory[8]. Moreover this L.C. field theory was used recently to give a proof[9] of the unitarity of the covariant first quantised bosonic string theory. It would therefore seem useful to extend the L.C. bosonic string field theory to the superstring case.

Such an extension has already been given by Green and Schwarz[10] where a purely cubic interaction term was proposed, the total L.C. Hamiltonian for the closed superstring case having the schematic form

$$H(\Psi) = \int [\Psi K \Psi + g V \Psi^3]$$

where K is the kinetic term, V the interaction $\delta$-function sewing the three strings together, g the string coupling constant and $\Psi$ the closed superstring L.C. field (depending on transverse bosonic and fermionic co-ordinates). But SUSY implies that H is positive definite, which cannot be valid for $H(\Psi)$ above due to the cubic term. This was pointed out by Greensite and Klinkhammer, both for the open and closed superstring[11]. The latter pointed out the need for at least a quartic addition to $H(\Psi)$ of contact form (in addition to the zipper diagram in the open case). They have suggested a form for this quartic term, but not discussed the closure of the super-Poincaré algebra beyond hinting at possible non-polynomiality in the complete $H(\Psi)$ (after the present work[12] was reported on at the meeting a further paper[13] was received by the authors giving more details of their proposed quartic interaction term; that is presently under consideration). The present account is to present preliminary results of our analysis[12] of the above situation, both for the open and closed case. In particular we give a suggested form of new quartic term in both the Hamiltonian and supersymmetry charges, and also indicate why we feel that there are no further terms in the closed case. This latter is important for the covariant first quantised approach since it would seem to lead to compactifications of moduli space outside the stable case considered up to now.

SUPERSTRINGS AT CUBIC ORDER

We use the notation of ref.10, so SO(8) $\rightarrow$ SU(4)$\times$U(1) description of the Grassmann valued string variables $S_\alpha$ as (for type II) $S_\alpha \rightarrow (\lambda^A, \theta_A), (\tilde{\lambda}^A, \tilde{\theta}_A)$, where $\lambda, \tilde{\lambda}$ are left and right moving momentum-type and co-ordinate type variables with A = 1 to 4, $\lambda^A$ transforming as $\bar{4}$, $\theta_A$ as 4 of SU(4). In the Wick-rotated space-time with $\rho = \tau + i\sigma$, $t = -i\tau$ denoting the real time, the free first quantized lagrangian is

$$L_0 = \lambda\, \partial_{\bar{\rho}} \theta + \tilde{\lambda}\, \partial_\rho \tilde{\theta}$$

The generators ($J_{\mu\nu}$, $Q_i^{-A}$, $\bar{Q}_i^{-A}$, $q_i^{-A}$, $\bar{q}_i^{-A}$, i = 1,2) may be written down explicitly[10], such as

$$J_{\mu\nu} = M_{\mu\nu} + S_{\mu\nu}$$

$$S_{ij} = -\tfrac{1}{2}i\int(\lambda\rho_{ij}\theta+\tilde{\lambda}\rho_{ij}\tilde{\theta})d\sigma$$

$$q_1^{-\bar{A}} = \int(\surd 2\ \rho^i\partial_\rho X^i\theta^{\bar{A}} + 2\pi\partial_\rho X^L\lambda^{\bar{A}})d\sigma$$

and the other quantities are constructed similarily. Then the second quantised generators of the super-Poincaré algebra are written as

$$(\hat{J}_2,\hat{Q}_2) = \int\Psi^+(J_{\mu\nu},Q^{-A,-\bar{A}},q^{-A,-\bar{A}})\Psi(\underline{X},\theta)$$

where $\underline{X}$ denotes the transverse bosonic modes and the suffix 2 denotes the number of string fields involved (where the type I superstring has supercharges $q_1+q_2$).

Interaction is introduced as is usual for L.C. bosonic string field theory[7] by a cubic interaction which has a $\delta$-function along the three fusing strings 1,2,3 which fuses, say 1 and 2 with 3. This latter, in the closed case, must therefore be self-intersecting at the assumed common junction or interaction point P of 1 and 2. In order to preserve the super-Poincaré group (SP) it is necessary to include insertion factors. For generators of SP outside the light-cone set (the only ones non-linearly realised) the cubic terms are of form for the open string denoted by the field $\Phi$,

$$\hat{Q}_3^{-\bar{A}} = c\int \theta^{\bar{A}}(P^1)\Delta(1,2,3)(\Phi^+\Phi\Phi + \text{h+ conj})$$

with $\Delta(1,2,3)$ being the three-string functional $\delta$-function for the string fusion. For closed superstrings the above is modified by the Hamiltonian insertion factor considered below for the right moving mode (and $L\leftrightarrow R$ for $\hat{Q}^{-A}$). The constant c is rather important due to the singular nature of the operator $\theta^A$ at the interaction point P, and $P^1$ is chosen to be close to P on one or other of the strings with $c \propto (P-P^1)^{\frac{1}{2}}$ (since $\theta$ transforms with conformal weight 1). A similar expression for $\hat{Q}_3^{-A}$ is available, with $c\theta$ replaced by $(c\theta)^3$, so leading[10], through

$$[\hat{Q}_2^{-(A},\hat{Q}^{-\bar{B})}]_+ = \delta^{A\bar{B}}H_3$$

to the expression for $H_3$ for open strings similar to that for $\hat{Q}_3^{-\bar{A},A}$ but now with the insertion factor

$$c[2^{-\frac{1}{2}}\partial_{\bar{\rho}}X^L - \partial_{\bar{\rho}}X^i c^2\theta\rho^i\theta + (\surd 2/3)\partial_{\bar{\rho}}X^R c^4\theta^4]$$

where $X^{L,R} = X^{7\pm i8}$ and $1 < i < 6$. For closed superstrings there is such an insertion factor for both L and R modes. These results were proved by using a detailed mode analysis in ref.10, but can be equally well shown valid by conformal operator techniques with

$$\langle \lambda^A(z)\theta^{\bar{B}}(w)\rangle \sim \delta^{A\bar{B}}(z-w)^{-1}, \quad \langle \partial_z X^I \partial_w X^J\rangle \sim \delta^{IJ}(z-w)^{-2}$$

In order to justify the SUSY algebra is satisfied it is necessary to show, for example, that

$$[\hat{Q}^{-\bar{A}},\hat{Q}^{-\bar{B}}]_+ = 0, \quad [\hat{Q}^{(A},\hat{Q}^{-\bar{B})}]\propto\delta^{A\bar{B}}$$

If

$$Q^{-\bar{A}} = Q_2^{-\bar{A}} + Q_3^{-\bar{A}}$$

this requires both

$$[Q_2^{-(\bar{A}},Q_3^{-\bar{B})}]_+ = 0$$

and

$$[Q_3^{-(\bar{A}}, Q_3^{-\bar{B})}]_+ = 0 \tag{1}$$

The former of these equations was proven true in ref.10. We turn to consider the validity of the latter in the next section.

## SUPERSTRINGS AT QUARTIC ORDER

It is possible to calculate the last expression $[Q_3^{-(\bar{A}},Q_3^{-\bar{B})}]_+$, and similar expressions involving pairs of non-linearly realised generators of SP at cubic order by making direct use of the equal time CCR's for the fields $\Phi$ or $\Psi$:

$$[\Psi[1],\Psi[2]] = \frac{1}{p_2^+}\,\delta(p_1^+ + p_2^+)\int_0^{\pi p_2^+} \frac{d\sigma_0}{\pi|p_2^+|}\,\Delta^{16}[z_1(\sigma)-z_2(\sigma+\sigma_0)]$$

where z denotes the $\underline{X}$ and $\theta,\tilde{\theta}$ variables of a particular string, and the integral removes the origin dependence. The results of these

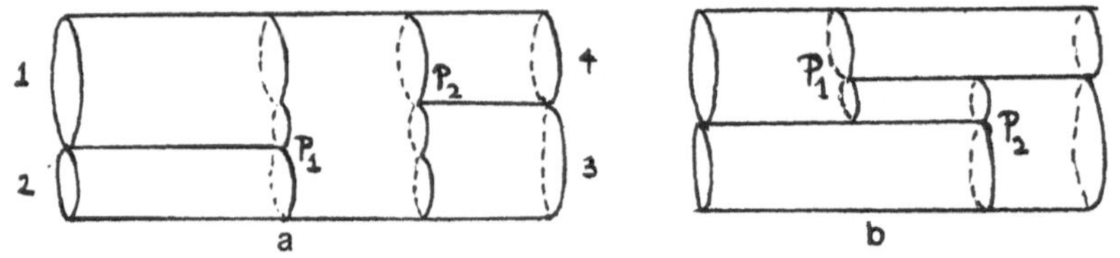


Figure 1

The two contributions to $[\hat{Q}_3^{-\bar{A}},\hat{Q}_3^{-\bar{B}}]_+$ for the closed superstring, giving a spurious time displacement between the interaction points $P_1$ and $P_2$ to show the relative order of the associated insertion factors at $P_1$ and $P_2$.

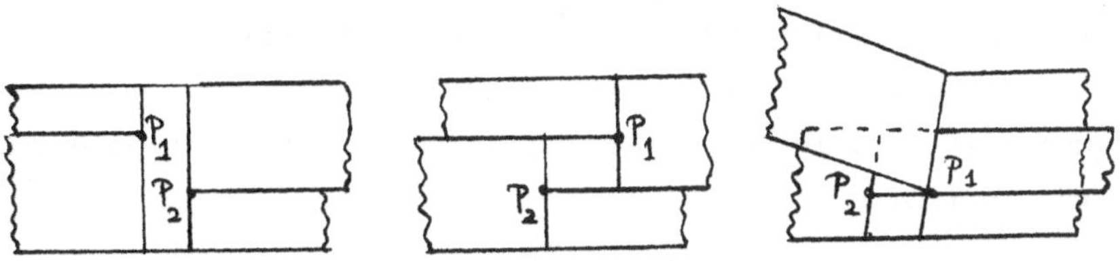


(a) (b) (c)

Figure 2

The three contributions to $[\hat{Q}_3^{-(\bar{A}}, \hat{Q}_3^{-\bar{B})}]_+$ for the open superstring.

calculations[12] show that, in particular, $[\hat{Q}_3^{-(A}, \hat{Q}_3^{-B)}]_+$ and $[\hat{Q}_3^{-(A}, \hat{Q}_3^{-B)}]_+$ have no contribution from string configurations other than for coinciding interaction points. This occurs rather immediately in the closed string case, as can be seen by the fact that the two contributions of fig.1,(a) and (b) combine to give an anticommutator of two Grassmann-valued insertion factors. A similar situation occurs for the open string, where care must be taken in following the Chan-Paton factors round the edges of the open-string analogue of figure 1. However there will be no contribution from (b) in the case when $P_1 = P_2$ and the internal string vanishes. Thus a non-zero contribution will be obtained to the l.h.s. of (1), contrary to what is required.

A similar situation occurs for the open superstring, as was originally pointed out in ref.10. The corresponding diagrams to fig.1, describing the ordering of the insertion factors, are given in fig.2. There are three types of diagram, the first two being planar, the third non-planar. In the equal time limit there is exact cancellation between the first two when the interaction points are distinct, whilst only the first and third contribute when $P_1 = P_2$. Since these two latter contribute to different topologies, for example only the first to the s-t channel in 2 string → 2 string amplitudes, then (1) does not cancel in the open superstring case.

Let us consider the possible contribution $\hat{Q}_4$ which is required by the above result. It is to be noted that the contributing term (a) in figure 2, in the limit $P_1 = P_2$, has a similar contact form to that in (a) of figure 1. In other words only the configuration $\underline{X}_1 = \underline{X}_4$, $\underline{X}_2 = \underline{X}_3$ is present, so that the form of any contribution $\hat{Q}_4^{-A,\bar{A}}$ required to cancel $[\hat{Q}_3^{-A}, \hat{Q}_3^{-B}]_+$ at quartic order in the string fields, by the additional term $[\hat{Q}_2^{-(A}, \hat{Q}_4^{-B)}]_+$, must be of the form

$$\hat{Q}_4 = \int D\, Z_1\, D\, Z_2\, \Psi^+(Z_1)\Psi(Z_1)\Psi^+(Z_2)\Psi(Z_2)\delta^{16}(Z_1(P)-Z_2(P))$$

to within some insertion factor which has to be determined, and a similar form for the open case. This is essentially a "straight through" term, only contributing to forward scattering, where the term $\hat{H}_4$ is generated from the equation

$$[\hat{Q}_3^{-A},\hat{Q}_3^{-\bar{B}}]_+ + [\hat{Q}_2^{-(A},\hat{Q}_4^{-\bar{B})}]_+ = \delta^{A\bar{B}}\hat{H}_4$$

We have found the form, to leading order in $(P_1-P_2)$,

$$\hat{Q}_4^{-\bar{A}} = \lim_{P_1\to P_2} \int c_{12}[\Theta^{\bar{A}}(P_1)V(X(P_2),\Theta(P_2))\Phi^+(Z_1)\Phi(Z_1) \times_{P_1\leftrightarrow P_2}$$

$$\times\ \Phi^+(Z_2)\Phi(Z_2)\delta(Z_1(P_1)-Z_2(P_1)) \qquad (2)$$

with V being the same function of its variables $X,\Theta$ as the insertion is in $\hat{H}_3$ of $\partial_{\bar{\rho}}X$ and $\Theta$. Similar forms for the corresponding quantities $\hat{Q}_4^{-\bar{A}}$, $\hat{H}_4$ have also been found to leading order in $(P_1-P_2)$ so that the SUSY algebra involving these operators closes correctly to quartic order in the fields, as does that involving $\hat{J}^{ij},\hat{J}^{Li},\hat{J}^{Ri}$. We have not finished investigating the term involving (c) of figure 2, though our methods allow us to do so. The higher order terms in $(P_1-P_2)$ involve multiple commutators of $\hat{H}$ with the above expressions, due to the time separation used to regularise when $P_1 = P_2$ in (1). For this regularisation requires the use of time translation to bring the field operators at the earlier time to the later time, and thus allow use of equal time CCR's. These extra terms can also be analysed; the results will be reported on elsewhere[12].

SUPERSTRINGS AT HIGHER ORDER

It is important to answer the question as to the necessity or otherwise of higher order terms in Q or H. Thus we write

$$\hat{Q}^{-\bar{A}} = \hat{Q}_2^{-\bar{A}} + \hat{Q}_3^{-\bar{A}} + \hat{Q}_4^{-\bar{A}} + \ldots$$

and consider the conditions

$$[\hat{Q}^{-\bar{A}},\hat{Q}^{-\bar{B}}]_+ = [\hat{Q}_2^{-(\bar{A}},\hat{Q}_3^{-\bar{B})}]_+ + \{[\hat{Q}_3^{-\bar{A}},\hat{Q}_3^{-\bar{B}}]_+ + [\hat{Q}_2^{-(\bar{A}},\hat{Q}_4^{-\bar{B})}]_+\}$$

$$+ \{[\hat{Q}_3^{-(\bar{A}},\hat{Q}_4^{-\bar{B}}]_+ + [\hat{Q}_2^{-(\bar{A}},\hat{Q}_5^{-\bar{B})}]_+\} + \ldots \qquad (3)$$

Then we note from (2) that only if

$$[\hat{Q}_3^{-(\bar{A}},\hat{Q}_4^{-\bar{B})}]_+ = 0 \qquad (4)$$

can we hope that the process of adding counter terms can terminate at the quartic term. Moreover the further condition

$$[\hat{Q}_4^{-\bar{A}}, \hat{Q}_4^{-\bar{B}}]_+ = 0 \tag{5}$$

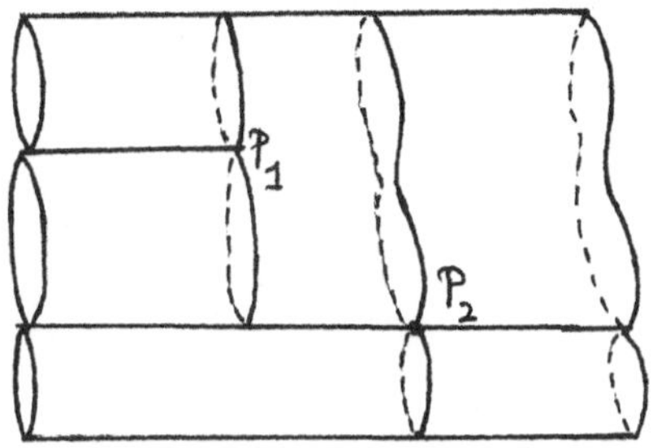


Figure 3

Contribution to $[\hat{Q}_3^{-(\bar{A}}, \hat{Q}_4^{-\bar{B})}]_+$ in terms of time-separated interaction points $P_1, P_2$.

must also be shown to be valid. We will not consider (4) and (5) in the open superstring case, where in any case there are a number of further three string interactions besides that of only open superstrings, as well as the quartic zipper interaction. The situation is much simpler when only closed strings are present. In that case the only contribution to (4) is of the form shown in figure 3 plus the term with $P_1$ and $P_2$ interchanged. A similar situation occurs for (5), so that the algebra appears to be at most quartic for the closed superstring.

It is important to remark here that the above cancellations depend crucially on the highly reduced form that the quartic term takes as seen by (2); in other words it is because the strings effectively go straight past each other that nonpolynomiality can be so easily avoided. In the open superstring case it is not clear, due to the presence of quartic counter-terms involved with fig.2(c), that such a simplification still occurs. This may indicate non-polynomiality for the open superstring interaction. However this has to be investigated more carefuly.

Such non-polynomiality could, in fact, be serious. The effect of contact terms on amplitudes can be seen to take the form indicated in figure 4. In (a) the contact interaction is seen to act as a vanishing dividing geodesic, whilst in (b) and (c) it acts as a vanishing handle geodesic. It might be possible that higher order terms arise from, say, terms of form

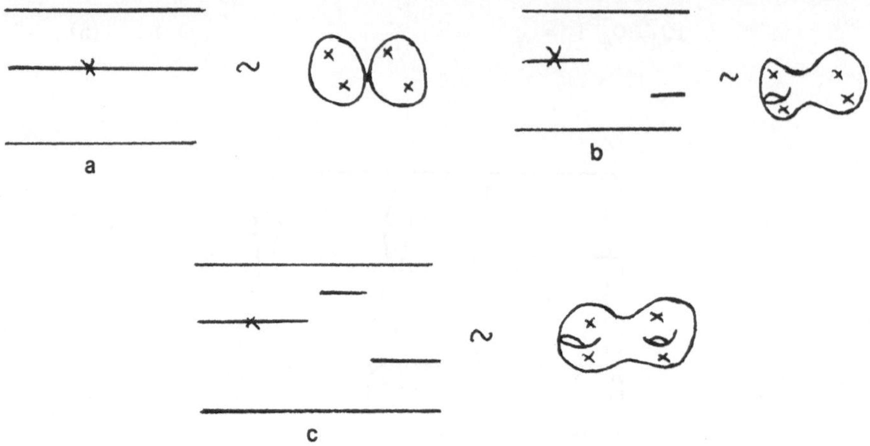


Figure 4

The representation of scattering amplitudes with contact insertions in terms of Riemann surfaces.

$$\int \prod_{r=1}^{N} \Phi^{+}(Z_r)\Phi(Z_r)\ D\ Z_r \ \cdot \prod_{r<s} \delta(Z_r(P)-Z_5(P))$$

Such terms would not be described by Riemann surfaces with stable nodes, but only as unstable ones. Since these have not been classified, compactification of moduli space to include these would be problematic. This does not seem to occur for the closed superstring, so leading to agreement with the Polyakov action.

Finally these contributions seem to be related to boundary terms arising from partial integration in previously constructed multi-loop amplitudes[14]. In the heterotic string they do not seem to affect the finiteness analysis given earlier[15]; the situation for Type II superstrings is presently under study.

ACKNOWLEDGEMENTS

I would like to thank the organisers, Peter Freund and K.T. Mahanthappaa, for organising such an excellent meeting in such a stimulating location.

REFERENCES


1. A.N. Schellekens, "Four Dimensional Strings", CERN preprint Th.4807/87 July 1987; J.H. Schwarz, "Recent Developments in Superstrings" in Superstrings, ed. K.T. Mahanthappa and P.G.O. Freund, Plenum Pub. Co. New York, (1988).
2. D. Friedan and S. Shenker, Phys. Lett. 175 287 (1986); ibid Nucl. Phys. B281 509 (1987).
3. L. Alvarez-Gaumé, C. Gomez and C. Reina, Phys. Lett. 190 55 (1987); ibid "New Methods in String Theory", CERN preprint TH.4775/87; E. Witten, "Quantum Field Theory, Grassmannians and Algebraic Curves", Princeton preprint PUPT-1057.
4. M. Bowick and S. Rajeev, Phys. Lett. 58 535 (1987); ibid "The Holomorphic Geometry of Closed Bosonic String Theory and Diff $S^1/S^1$" MIT preprint CTP#1450 (1987).
5. E. Witten, Nucl. Phys. B268 253 (1986); also ibid B276 2 (1987).
6. For a discussion of this and other approaches see M. Nouri-Moghadam and J.G. Taylor, "A Review of String Field Theory", to appear in Proc. A.M.S. Workshop on Theta Functions.
7. M. Kaku and K. Kikkawa, Phys. Rev. D10 1110, 1832 (1974).
8. H. Hata, K. Itoh, T. Kugo, H. Kunitomo and H. Ooguri, Phys. Lett. 172B 186, 195 (1986); ibid Phys. Rev. D34 2360 (1986); ibid Phys. Rev. D35 1318 (1987); A. Neveu and P.C. West, Phys. Lett. 168B 192 (1986); Nucl. Phys. B278 601 (1986).
9. S.B. Giddings and E. D'Hoker, "Unitarity of the Polyakov Bosonic String", Princeton preprint 1987.
10. M.B. Green and J.H. Schwarz, Nucl. Phys. B243 536 (1984).
11. J. Greensite and F.R. Klinkhammer, Nucl. Phys. B281 269 (1987); ibid B291 557 (1987).
12. N. Linden and J.G. Taylor, in preparation.
13. J. Greensite and F.R. Klinkhammer, "Superstring Amplitudes and Contact Interactions", Niels Bohr preprint NBI-HE-87-58, August 1987.
14. A. Restuccia and J.G. Taylor, Phys. Rev. D, July 15, 1987.
15. A. Restuccia and J.G. Taylor, Comm. Math. Phys. (to appear).

# Chapter III
# String Geometry and Arithmetic

# KÄHLER GEOMETRY OF STRING THEORY

S. G. Rajeev

Department of Physics and Astronomy
University of Rochester
Rochester, NY 14627

## INTRODUCTION

General relativity describes gravitation in terms of the Riemannian geometry of space-time. This leads one to expect that any more accurate theory of gravity should also describe the geometry of some manifold.

Space-time may be thought of the set of all instantaneous states of a point-particle. Its evolution is then described by a world-line. The analogous object in string theory is the manifold of all loops in space-time. We should expect that a string theory of gravity describes the geometry of loop space.

But this cannot be the whole story. A theory in loop space presupposes the existence of an underlying space-time manifold. What we should do is to study the loop space of a Riemannian manifold and then postulate axioms for a geometry generalizing this. In special cases, or at low energies, this new geometry must reduce to that of the loop space of a Riemannian manifold. Furthermore, solutions to Einstein's equations must have corresponding approximate solutions to the field equations of this theory.

I have proposed such a theory in collaboration with M. Bowick [3]. It is based on infinite dimensional Kähler geometry. There are solutions to our theory that correspond to loop spaces. But it should be emphasized that there could also be other solutions which have no interpretation as loop spaces.

It is not clear what sort of string model our theory corresponds to. The closest two-dimensional field theories seem to be the chiral bosonic systems [5,10,12]

## THE LOOP SPACE OF A RIEMANN MANIFOLD

In order to learn what sort of geometry is relevant, let us first consider the loop space of a Riemann manifold.

Let L be the set of all maps from $S^1$ to a Riemann manifold (M,g). [We will leave aside the question of which precise class of maps are allowed. For various proposals, see [6,3].

The first thing we note about L is that it allows an action of the group of diffeomorphisms of $S^1$, $\mathrm{Diff}S^1$:

$$x(\bullet) \in L \quad , \quad \rho \in \mathrm{Diff}S^1 \quad , \quad \theta \in S^1$$
$$[\rho\bullet x](\theta) = x(\rho^{-1}(\theta)) \tag{2.1}$$

We look for structures on L invariant under the connected component of $\mathrm{Diff}S^1$. (Our theory is in fact not invariant under the orientation reversing diffeomorphisms. From now on, $\mathrm{Diff}S^1$ denotes the connected component).

Since M is Riemannian it is natural to look first for a Riemann metric on L. If u and v are vectors in L at the point $x(\bullet) \in L$, they can be identified as vector fields defined along the loop $x(\bullet)$ in M. A Riemann metric on L is easily obtained.

$$\gamma(u,v) = \int d\theta \; g_{\mu\nu}(x(\theta))u^{\mu}(\theta)v^{\nu}(\theta) \tag{2.2}$$

However, this is not invariant under $\mathrm{Diff}S^1$. Notice however that if we consider instead,

$$\omega(u,v) = \int d\theta \; g_{\mu\nu} \frac{dx^{\rho}}{d\theta} u^{\mu} \nabla_{\rho} u^{\nu} \tag{2.3}$$

with $\nabla$ being the Riemannian connection, we get a $\mathrm{Diff}S^1$-invariant object. But $\omega$ is antisymmetric under the interchange of u,v: it is a 2-form in $\Gamma$. In fact, it can be shown that it is a closed 2-form:

$$d\omega = 0 \quad . \tag{2.4}$$

$\omega$ is degenerate in general. If there is a solution to

$$x^{\cdot\rho}\nabla_{\rho} \; v = 0 \tag{2.5}$$

along $\Gamma$, we have a zero mode. (2.5) describes a vector that is covariantly constant along a closed loop. If M is $R^d$, there are d linearly independent solutions to (2.5). In general there will never be more than d independent solutions.

So $\omega$ is a closed 2-form with at most a finite dimensional kernel. If it had been non-degenerate, $(L,\omega)$ would be an infinite dimensional symplectic manifold [1]. By an abuse of language we will continue to refer to this as a symplectic manifold even though the term "contact" manifold may be more appropriate.

## SYMPLECTIC GEOMETRY

We might now propose that our theory is based on the following:

I. An infinite dimensional manifold $\Gamma$.

II. An action of $\mathrm{Diff}S^1$ on $\Gamma$.

III. A closed 2-form $\omega$ on $\Gamma$ invariant under $\mathrm{Diff}S^1$ and having a finite dimensional kernel.

However, we do not have enough structure to produce a field equation. In fact, we saw above that the loop space of any Riemann manifold is a solution to our scheme.

We propose that the additional structure required is a "polarization" in the language of geometric quantization [14]. In particular we chose a Kähler polarization because of its elegance, though strictly speaking we can choose others.

But before pursuing this, let me interpret what I described in terms of a more physical (rather than geometric) language. Recall that a symplectic manifold can be thought of as the phase space of some system in classical physics [1]. The symmetries of $\omega$ are the canonical transformations. In fact, if a vector field leaves $\omega$ invariant,

$$L_u \omega = 0 \quad , \quad \text{then} \quad \tilde{u}_\mu = \omega_{\mu\nu} u^\nu \text{ satisfies} \tag{3.1}$$

$$\partial_\mu \tilde{u}_\nu - \partial_\nu \tilde{u}_\mu = 0 \tag{3.2}$$

So, such transformations are described by a function,

$$\tilde{u}_\mu = \partial_\mu f$$

f is the generator of the canonical transformation. Notice that even when $\Gamma$ is finite dimensional, the symmetry group of $\omega$ is infinite dimensional. Notice the contrast with a Riemann manifold, where

$$L_\xi g = 0 \quad \Rightarrow \quad \partial_\mu \tilde{\xi}_\nu + \partial_\nu \tilde{\xi}_\mu = 0 \quad \text{with} \quad \tilde{\xi}_\mu = g_{\mu\nu} \xi^\nu \quad . \tag{3.3}$$

If $\Gamma$ has finite dimension, there are only a finite number of independent solutions.

Let the vector field $X_f$ correspond to the function f on $\Gamma$. Then we can show that

$$\left[X_f, X_{\tilde{f}}\right] = X_{\{f,\tilde{f}\}} \tag{3.4}$$

where the new function $\{f,\tilde{f}\}$ is the Poisson-bracket of f and $\tilde{f}$.

In classical mechanics there is a preferred function, the Hamiltonian H. It determines a vector field $X_H$. The integral curves of this field satisfy

$$\frac{dx}{dt} = X_H \tag{3.5}$$

This is just the equation of motion.

Applying to our situation, we know there are vector fields $X_n$ an $\Gamma$ implementing the action of the infinitesimal diffeomorphisms of $S^1$, $e^{in\theta}\partial/\partial\theta$. These leave $\omega$ invariant:

$$L_{X_n}\omega = 0 \tag{3.6}$$

There are then functions $\lambda_n$ which are the generators of these canonical transformations. They satisfy the algebra

$$\{\lambda_m,\lambda_n\} = (m-n)\,\lambda_{m+n} \qquad m,\, n \in Z \tag{3.7}$$

under Poisson bracket. Note that they also satisfy a reality condition

$$\bar{\lambda}_m = \lambda_{-m} \tag{3.8}$$

Let us now see which classical system corresponds to the particular example of the loop space.

A system whose phase space is L would have an equation of motion whose initial data is just a function $x: S^1 \to M$.

It can be verified that with the choice for the Hamiltonian

$$H = \frac{1}{2}\int g_{\mu\nu}\frac{dx^\mu}{d\theta}\frac{dx^\nu}{d\theta}\,d\theta \tag{3.9}$$

the symplectic form (2.3) yields the equation of motion

$$\frac{\partial x^\mu}{\partial t} = \frac{\partial x^\mu}{\partial\theta} \tag{3.10}$$

This is just the field equation of a chiral bose field in 2-dimensional space. Note that the equation of motion is independent of $g_{\mu\nu}$ even though $\omega$ and H depend on it. In fact the equation of motion makes use only of the differential topology of the target manifold M.

Another realization of our axioms of section 3 is the coset space $M = \mathrm{Diff}S^1/S^1$. There is a 2-parameter family of invariant closed 2-forms on M. [3,8] This is in fact a special case of the fact that a co-adjoint orbit of a Lie group is a homogeneous symplectic manifold. [13,7] We are studying the geometry of M in more detail and relate it to the representation theory of $\mathrm{Diff}S^1$. See also [2,15] We will return to $\mathrm{Diff}S^1/S^1$ later in connection with ghosts.

We saw in section 3 that additional structure is needed on $\Gamma$ to obtain field equations. We now argue that what we need is a complex structure J such that $(\Gamma,\omega,J)$ is a Kähler manifold. (Like $\omega$, J is also allowed to have zero modes. It satisfies $J^4 = -J^2$ rather than $J^2 = -1$; we again require there to be only a finite number of zero modes for J).

Just as $(\Gamma,\omega)$ has an interpretation in terms of classical mechanics, $(\Gamma,\omega,J)$ has one in terms of quantum mechanics. If we require,

IV. $\omega$ is an integral element of $H_2(\Gamma)$.

There is a line bundle $\Lambda$ an $\Gamma$ whose Chern class is $[\omega]$. Furthermore, there is a connection on it with curvature $\omega$. [These statements constitute the Weyl theorem. See e.g. [7]]. Given the Kähler structure, we can consider the space of holomorphic sections of this line bundle. It is possible to identify this (after completion with a suitable inner product) as the Hilbert space B of a quantum system whose classical limit is $(\Gamma,\omega)$.

It is clear that there could be many quantum systems corresponding to the same classical system. Some data other than $(\Gamma,\omega)$ is needed to specify a quantum system. From the current point of view (called "geometric quantization") this new structure is J.

Now we could run into trouble. Symmetries of $(\Gamma,\omega)$ need not be symmetries of $(\Gamma,\omega,J)$. If a function f generates the vector field $X_f$ which leaves J invariant, it is not difficult to construct an operator representing this on the above Hilbert space. I will confine myself to quoting the formula [14]

$$r(f) = \frac{\hbar}{i} \nabla_{X_f} + f \tag{4.1}$$

$\nabla$ being the connection on $\Lambda$. A holomorphic section of $\Lambda$ satisfies

$$\nabla_X \psi = 0 \quad \text{if} \quad X \quad \text{is type} \quad (0,1) \tag{4.2}$$

with respect to J. So it is clear that if $X_f$ leaves J invariant,

$$L_{X_f} J = 0 \tag{4.3}$$

holomorphic sections are mapped to holomorphic sections.

In general however $X_f$ may not leave J invariant. Even then, it is often possible to extend the construction above and produce an operator in H.

But sometimes a group of symmetries of the classical system cannot be extended to the quantum system. This is the signature of an anomaly. It is useful to find a criterion for when an anomaly occurs. It is not enough that $L_X J \neq 0$. With M. Bowick, I have introduced a procedure [4] to identify such anomalies at the curvature of a complex vector bundle.

Let $(\Gamma,\omega)$ be invariant under G. Let H be subgroup leaving J invariant. To each point in G/H, I can associate a complex structure in $\Gamma$. So at each point on G/H I have a complex Hilbert space H. The operators (4.1) can be used to provide a connection D on this vector bundle.

$$\begin{array}{ccc} B & \to & V \\ & & \downarrow \\ & & G/H \end{array} \qquad (4.4)$$

In fact it turns out that V is naturally holomorphic vector bundle, but I will not use this fact now.

The group G can be realized on B, if the connection D is flat.

The above is the outline of a quite general framework, applicable not just to string theory. In fact in [4] we have shown that all anomalies in two-dimensional theories can be interpreted as curvatures this way. [The extension to four dimensions requires some mathematical developments, see [11].]

Let us return to string theory. In addition to axioms I-IV we now introduce J:

V. There exists a tensor satisfying $J^4 = -J^2$ and the integrability condition.

What we mean by the integrability condition is that the equation for holomorphic functions,

$$\partial_X \psi = 0 \quad \text{for} \quad X \quad \text{of type} \quad (0,1)$$

has <u>locally</u> well-defined solutions. When $J^2 = -1$, this requires that the Nijenhuis tensor vanishes.

Motivated by special cases we postulate also

VI. The subgroup $S^1$ of Diff$S^1$ leaves J invariant.

A realization is the loop space of $R^d$. A curve $x(\theta)$ can be Fourier analyzed,

$$x(\theta) = \sum_{n\in Z} x_n \, e^{in\theta} \quad ; \quad \bar{x}_n = x_{-n} \qquad (4.5)$$

and J is defined as

$$Jx = \sum i \, \epsilon(n) \, x_n \, e^{in\theta} \qquad (4.6)$$

$$\begin{aligned} \text{where } \epsilon(n) &= 0 \quad \text{for} \quad n = 0 \\ &= \pm 1 \quad \text{for} \quad n \quad 0 \,. \end{aligned} \qquad (4.7)$$

We can now construct the vector bundle

$$
\begin{array}{ccc} B & \to & V \\ & & \downarrow \\ & & \mathrm{Diff}S^1/S^1 \end{array} \tag{4.8}
$$

To the functions $\lambda_m$ an $\Gamma$, we can associate operators $r(\lambda_m)$ as in (4.1). This determines a homogeneous connection on 4.8 whose curvature is

$$
F_{m,n} = [r(\lambda_m), r(\lambda_n)] - r(\{\lambda_m, \lambda_n\}) \tag{4.9}
$$

This depends on the choices, $(\Gamma, \omega, J, X_m)$ made so far. In the example above of loops on flat space, we will get

$$
F_{m,n} = \frac{d}{12}\,(m^3 + \beta m) \tag{4.10}
$$

$\beta$ being a parameter depending on the ordering chosen to define $r(\lambda_m)$.

I now present the field equation for the variables, $(\Gamma, \omega, J, X_m)$

VII. $$F_{m,n} + \left[-\frac{26}{12}\,m^3 + \frac{1}{6}\,m\right]\delta_{m,-n} = 0 \quad . \tag{4.11}$$

The motivation for this comes from noting that the second term is the curvature of the canonical line bundle on $\mathrm{Diff}S^1/S^1$, and that this can be viewed as the contribution of the ghosts to the anomaly. I will not describe the geometry of $\mathrm{Diff}S^1/S^1$ since a paper devoted exclusively to it is in preparation.

It is clear that our field equations just require the cancellation of the $\mathrm{Diff}S^1$ anomaly. It can be viewed this way whenever $(\Gamma, \omega)$ is the phase space of a classical system for which $\mathrm{Diff}S^1$ acts as canonical transformation.

There are many such systems. We already described the chiral boson. Other examples are the KdV system [2] or the Wess-Zumino model [9].

I mention in particular that we have shown [3] the loop space of a Riemann manifold to be a solution to 4.11 in the low energy limit, if

$$
R_{\mu\nu}(g) = 0 \tag{4.12}
$$

It would be interesting to find a solution to (4.11) that approached a black-hole solution to (4.12) at long distances.

## REFERENCES


1. V. I. Arnold, "Mathematical Methods of Classical Mechanics", (Springer-Verlag, NY 1978).
2. I. Bakas, University of Texas at Austin preprint (1987).
3. M. Bowick and S. G. Rajeev, Phys. Rev. Lett. 58 (1987), 535, MIT preprint No. 1450 (1987) (to be published: Nucl. Phys. B).
4. M. Bowick and S. G. Rajeev, MIT preprint No. 1449 (1987), to appear in Nucl. Phys. B.
5. R. Floreanini and R. Jackiw, MIT preprint No. 1512 (1987).
6. G. Horowitz and D. M. Witt, Santa Barbara preprint UCSB-TH-8736.

7. A. Kirillov, "Elements of the Theory of Representations", Springer-Verlag (1976).
8. A. Kirillov and D. V. Yur'ev, Funkt. Anal. Pril. 20, (1986) 79.
9. J. Mickelsson, MIT preprint No. 1448 (1987) to appear in Comm. Math. Phys.
10. L. Mezineseu and R. Nepomechie, University of Miami preprint UMTG-40 (1987).
11. J. Mickelsson and S. G. Rajeev, MIT preprint No. 1482 (1987) submitted to Comm. Math. Phys.
12. W. Siegel, Nucl. Phys. B238 (1984) 307.
13. G. Segal, Comm. Math. Phys. 80 (1981) 201.
14. N. J. M. Woodhouse, "Geometric Quatization" (Clarendan Press, Oxford, UK 1980).
15. E. Witten, Princeton preprint PUTP-1061.

NUMBERS AND STRINGS*

Peter G. O. Freund

Enrico Fermi Institute and Department of Physics
The University of Chicago
Chicago, IL 60637

Compact Riemann surfaces figure prominently in string theory where they appear as string world sheets. Compact Riemann surfaces are "algebraic curves" (curves, in that their complex dimension is one while their real dimension is two), and their algebraic geometry has strong arithmetic (number-theoretic) flavor. Some of this arithmetic geometry has already surfaced in discussions of higher loop diagrams in string theory. It is well known that in algebraic geometry it pays to be flexible concerning the nature of the number-field over which the algebraic curves (or varieties) are defined. Could one be equally flexible in physics, and envisage strings over the still locally compact but non-archimedean field of p-adic numbers? Could such non-archimedean strings shed light on the physically interesting archimedean string over the field R of real numbers? As we shall see [1),2)] this is indeed the case and the extremely simple non-archimedean strings can be viewed, essentially, as the basic building blocks of the much richer physical (archimedean) string. Before constructing such non-archimedean strings one has to confront three crucial decisions.

(A) Are quantum amplitudes to be valued in the field C of complex numbers in the time-honored fashion, or in some p-adic field $Q_p$?

(B) Is space-time to be an ordinary manifold with charts into $R^n$ or a p-adic manifold with charts into $Q_p^n$?

(C) Is the string world sheet to be an ordinary or a p-adic manifold?

Depending on how one answers these three questions one encounters eight possibilities, the usual archimedean theory and seven partially or totally non-archimedean alternatives. We shall see that the well-known form of the archimedean string tree amplitudes virtually forces upon us[1),2)] a unique non-archimedean generalization which uniquely answers the questions (A)-(C).

We therefore start from the ordinary archimedean bosonic open string without Chan-Paton rules (or equivalently with "SO(1)" as "internal

*The actual talk I gave at the Boulder Workshop under the title "Non-archimedean Strings" presented only the work[1)] on closed non-archimedean strings by M. Olson and myself. The progress registered since, has made me opt here for a presentation centered on the simpler open non-archimedean string based on work by E. Witten and myself,[2)] and including the adelic string constructed in reference 2). This work was started after the Boulder Workshop and causality prevented its presentation there. The present work was supported in part by the NSF Grant No. PHY 85-21588.

symmetry"). In terms of the Euler beta function

$$B(a,b) = \int_0^1 dx\, x^{a-1}(1-x)^{b-1} \tag{1}$$

the open string tachyon-tachyon scattering amplitude $A_4^{(\infty)}(s,t,u)$ is given by[3)]

$$A_4^{(\infty)}(s,t,u) = g_\infty^2[B(-\alpha(s),-\alpha(t)) + B(-\alpha(t),-\alpha(u)) + B(-\alpha(u),-\alpha(s))] \tag{2}$$

i.e., by a sum of three terms corresponding to the s-t, t-u, and u-s pairs of channels. Here[3)] $s,t,u$ are the standard Mandelstam kinematical variables, $\alpha(s) = 1 + \frac{1}{2}s$, so that $\alpha(0) = 1$, $\alpha(-2) = 0$, $\alpha' = \frac{1}{2}$. The tachyon mass is then $m^2 = -2$ in these units and we have $\alpha(s) + \alpha(t) + \alpha(u) = -1$. As it stands, equation (2) has no simple p-adic counterpart, as the concept of integrating from zero to one is not p-adically natural. But, in the absence of Chan-Paton factors, equation (2) can be recast in the form[4)]

$$A_4^{(\infty)}(s,t,u) = g_\infty^2 B^{(\infty)}(-\alpha(s),-\alpha(t)) \tag{3a}$$

$$B^{(\infty)}(-\alpha(s),-\alpha(t)) = \int_R dx\, |x|^{-\alpha(s)-1}|1-x|^{-\alpha(t)-1} \tag{3b}$$

where the integral runs from $-\infty$ to $+\infty$, or, in other words, over the full field $R$ of real numbers. The three terms in equation (2) correspond in equation (3) respectively to integration from 0 to 1, from 1 to $\infty$, and from $-\infty$ to 0. Using the customary notations[5),6)] $R \equiv Q_\infty$ and $|\;| \equiv |\;|_\infty$ for the field of real numbers and its standard norm: the absolute value, we rewrite equation (3b) in the suggestive form

$$B^{(\infty)}(-\alpha(s),-\alpha(t)) = \int_{Q_\infty} dx\, |x|_\infty^{-\alpha(s)-1}|1-x|_\infty^{-\alpha(t)-1} \; . \tag{4}$$

This equation shows clearly that $B^{(\infty)}$ is an integral over the full field $Q_\infty$ of the product of the multiplicative characters[7)] $|x|_\infty^{-\alpha(s)-1}$ and $|1-x|_\infty^{-\alpha(t)-1}$ on $Q_\infty$. The integration measure $dx$ used in (4) is the additive Haar measure on $Q_\infty$ ($\equiv R$). The multiplicative characters and the additive Haar measure dx, both have immediate p-adic generalizations[7)] and one is unavoidably led to the non-archimedean (p-adic) scattering amplitude (p = prime number)

$$A_4^{(p)}(s,t,u) = g_p^2 B^{(p)}(-\alpha(s),-\alpha(t)) \tag{5a}$$

$$B^{(p)}(-\alpha(s),-\alpha(t)) = \int_{Q_p} dx\, |x|_p^{-\alpha(s)-1}|1-x|_p^{-\alpha(t)-1} \; . \tag{5b}$$

$B^{(p)}$ is the Gelfand-Graev beta function[7),8)] on $Q_p$, just as $B^{(\infty)}$ is the Gelfand-Graev—not Euler!—beta function on $Q_\infty \equiv R$. In the familiar archimedean case, $B^{(\infty)}$ can be expressed in terms of gamma functions, which, of course, are special functions. Specifically, making use of the kinematics of the tachyon scattering and of the functional equation obeyed by the Euler gamma function

$$\Gamma(z)\Gamma(1-z) = \pi/\sin \pi z \tag{6}$$

we can recast $B^{(\infty)}$ as given in equation (2), in the form[2)]

$$B^{(\infty)}(-\alpha(s),-\alpha(t)) = (2\pi)^{-1} \prod_{x=s,t,u} \Gamma^{(\infty)}(-\alpha(x)) \tag{7}$$

$$\Gamma^{(\infty)}(z) = 2\Gamma(z)\cos\frac{\pi z}{2} \; .$$

Here $\Gamma^{(\infty)}$ is the Gelfand-Graev gamma function on $Q_\infty$. The manipulations that lead from equation (2) or more conveniently equation (4) to equation (7) are independent of the details of the local number field $Q_\infty$ and have direct p-adic counterparts,[7),8)] so that we also have (e.g., by explicitly carrying out the p-adic integration in equation (5b))

$$B^{(p)}(-\alpha(s),-\alpha(t)) = \prod_{x=s,t,u} \Gamma^{(p)}(-\alpha(x))$$

$$\Gamma^{(p)}(z) = \frac{1 - p^{z-1}}{1 - p^{-z}} \qquad \alpha(x) = 1 + \frac{x}{2} \ . \tag{8}$$

The novelty here is that, unlike $\Gamma^{(\infty)}$, the p-adic gamma functions $\Gamma^{(p)}$ are not special functions! The non-archimedean case is considerably simpler than its archimedean counterpart.

At this point already we can see the answers given to all three questions (A)-(C) mentioned in the beginning. The p-adic quantum amplitudes (8) are complex valued as in the archimedean case. The momenta enter these amplitudes in the forms of the exponents $\alpha(x)$ of the multiplicative characters in the integrand. These characters are complex valued, hence the Mandelstam variables must be valued in C, hence the momenta for physical processes must be real as in the ordinary case. Their canonical conjugates, the coordinates of charts of the embedding space-time manifold, must also be real. Space-time is thus seen to remain an ordinary manifold with charts to ordinary euclidean space. What has been rendered p-adic is only the world sheet, now a p-adic surface. The physically directly accessible space-time and quantum amplitudes have not been tampered with. Only the nature of the world sheet has been altered. But by the time the integration over the world sheet is carried out all explicit p-adic dependence is gone. The only memory the amplitudes retain of their p-adic origin is in the explicit appearance of the prime number p in the physical amplitudes, as is evident in equation (8).

A more radical proposal has been advanced by Volovich[9)] and Grossmann.[10)] They choose the p-adic option on each of the three questions (A)-(C). In particular, their space-time is a p-adic manifold and their quantum amplitudes are p-adically valued. Though rather different from the proposal of references (1) and (2), the proposal of Volovich and Grossmann should be further pursued on its own merits.

Now let us explore in more detail the non-archimedean string whose 4-point amplitudes are given by equation (8). To that effect we first briefly consider the general N-point amplitudes.[1),2)]

In the Koba-Nielsen form, the archimedean N-point function is given as a (N-3)-fold integral over $(Q_\infty)^{N-3}$ of an integrand which is again built out of multiplicative characters. This again can be replicated p-adically.[1),2)] We thus obtain a full set of tree amplitudes for the p-adic string. For these tree-amplitudes to be acceptable they must obey the following five conditions:

i) they must be meromorphic;
ii) they must factorize;
iii) they must be p-adic Möbius invariant;
iv) they must be crossing symmetric;
v) they should have no simultaneous poles in "incompatible" channels.

All but the second of these conditions receive non-trivial tests already at the level of the 4-point amplitude (7). From equation (8) it is evident that $B^{(p)}$ is a periodic function of the Mandelstam variables s, t, and u, with purely imaginary period $i\,4\pi/\ln p$ (p is a prime number, not a momentum!). It is also evident that $B^{(p)}$ is meromorphic, its only pole at the tachyon position and its periodic repeats. Specifically $B^{(p)}$ has poles at $s = m_n^2 = -2 + i\,4\pi n/\ln p$ whose residues are equal constants (independent of the other two Mandelstam variables) as expected for spin zero. Of course $B^{(p)}$ also has poles at $t = m_n^2$ again with constant residues, and at $u = m_n^2$ the same way, as required by crossing symmetry. Thus $B^{(p)}$ obeys conditions i) and v), and the manifest crossing symmetry of equation (8) shows that iv) also holds. The argument that yields the well-known $SL(2,Q_\infty)$-Möbius invariance for the integral (4), can be followed step for step to provide the $SL(2,Q_p)$-Möbius invariance of (8). This spectrum and these considerations generalize to the N-point functions, and one readily checks that factorization ii) holds as well. All conditions i)-v) are thus met. The surprise is the apeparance of the pairs of complex conjugate poles at $m_n^2 = 2 + i\,4\pi n/\ln p$ $n = 1,2,3,\ldots$ in addition to the tachyon pole at -2. These pairs of complex conjugate poles on the physcial sheet signal a breakdown of locality having to do with an elementary length of the order of the Planck length times $\sqrt{\ln p}$. This is the one feature which may be undesirable. What this means is that we should not settle "phenomenologically" on some prime number p, for that entails the presence of these complex poles. Rather we should opt for an adelic point of view which treats the archimedean and the non-archimedean amplitudes corresponding to each prime on an equal footing. After all, it would have been most peculiar if access to the laws of nature had to be gained by using as a key, a prime number of all things. An adelic[11] treatment of archimedean and non-archimedean amplitudes has been given by Witten and myself.[2] Making use of the functional equation of the Riemann zeta-function, we found that

$$B^{(\infty)}(-\alpha(s),-\alpha(t)) \prod_p B^{(p)}(-\alpha(s),-\alpha(t)) = 1 \,. \qquad (9)$$

This equation shows that the Veneziano amplitude $B^{(\infty)}$ factors as an infinite product over the inverses of the non-archimedean amplitudes for all primes. It may be surprising at first that inverses of $B^{(p)}$'s rather than the $B^{(p)}$'s themselves appear in the expression of $B^{(\infty)}$, but the adelic "equipartition" of archimedean ($\infty$) and non-archimedean (p) places then somewhat cuts the surprise.

Closed strings can be treated[1] along similar lines with a quadratic extension $Q_p(\sqrt{\tau})$ of the p-adic field. In that case as well there is[2] an adelic formula of the type (9).

How is one to interpret such adelic formulas? In some sense one has factored the Veneziano amplitude with its rich spectrum and without complex poles, into an infinite product of incomparably simpler amplitudes with complex poles ($1/B^{(p)}$ has complex poles and zeros). These factors are essentially one-state "strings" with periodicity in the complex planes of the kinematic variables. Maybe it is not too far-fetched to see an analogy with "factorization" of a baryon into three quarks. The string's non-archimedean building blocks can be studied to all orders much more simply, and the results of such a study hopefully adelically restructured into the archimedean string which we are ultimately interested in.

This opens up an entire host of problems. First of all, can one extend the adelic treatment to N-point functions? Then to p-adic algebraic curves of higher genus? Can one incorporate Chan-Paton rules in the p-adic (open string) case? Then supersymmetry? Can one write down a non-archimedean string lagrangian? Can one get this way nonperturbative results in string dynamics? "Stay tuned," as they say.

REFERENCES


(1) P.G.O. Freund and M. Olson, Phys. Letters (in press).
(2) P.G.O. Freund and E. Witten, Phys. Letters (in press).
(3) M.B. Green, J.H. Schwarz, and E. Witten, Superstring Theory, Cambridge University Press, Cambridge, 1987.
(4) M. Ademollo, A. d'Adda, R. d'Auria, E. Napolitano, P. di Vecchia, F. Gliozzi, and S. Sciuto, Nucl. Phys. B77 (1974) 189.
(5) P.G.O. Freund and M. Olson, Nucl. Phys. B (in press).
(6) N. Koblitz, p-adic Numbers, p-adic Analysis, and Zeta Functions 2nd edition, Springer, Berlin, 1984.
(7) I.M. Gel'fand, M.I. Graev, and I.I. Pyatetskii-Shapiro, Representation Theory and Automorphic Functions, Saunders, London, 1966.
(8) P.J. Sally and M.H. Taibleson, Acta Math. 116 (1966) 279.
(9) I.V. Volovich, Classical and Quantum Gravity 4 (1987) L83.
(10) B. Grossmann, Rockefeller University preprint, 1987.
(11) A. Weil, Basic Number Theory, Springer, Berlin, 1967.

# Chapter IV
# Compactification and Phenomenology

# TYPE II SUPERSTRINGS IN FOUR DIMENSIONS

John H. Schwarz

Lauritsen Laboratory 452-48

California Institute of Technology

Pasadena, California 91125

## INTRODUCTION

Until recently it was believed that compactification of type II superstring theory could not give rise to interesting Yang–Mills gauge symmetries, and thus this theory was not promising for realistic phenomenology. It has now been realized that if one treats the internal degrees of freedom required to saturate the conformal anomaly in four dimensions abstractly, rather than insisting that they describe a compact six-dimensional manifold, many new possibilitiies arise. In particular, it is possible to achieve a non-Abelian gauge symmetry with as many as eighteen generators. In this talk I will show how this can be done and then present a rather general scheme for classifying potential models of this type in terms of the unitary representations of super Kac–Moody algebras. The conclusion appears to be that it is possible to come rather close to achieving a realistic scheme, but that fully realistic fermion representations are not possible. Of course, this result (first presented by Dixon, Kaplunovsky, and Vafa) requires making certain assumptions. It will be interesting to see whether any of them can be modified so as to evade their conclusion.

## CLASSIFICATION OF STRING THEORIES

Superstring theories are attractive candidates for unifying all fundamental forces, because they can give a consistent quantum theory containing gravity[2]. Another attractive feature is that there are very few possible theories, so that it is conceivable that we could be led more or less uniquely to realistic phenomenology purely from considerations of mathematical consistency. In assessing the extent to which this is the case, it is desirable to make a distinction between "theories" and "solutions." For example, it appears that there is just one heterotic string theory, but it has many classical solutions. These include ones with ten-dimensional Poincaré symmetry and a gauge symmetry $E_8 \times E_8$, $SO(32)$, or $SO(16) \times SO(16)$, as well as very many others with $D < 10$.*

* We do not distinguish $SO(n)$ from spin$(n)$ here.

Strings can occur in only two distinct topologies, called open and closed. Open strings have free ends, whereas closed strings have the topology of a circle. In each case, the strings may or may not carry an intrinsic orientation. All theories that contain open strings necessarily contain closed strings as well. The reason is that the same basic interaction that allows two open strings to join ends and give one open string, also allows a single open string to join its ends and give a closed string. In such theories the closed strings are oriented if and only if the open strings are. Theories that contain closed strings need not contain open strings, however.

Associated with any string theory ground state there is a two-dimensional quantum field theory that describes the dynamics of the strings. In other words, it is formulated on the two-dimensional world sheet that is swept out by the motion of the string through space-time. The possible two-dimensional theories that one needs to consider are severely limited by the fundamental requirement of conformal invariance. I cannot go into the details here, but the bottom line is that only conformally invariant two-dimensional theories can be consistently interpreted as string theories. Two-dimensional field theories can be supersymmetric. In fact, the fundamentally distinct possibilities for string theories can be characterized by the number of local supersymmetries that the conformally invariant two-dimensional action possesses.

## Superstring Theories

String theory is not understood well enough yet to assert with certainty in all cases when two configurations are classical solutions of the same or distinct string theories. I find it convenient to use a classification scheme that presupposes a particular answer to this question. This scheme characterizes string theories by the local supersymmetry and topology of their world sheets. In this classification scheme one has the following possibilities:

1) *Type I superstring theory*[3–5] — this theory has $N = 1$ world-sheet supersymmetry for both left- and right-moving modes. The strings are unoriented and can be open or closed. Thus, both oriented and unoriented world sheets with arbitrary boundaries are included in the sum (path integral) over conformally inequivalent metrics. This theory, originally constructed in 1971[3], was shown to have space-time supersymmetry as well, when properly interpreted[4]. At the classical level it can accommodate the same gauge groups as the Veneziano model. However, these theories have chiral fermions, and therefore one-loop effects can give rise to anomalies that break the gauge symmetry rendering the theory inconsistent. There is a unique anomaly-free solution for $D = 10$ that has $N = 1$ space-time supersymmetry and $SO(32)$ gauge symmetry[5]. The possibilities in lower dimensions have not been explored extensively. A recent work by Harvey and Minahan[6] suggests that it is not out of the question that a realistic four-dimensional solution could be obtained. Nonetheless, this theory will not be considered any further here.

Let us now consider closed-string theories. In these theories the normal modes consist of left-moving and right-moving waves, in other words excitations that travel in one direction or the other on the string. The reason for this is very simple. The two-dimensional wave equation

$$\left(\frac{\partial^2}{\partial \tau^2} - \frac{\partial^2}{\partial \sigma^2}\right) X(\sigma, \tau) = 0 \tag{1}$$

has the general solution

$$X(\sigma,\tau) = X_R(\tau-\sigma) + X_L(\tau+\sigma), \tag{2}$$

where $X_R$ and $X_L$ describe right-moving and left-moving modes, respectively. By a Wick rotation $\tau \to i\tau$ it is possible to Euclideanize the world sheet, which can then be regarded as a Riemann surface and studied using methods of complex analysis. In particular, $X_R$ becomes a holomorphic function (i.e., it depends only on the complex coordinate $z$) and $X_L$ becomes an antiholomorphic function (it depends only on $\bar{z}$).

The classification of open-string theories is immediately applicable to the study of closed strings. The reason for this is that left- or right-moving modes of closed strings are described by essentially the same mathematics as the modes of open strings (standing waves). Therefore we can associate a closed-string theory to any pair of open-string theories by the rule

$$\mathrm{CLOSED} \sim (\mathrm{OPEN\#1})_{\mathrm{L}} \otimes (\mathrm{OPEN\#2})_{\mathrm{R}}. \tag{3}$$

For each open-string factor we can consider using the $N = 0$ bosonic string or the $N = 1$ superstring. If both factors have $N = 0$ one obtains the bosonic Shapiro–Virasoro model[7], which we reject for reasons discussed below. Choosing both to have $N = 1$ gives the type II superstring theory and choosing one to have $N = 0$ and the other to have $N = 1$ gives the heterotic string theory.

2) *Type II superstring theory*[16] — this theory also has $N = 1$ local world-sheet supersymmetry for both left- and right-moving modes. However, it only contains oriented closed strings, so that the relevant world sheets are orientable surfaces without boundary. This theory has two classical $D = 10$ solutions. One, called IIA, has two space-time supersymmetries of opposite chirality and no gauge group. The other, called IIB, has two space-time supersymmetries of the same chirality and no gauge group. The possibilities for $D < 10$ is one of the main topics we will discuss.

3) *Heterotic string theory* — this theory has $N = 1$ local world-sheet supersymmetry for right-moving modes, but no supersymmetry for left-moving modes. Like the type II theory, it only contains oriented closed strings. As was already remarked, there are three tachyon-free $D = 10$ solutions. The ones with $E_8 \times E_8$ and $SO(32)$ gauge symmetry have $N = 1$ space-time supersymmetry[9]. The one with $SO(16) \times SO(16)$ gauge symmetry has no space-time supersymmetry[10]. Many solutions with $D < 10$ have been obtained by compactifying six dimensions on Calabi–Yau spaces[11] or orbifolds[12]. One is even very close to being realistic[13].

In addition to these theories, there are also the bosonic string theories that have no world-sheet supersymmetry. The open string theory with $N = 0$ and $D_c = 26$, often referred to as the Veneziano model, was developed in the late 1960's. By attaching suitable "charges" to the ends of the open strings it can accommodate any classical group as a local gauge symmetry. The choices $SO(n)$ or $Sp(n)$ correspond to theories of unoriented strings, whereas $U(n)$ corresponds to oriented strings. These theories have several serious problems. First of all, their spectra do not contain any fermions; they are purely bosonic theories. Secondly, in flat 26-dimensional Minkowski space the classical spectrum contains a tachyon. A ground state that allows tachyonic excitations cannot be stable as a quantum theory. Either there is another stable vacuum (as in a Higgs theory) or the theory does not exist. A possible answer is suggested by a recent work of Orland[14], which studied the $U(n)$ theory for large $n$. It concluded that for $n \to \infty$ the theory is free, and suggested that this might be the case in general. I find that conjecture quite attractive.

Some authors have suggested[15] that the type II and heterotic strings can be identified as specific classes of solutions of the closed oriented bosonic string. If this were true, it would shorten the list of candidate theories. However, the analysis requires assumptions that seem difficult to justify dynamically.

To summarize, there are three string theories that are potentially realistic. They are the SO(32) type I superstring, the type II superstring, and the heterotic string. Of these the heterotic theory is generally considered the most promising for phenomenology, but the possibilities for constructing ground states are not understood well enough to say that the other two possibilities are excluded. The discussion in the remainder of this paper will be restricted to the type II and heterotic theories. The obvious problem in each case is to characterize all possible solutions and to determine which ones, if any, could be consistent with known phenomenological requirements. Of course, finding the answer is not so straightforward. New possibilities for solutions are being discovered at a rapid pace, and the end of the search does not seem to be in sight.

We have narrowed our search to just three theories, each of which is completely free from adjustable dimensionless parameters. However, to know a theory is to know the equations and not the solutions of the equations. Ideally the solution of lowest energy, called the "vacuum" or "ground state" of the theory, would be unique and would possess the symmetries observed in nature such as $D = 4$ Poincaré invariance and $SU(3)_c \times U(1)_{\mathrm{EM}}$. At our present level of understanding, we can only look for solutions to the classical string equations. There appear to be an enormous number that have unbroken supersymmetry and zero energy. (Zero energy means zero cosmological constant, which is the desired result. The real challenge is to obtain this result for a solution with spontaneously broken supersymmetry.) The properties of these classical solutions can be studied in perturbation theory, where, for the most part, they remain consistent (perhaps with small corrections) to every order. This vacuum degeneracy problem is a real threat to our most ambitious goal – a parameter-free explanation of the particle physics data. This requires not only a unique theory, but a unique vacuum as well. The hope (which is mostly based on wishful thinking) is that when nonperturbative effects are taken into account, the theory will resolve the degeneracy problem.

The techniques that are being used to construct new solutions can be separated into two broad categories — bosonic and fermionic. As is well-known by now, two-dimensional field theories can be equivalently described in terms of bosonic or fermionic fields whenever the bosonic fields take their values in a compact domain. As a result, all the internal (non space-time) degrees of freedom in any particular solution can be equivalently described in terms of bosonic or fermionic degrees of freedom. Some cases are easier to describe using bosonic variables and others using fermionic variables. It is also possible to use some of each, and there are undoubtedly cases where that would prove useful.

The simplest example of a bosonic solution is one in which the coordinates of "extra dimensions" describe a torus[16]. This can be equivalently described in terms of fermionic coordinates. However, free fermions with simple boundary conditions corresponds to special values of the compactification radii. In general, it is necessary to introduce either four-fermi interactions or twisted boundary conditions in the equivalent fermi theory[17].

Compactification

It is customary to describe the heterotic and type II theories as having ten space-time dimensions, but admitting solutions in which some of them are "compactified." However, some of the constructions considered during the past year or so are sufficiently intricate that it is doubtful whether they admit a straightforward interpretation of this type. Even though this is a tempting way to describe the situation in popular or semipopular presentations, something of which I am guilty myself, it does not seem to be valid in general. It is probably better to say that the dimension of space-time is not an intrinsic property of the theory itself, but rather of any particular solution, and that any value $D \leq 10$ can be obtained. In each case there are a certain number of internal degrees of freedom (which can be taken to be bosonic or fermionic) that are required to saturate the conformal anomaly.

There is by now a vast literature on solutions using bosonic and fermionic formulations. It is not a manageable task for me to review all of them here. Therefore, I will briefly discuss here a few highlights of the bosonic approach, and emphasize fermionic approaches in subsequent sections. Both approaches were used in the original heterotic string papers[9], where it was shown that the left-moving degrees of freedom (for a $D = 10$) solution consist of ten space-time coordinates, plus additional degrees of freedom that could be taken to be either 16 free bosonic variables parametrizing a certain 16-dimensional torus, or 32 free fermionic variables with appropriate boundary conditions.

The bosonic formulation of the heterotic string was generalized to the case of $D < 10$ in a seminal paper of Narain[18]. To define a heterotic string ground state with $D = 10 - n$, let us compactify $16 + n$ left-movers and $n$ right-movers in a $(16 + 2n)$-dimensional torus. This basically means that each dimension forms a circle, although the circles are not required to be orthogonal to one another. The momentum quantum number for a periodic coordinate is quantized in integer multiples of a fundamental unit. Thus the compactified dimensions have discrete momenta $(k_1, k_2, \ldots, k_{16+n}\,; l_1, l_2 \ldots l_n)$ associated with them. The set of all allowed momenta form a $(16 + 2n)$-dimensional lattice of points, which Narain showed must be self dual and even in order to have modular invariance. In carrying out the construction it is necessary to use a "Lorentzian" metric. Thus, for example, $(k\,; l)^2 \equiv k^2 - l^2$. The statement that the lattice is even means that this is an even integer for all allowed $(k\,; l)$. The set of all such self-dual lattices that are equivalent is described by $(16 + n)n$ parameters and can be put in correspondence with the coset space

$$\frac{SO(16+n,n)}{SO(16+n)\times SO(n)}. \tag{4}$$

Many different gauge symmetries can be obtained depending on which point of this space is selected. They all have rank $16 + 2n$, however. These results were reinterpreted in terms of background fields in ref. 19. Subsequently, various generalizations involving twists[20–23], odd self-dual lattices[24], and asymmetric orbifolds[25] have been introduced. This is a very interesting and active area that would deserve a separate review.

Until recently it was generally assumed that the type II theory could not have any realistic solutions. In particular, one saw no possibility of achieving chiral asymmetry or realistic gauge groups for $D = 4$. This situation has changed as a result of a number of remarkable developments. First, it was discovered that two-dimensional supersymmetry could be represented nonlinearly in a theory consisting entirely of free fermions[26,27]. This representation was utilized to obtain new solutions to type II and heterotic string theories by three different groups[28–30]. In particular, these papers demonstrated that gauge groups such as $SU(3)$ and $SU(2)$ could arise in $D = 4$ solutions of the type II theory. The method of construction seemed to allow for chiral asymmetry, although none of the groups gave such an example. Recently, Dixon, Kaplunovsky, and Vafa[31] have found chiral examples and examples with $SU(3) \times SU(2) \times U(1)$ gauge symmetry. However, they claim that it is not possible to obtain a realistic spectrum of fermions and are therefore proclaiming the death of the type II string. They may well be right, but I would prefer to reserve final judgment.

## LOCAL SYMMETRIES ON THE WORLD SHEET

### Conformal Symmetry

Local symmetries of the world-sheet action have been recognized for a long time to play a fundamental role in string theory. Reparametrization invariance (two-dimensional general coordinate invariance) was first emphasized by Nambu[32]. This was recast using a two-dimensional metric tensor and extended to local supersymmetry on the world sheet by two groups[33]. This formulation was used by Polyakov, who drew attention to local conformal symmetry and the necessity of canceling the conformal anomaly[34]. The understanding of the fundamental role of conformal symmetry was developed further by Friedan[35], Alvarez[36], Belavin, Polyakov, and Zamolodchikov[37], Friedan, Martinec, and Shenker[38], and others.

The upshot of these works is that the reparametrization invariance allows one to choose a conformally flat gauge in which the two-dimensional metric takes the form

$$g_{\alpha\beta}(\sigma,\tau) = \phi(\sigma,\tau)\eta_{\alpha\beta} \quad . \tag{5}$$

Conformal invariance ensures that the mode $\phi$ decouples from the theory as well. It is convenient to Wick rotate the world-sheet action to Euclidean metric so that the surface can be regarded as a Riemann surface described (in coordinate patches) by a holomorphic variable $z$ and its antiholomorphic conjugate $\bar{z}$. The energy–momentum tensor $T_{\alpha\beta}$ plays a fundamental role. Scale invariance ensures that its trace $T_{z\bar{z}} = 0$. Conformal transformations $z \to f(z)$ are then generated by $T_{zz}$ and $\bar{z} \to \bar{f}(\bar{z})$ by $T_{\bar{z}\bar{z}}$. More precisely, $z \to z + f(z)$, for infinitesimal $f$, is generated by

$$T_f = \oint \frac{dz}{2\pi i} f(z) T_{zz}(z) \quad . \tag{6}$$

Henceforth, we simply write $T(z)$ for $T_{zz}(z)$ and do not mention $T_{\bar{z}\bar{z}}(\bar{z})$, which has analogous properties.

A holomorphic (right-moving) field $\phi(z)$ is said to have conformal dimension $h$ if as $z$ approaches $w$

$$T(z)\phi(w) \sim \frac{h\phi(w)}{(z-w)^2} + \frac{\partial_w \phi(w)}{z-w} + \text{finite} \quad , \tag{7}$$

which is the operator product expansion (OPE) version of the statement that

$$\phi(z) \to \left(\frac{\partial z'}{\partial z}\right)^h \phi(z') \tag{8}$$

under the conformal transformation $z \to z'(z)$. Introducing Laurent expansions

$$T(z) = \sum_{-\infty}^{\infty} \frac{L_n}{z^{n+2}} \tag{9}$$

$$\phi(z) = \sum_{-\infty}^{\infty} \frac{\phi_n}{z^{n+h}} \quad , \tag{10}$$

the OPE becomes equivalent to the commutation relations

$$[L_m, \phi_n] = ((h-1)m - n)\phi_{m+n} \quad . \tag{11}$$

$T(z)$ is essentially an operator of conformal dimension two except that in general there is a conformal anomaly given by the first term in the OPE

$$T(z)T(w) \sim \frac{c}{2(z-w)^4} + \frac{2T(w)}{(z-w)^2} + \frac{\partial_w T(w)}{z-w} + \text{finite} \tag{12}$$

or, equivalently,

$$[L_m, L_n] = (m-n)L_{m+n} + \frac{c}{12}(m^3 - m)\delta_{m+n} \quad , \tag{13}$$

where $\delta_k$ is one if $k = 0$ and zero otherwise. A necessary requirement for conformal invariance is that the conformal anomaly cancels when all contributions to $T(z)$ (and $\bar{T}(\bar{z})$) are included.

A free scalar field with $h = 0$ has OPE

$$\phi(z)\phi(w) \sim -\ell n(z-w) + \ldots \tag{14}$$

and contributes a piece $-\frac{1}{2} : \partial_z\phi\partial_z\phi :$ to $T(z)$. By definition, the dots mean removing the singular part as follows

$$: \partial_z\phi\partial_z\phi := \lim_{w\to z}\left(\partial_z\phi\partial_w\phi + \frac{1}{(z-w)^2}\right) . \tag{15}$$

They will be implicit in subsequent formulas. Such an $h = 0$ field gives a contribution of 1 to the conformal anomaly $c$. An important example is the space-time coordinate $X^\mu(z,\bar{z})$, which can be written $X^\mu(z) + \bar{X}^\mu(\bar{z})$ in conformal gauge. In dimension $D$ it gives a contribution of $D$ to $c$ and $\bar{c}$.

Similarly, a free fermi field $\psi(z)$ with $h = \frac{1}{2}$ and OPE

$$\psi(z)\psi(w) \sim \frac{1}{z-w} \tag{16}$$

gives a contribution $-\frac{1}{2}\psi\partial_z\psi$ to $T(z)$ and $\frac{1}{2}$ to $c$. When bose fields taking values on a compact domain are replaced by fermi fields or vice versa, it is necessary to associate two fermi fields to each bose field in order to preserve the conformal anomaly.

In the Faddeev–Popov analysis of the path integral the choice of conformal gauge (or superconformal gauge in the $N = 1$ case) results in the introduction of pairs of ghost fields on the world sheet. Generically, a pair of ghost fields $b$ and $c$, with conformal dimension $\lambda$ and $1 - \lambda$, respectively, have an OPE

$$c(z)b(w) \sim \frac{1}{z - w} \quad . \tag{17}$$

Moreover, they give a contribution to $T$ of the form

$$T = -\lambda b\partial c + (1 - \lambda)(\partial b)c \tag{18}$$

and a contribution to the conformal anomaly

$$c(\epsilon, \lambda) = -2\epsilon(6\lambda^2 - 6\lambda + 1), \tag{19}$$

where $\epsilon = +1$ if they satisfy fermi statistics and $\epsilon = -1$ if they satisfy bose statistics. If the holomorphic modes have no supersymmetry ($N = 0$) then there is a single pair of ghosts (associated with reparametrization invariance) satisfying $\epsilon = 1$ and $\lambda = 2$. Thus $c^{gh} = -26$ in this case, and the conformal anomaly from all other sources must total $+26$ in order to give $c = 0$ altogether. For example, 26 spatial coordinates $X^\mu$ would work. In the case of $N = 1$ world-sheet supersymmetry, the choice of superconformal gauge (in which one also puts the gravitino field $\chi_\alpha = \rho_\alpha \eta$) gives an additional pair of ghost fields with $\epsilon = -1$ and $\lambda = \frac{3}{2}$. Since $c(-1, \frac{3}{2}) = 11$, the total ghost contribution to the conformal anomaly in this case is $c^{gh} = -15$. This must again be balanced by other contributions. For example, ten pairs $(X^\mu, \psi^\mu)$ would work.

Kac–Moody Algebras

Sometimes we are also interested in local Lie group symmetries on the world sheet as well. Holomorphic currents with conformal dimension $h = 1$ satisfy the OPE*

$$J^a(z)J^b(w) \sim \frac{k\delta^{ab}}{(z - w)^2} + \frac{if^{abc}J^c(w)}{z - w} + \ldots \tag{20}$$

where $k$ is called the "level" of the representation and $f^{abc}$ are the structure constants of the Lie algebra. This OPE is equivalent to the Kac–Moody algebra $\hat{g}$

$$[J^a_m, J^b_n] = if^{abc}J^c_{m+n} + km\delta^{ab}\delta_{m+n} \quad . \tag{21}$$

The generators $J^a_0$ give the associated Lie algebra $g$. (For a review of Kac–Moody and Virasoro algebras see ref. 39.) An important class of examples can be constructed as bilinears in fermi fields. Specifically,

$$J^a(z) = \frac{i}{2}\lambda^a_{\alpha\beta}\psi^\alpha(z)\psi^\beta(z) \quad , \tag{22}$$

where $\lambda$ is a real representation $R$ of $g$. It satisfies

$$[\lambda^a, \lambda^b] = f^{abc}\lambda^c \tag{23}$$

and

$$tr(\lambda^a\lambda^b) = -c_R\delta^{ab} \quad , \tag{24}$$

giving a KM algebra $\hat{g}$ with $k = c_R/2$. In particular, we will be interested in the adjoint

* Many authors use normalization conventions for $c_R$, $f^{abc}$, etc. that differ by factors of two from those used here. Our conventions agree with ref. 39.

representation given by $\lambda^a_{bc} = -f^{abc}$, which gives $k = c_A/2$, where

$$f^{abc}f^{a'bc} = c_A\delta^{aa'} \quad . \tag{25}$$

Superconformal Symmetry

In theories with world-sheet supersymmetry in superconformal gauge, there is a conserved $h = \frac{3}{2}$ supercurrent, whose holomorphic part we denote $T_F(z)$. Together with the energy–momentum tensor, which we now denote $T_B(z)$, it forms a superconformal algebra with OPE

$$T_F(z)T_F(w) \sim \frac{\hat{c}}{4(z-w)^3} + \frac{T_B(w)}{2(z-w)} + \dots \quad . \tag{26}$$

In terms of components defined by Laurent expansions (9) and

$$T_F(z) = 2\sum_{-\infty}^{\infty} G_n z^{-n-3/2} \; , \tag{27}$$

this corresponds to the anti-commutation relations

$$\{G_m, G_n\} \;=\; 2L_{m+n} + \frac{\hat{c}}{2}(m^2 - \frac{1}{4})\delta_{m+n} \quad . \tag{28}$$

One also has

$$[L_m, G_n] \;=\; \left(\frac{m}{2} - n\right) G_{m+n} \, . \tag{29}$$

$$[L_m, L_n] \;=\; (m-n)L_{m+n} + \frac{\hat{c}}{8}(m^3 - m)\delta_{m+n} \quad , \tag{30}$$

so that $c = \frac{3}{2}\hat{c}$.

An explicit representation is given by

$$T_B \;=\; -\frac{1}{2}(\partial X^\mu)^2 - \frac{1}{2}\psi_\mu\partial\psi^\mu - 2b\partial c - (\partial b)c - \frac{3}{2}\beta\partial\gamma - \frac{1}{2}(\partial\beta)\gamma \tag{31}$$

$$T_F \;=\; \frac{i}{2}\psi_\mu\partial X^\mu + \frac{1}{2}b\gamma - (\partial\beta)c - \frac{3}{2}\beta\partial c \quad , \tag{32}$$

which gives $\hat{c} = D - 10$. Thus the superconformal anomaly cancels for $D = 10$.

It is sometimes convenient to use a superspace formalism involving a single Grassmann parameter $\theta$[40,41]. One can then combine $T_F$ and $T_B$ into a single expression

$$T(z,\theta) \;=\; T_F(z) + \theta T_B(z) \tag{33}$$

whose OPE is

$$T(z_1,\theta_1)T(z_2,\theta_2) \sim \frac{\hat{c}}{4z_{12}^3} + \frac{3\theta_{12}}{2z_{12}^2}T(z_2,\theta_2)$$

$$+\frac{D_2T(z_2,\theta_2)}{2z_{12}} + \frac{\theta_{12}}{z_{12}}\partial_2 T(z_2,\theta_2) + \dots \quad , \tag{34}$$

where

$$z_{12} = z_1 - z_2 - \theta_1\theta_2 \tag{35}$$

$$\theta_{12} = \theta_1 - \theta_2 \tag{36}$$

$$D = \frac{\partial}{\partial\theta} + \theta\partial_z \quad . \tag{37}$$

This describes the entire superconformal algebra. Note that $\theta_{12}$ and $z_{12}$ are invariant under the supersymmetry transformations $\delta\theta_i = \epsilon$, $\delta z_i = \theta_i\epsilon$. A superfield $\phi(z,\theta)$ with components of conformal dimension $h$ and $h+\frac{1}{2}$ satisfies

$$T(z_1,\theta_1)\phi(z_2,\theta_2) \sim h\frac{\theta_{12}}{z_{12}^2}\phi(z_2,\theta_2) + \frac{1}{2z_{12}}D_2\phi + \frac{\theta_{12}}{z_{12}}\partial_2\phi + \ldots \quad . \tag{38}$$

<u>Super Kac–Moody Algebras</u>

It is natural to try to extend the Kac–Moody algebra to a super Kac–Moody algebra generated by a supercurrent

$$J^a(z,\theta) = J_F^a(z) + \theta J_B^a(z) \quad , \tag{39}$$

with $h = \frac{1}{2}$. $J_B^a(z)$ is the $h = 1$ current introduced earlier. The appropriate OPE is

$$J^a(z_1,\theta_1)J^b(z_2,\theta_2) \sim k\frac{\delta^{ab}}{z_{12}} - i\frac{\theta_{12}}{z_{12}}f^{abc}J^c(z_2,\theta_2) + \ldots . \tag{40}$$

For the case of a representation based on adjoint representation fermi fields

$$J_B^a = -\frac{i}{2}f^{abc}\psi^b\psi^c \quad , \tag{41}$$

we saw earlier that $k = c_A/2$. The only reasonable guess for $J_F^a$ (an adjoint fermi field with $h = \frac{1}{2}$) is that it be proportional to $\psi^a$. A short calculation verifies that the algebra is satisfied for the normalization

$$J_F^a = \sqrt{k}\,\psi^a \quad . \tag{42}$$

If we are to utilize the super Kac–Moody algebra described above in string theory, we also require a representation of $T(z,\theta)$ based on $\psi^a$. We have already seen that we can take $T_B = -\frac{1}{2}\psi^a\partial\psi^a$, obtaining $c = N/2$ (or $\hat{c} = N/3$). Now we also need to make an $h = 3/2$ fermionic operator that transforms as a singlet of the group out of $\psi^a$. The only reasonable thing to try is a constant times $f_{abc}\psi^a\psi^b\psi^c$. It was discovered in refs. [26,27] that this does, in fact, work for the normalization

$$T_F = -\frac{i}{12\sqrt{k}}f^{abc}\psi^a\psi^b\psi^c \quad . \tag{43}$$

Thus, remarkably, the entire combined super Kac–Moody superconformal algebra can be represented in terms of free fermions $\psi^a$ in the adjoint representation of a Lie group. It is essential that the group be semisimple (i.e., no $U(1)$ factors), since the closure of the superconformal algebra requires that $f^{abc}f^{a'bc} \propto \delta^{aa'}$ as well as the Jacobi identity. This discovery paves the way to new solutions of superstring theories based on a purely fermionic description of the internal degrees of freedom. Some possibilities will be explored in subsequent sections.

As we have already emphasized, a necessary requirement for a consistent solution of a string theory is the cancellation of the conformal (or superconformal) anomaly. When there is no world-sheet supersymmetry (as for the left movers of the heterotic string) the $c = -26$ conformal anomaly of the ghosts must be canceled by other contributions. If there are $D$ space-time dimensions $X^\mu(z, \bar{z})$ and we choose to represent the remaining left-moving degrees of freedom in terms of fermionic variables $\lambda^A(\bar{z})$, we must require that the index $A$ takes $2(26 - D)$ values (32 for $D = 10$ or 44 for $D = 4$).

Similarly, if there is $N = 1$ supersymmetry on the world sheet, as in the case of the right movers of the heterotic string or both left and right movers of the type II superstring, the superconformal anomaly $\hat{c} = -10$ must be canceled by other contributions. If we represent the superconformal algebra of $D$ space-time dimensions in terms of fields $(X^\mu, \psi^\mu)$ and the remaining internal degrees of freedom in terms of fermionic coordinates $\psi^a$ belonging to the adjoint representation of a semisimple Lie algebra, then the analysis given above shows that the algebra must have $3(10 - D)$ generators in order to cancel the superconformal anomaly. In the case of $D = 4$, we require a semisimple group with eighteen generators. There are only a few possibilities, namely $[SU(2)]^6$, $SU(4) \times SU(2)$, and $SU(3) \times SO(5)$. In each case the coupling constants of the various factors are related. We thus have the exciting prospect of obtaining interesting gauge groups that are effectively unified in type II superstring theory. This possibility was pointed out in refs. [28–30].*

Other possibilities can be achieved by describing some of the internal degrees of freedom with bosonic fields. For example, there is a $\hat{c} = \frac{2}{3}$ representation of the superconformal algebra based entirely on a scalar field.† (This is one of the discrete series of unitary representations in the superconformal case[44].) Thus it can be used in place of a pair of fermi fields. For example, $SU(3) \times SU(3)$ becomes possible if one pair of fermi fields are replaced, $G_2$ becomes possible if two pairs are replaced, and so forth.

The super Kac–Moody construction given above can be generalized as follows. Suppose the KM algebra $\hat{g}$, for which we have given a level $c_A/2$ representation in terms of adjoint fermions $\psi^a$, also has an independent level $\tilde{k}$ representation with currents $\tilde{J}^a$. This could be provided, for example, by a group manifold described by a nonlinear sigma model with Wess–Zumino term[45]. These sigma models have supersymmetric extensions[46], which can be understood without describing their explicit construction. Just as with the addition of angular momentum, one can form a representation of the KM algebra with level

$$k = \tilde{k} + c_A/2 \tag{44}$$

by forming the sum

$$J_B^a = \tilde{J}^a - \frac{i}{2} f^{abc} \psi^b \psi^c \quad . \tag{45}$$

This can be extended to a super KM algebra by simply taking[47]

$$J_F^a = \sqrt{k} \psi^a \quad . \tag{46}$$

The next question is whether we can extend the representation of the superconformal algebra to this case. Consider first the conformal symmetry current $\tilde{T}(z)$ that

---

⋆ The $[SU(2)]^6$ model was obtained previously by means of a toroidal compactification in ref. 42.
† In fact, such a field can even be used to represent the $N = 2$ superconformal algebra[43].

accompanies the current $\tilde{J}^a(z)$. This is given by the Sugawara construction[48,39]

$$\tilde{T}(z) = \frac{1}{2\tilde{k}} : \tilde{J}^a(z)\tilde{J}^a(z) : \quad . \tag{47}$$

Note that the coefficient involves the total level given in (44). The central charge associated with this conformal current is

$$\tilde{c} = \frac{\tilde{k}N}{k} \quad , \tag{48}$$

where $N$ is the dimension of the group as before. Therefore the central charge of the sum

$$T_B(z) = -\frac{1}{2}\psi^a(z)\partial\psi^a(z) + \tilde{T}(z) \tag{49}$$

is

$$c = \frac{N}{2} + \tilde{c} = \left(\frac{3}{2} - \frac{c_A}{2k}\right) N \quad . \tag{50}$$

or

$$\hat{c} = \left(1 - \frac{c_A}{3k}\right) N \quad . \tag{51}$$

To complete the analysis we still need to form the fermionic current $T_F(z)$. Requiring that it satisfy the OPE

$$T_F(z_1)J_F^a(z_2) \sim \frac{1}{2(z_1 - z_2)} J_B^a(z_2) + \ldots \tag{52}$$

motivates the guess

$$T_F(z) = -\frac{i}{12\sqrt{k}} f^{abc}\psi^a\psi^b\psi^c + \frac{1}{2\sqrt{k}}\psi^a \tilde{J}^a \quad . \tag{53}$$

With somewhat more work it can be verified that this choice satisfies the superconformal algebra as well. This procedure thus gives a consistent superconformal current algebra. In fact, all unitary representations can be obtained in this way[49].

To illustrate the above we note that $SU(n)$ has $c_A = 2n$. Thus for $SU(2)$ we have

$$\hat{c} = 3 - \frac{4}{2 + \tilde{k}} = 1,\ 5/3,\ 2, \ldots \tag{54}$$

and for $SU(3)$

$$\hat{c} = 8 - \frac{16}{3 + \tilde{k}} = 8/3,\ 4,\ 24/5, \ldots \quad . \tag{55}$$

Therefore $\hat{c} = 6$ can be achieved for $SU(3)$ with $\tilde{k} = 5$. Also, $SU(3) \times SU(2)$ can give $\hat{c} = 6$ if we choose $\tilde{k} = 1$ for the $SU(3)$ factor and $\tilde{k} = 2$ for the $SU(2)$ factor. A third possibility is to achieve $\hat{c} = 6$ for $SU(2) \times SU(2) \times SU(2)$ by choosing $\tilde{k} = 2$ for each $SU(2)$ factor.

We can now describe the essential elements of the arguments of ref. 31, which claims that it is not possible to construct a realistic four-dimensional solution of type II superstring theory. A realistic solution should contain $SU(3) \times SU(2) \times U(1)$ or a larger algebra. Also, the spectrum of massless fermions should contain some triplets of $SU(3)$ and doublets of $SU(2)$. The argument has two key steps. The first is that all the group theory should arise entirely from left-movers. Accepting that, it is then necessary to find a super Kac–Moody algebra with $\hat{c} \leq 6$. To obtain doublets of $SU(2)$ and triplets of $SU(3)$ requires that $\tilde{k} \geq 1$ in each case. Thus, in view of (51), the contribution of $SU(3)$ with $\tilde{k} \geq 1$ is $\hat{c} \geq 4$ and of $SU(2)$ with $\tilde{k} \geq 1$ is $\hat{c} \geq 5/3$. In addition, $U(1)$ requires $\hat{c} = 1$, and thus the total necessarily exceeds six. Considering larger groups that contain these as subgroups only makes matters worse. For example, the $\tilde{k} = 0$ representation of $SU(5)$ has $\hat{c} = 8$.

Even though a representation of the desired type does not exist, it is tantalizing how close we can come. One way of achieving $\hat{c} = 6$, for instance, is to choose $SU(3) \times SU(2) \times U(1)$ with $\tilde{k} = 1$ for the $SU(3)$ factor and $\tilde{k} = 0$ for the $SU(2)$ factor. Another possibility is to choose $\tilde{k} = 0$ for the $SU(3)$ factor and $\tilde{k} = 2$ for the $SU(2)$ factor giving a total of 16/3. The remaining 2/3 can be represented by an additional bosonic field.

## CONCLUSION

Much interesting work has been done in string theory during the last two years. Thanks to the concerted efforts of many clever people, the pace of progress is very rapid. Nothing that has been learned makes us less optimistic than before that the ultimate goal of achieving an understanding of the fundamental principles of elementary particle physics is possible. Still, it is clear that we need to learn much more about the fundamentals of string theory, the role of nonperturbative effects, and the mechanisms by which a vacuum configuration is determined. It would be dumb luck if significant contact with experimental results were made prior to achieving such understandings, which could require another 10–20 years.

We have reviewed some interesting recent progress in the construction of solutions to type II string theories. The number of possibilities, just within the framework we have described, is enormous. Moreover, it is clear that there are various extensions and generalizations of the techniques described that could lead to even more solutions consistent with conformal symmetry, modular invariance, and world-sheet supersymmetry.

There are a number of questions that deserve to be investigated further. First, the argument of ref. 31 that a realistic gauge group and fermion spectrum is not possible for a type II solution needs to be checked carefully. That conclusion would be somewhat disappointing, since a type II solution typically gives groups of about the right size and a reasonable number of massless fermions, whereas heterotic solutions typically give groups of high rank and many more massless fermions. Of course, we know from ref. 13, for example, that the heterotic theory compactified on an appropriate Calabi–Yau space can give realistic groups and representations. Since $\hat{c} = 6$ for the internal degrees of freedom of a type II string is too restrictive to allow realistic symmetries and fermion representations, one might be tempted to try modifying the theory. In particular, one could try to keep (1,1) supersymmetry on the world sheet while allowing for a larger total $\hat{c}$. A couple of ideas have been discussed in the recent literature that might conceivably make this possible, although each still requires further checking to make sure it does not lead to inconsistencies. One such proposal is to introduce vector superfields on the world sheet. Since quantization of the vectors leads to additional ghosts, the conformal anomaly required for the physical degrees of freedom would be altered. A second possibility that also results in an altered value of the central charge is to introduce quartic couplings of the ghost fields in the world-sheet action[50]. It is not known what this implies for the action prior to gauge fixing, although analogous structures have been uncovered in supergravity theories by canonical procedures.

It is implicit in the program described here that a classical solution bears sufficient resemblance to the full nonperturbative result that it makes sense to look for realistic groups and representations at this stage of the analysis. Until we understand the mechanism that breaks supersymmetry, while maintaining a vanishing cosmological constant, we cannot be confident about these matters. (An intriguing proposal for achieving this has been made by Moore[51].) Maybe then we will also have a better theoretical understanding of how a particular vacuum configuration is selected.

I am grateful to M. Douglas and L. Dixon for discussions.

## REFERENCES

1. J.H. Schwarz, *Superstrings, The First Fifteen Years of Superstring Theory*, two volumes (World Scientific, 1985); M. Green and D. Gross, eds., *Unified String Theories*, (World Scientific, 1986); M.B. Green, J.H. Schwarz, and E. Witten, *Superstring Theory*, two volumes (Cambridge University Press, 1987).
2. J. Scherk and J. H. Schwarz, Nucl. Phys. B81 (1974) 118; T. Yoneya, Prog. Theor. Phys. 51 (1974) 1907.
3. P. Ramond, Phys. Rev. D3 (1971) 2415; A. Neveu and J.H. Schwarz, Nucl. Phys. B31 (1971) 86.
4. F. Gliozzi, J. Scherk, and D. Olive, Nucl. Phys. B122 (1977) 253.
5. M. B. Green and J.H. Schwarz, Phys. Lett. 149B (1984) 117; Phys. Lett. 151B (1985) 21.
6. J.A. Harvey and J.A. Minahan, Phys. Lett. 188B (1987) 44.
7. J. Shapiro, Phys. Lett. 33B (1970) 361; M. Virasoro, Phys. Rev. 177 (1969) 2309.

8. M.B. Green and J.H. Schwarz, Phys. Lett. 109B (1982) 444.

9. D.J. Gross, J.A. Harvey, E. Martinec, and R. Rohm, Nucl. Phys. B256 (1985) 253; Nucl. Phys. B267 (1986) 75.

10. L.J. Dixon and J.A. Harvey, Nucl. Phys. B274 (1986) 93; L. Alvarez-Gaumé, P. Ginsparg, G. Moore and C. Vafa, Phys. Lett. 171B (1986) 155.

11. P. Candelas, G.T. Horowitz, A. Strominger, and E. Witten, Nucl. Phys. B258 (1985) 46; E. Witten, Nucl. Phys. B258 (1985)75; P. Candelas, A.M. Dale, C.A. Lütken, and R. Schimmrigk, "Complete intersection Calabi–Yau manifolds," Univ. of Texas preprint UTTG-10-87.

12. L. Dixon, J.A. Harvey, C. Vafa, and E. Witten, Nucl. Phys. B261 (1985) 651; Nucl. Phys. B274 (1986) 285.

13. B.R. Greene, K.H. Kirklin, P.J. Miron, and G.G. Ross, Phys. Lett. 180B (1986) 69; Nucl. Phys. B278 (1986) 667.

14. P. Orland, Nucl. Phys. B278 (1986) 790.

15. P. Freund, Phys. Lett. 151B (1985) 387; A. Casher, F. Englert, H. Nicolai, and A. Taormini, Phys. Lett. 162B (1985) 121; F. Englert, H. Nicolai, and A. Schellekens, Nucl. Phys. B274 (1986) 315. D. Lüst, Nucl. Phys. B292 (1987) 381.

16. M.B. Green, J.H. Schwarz, and L. Brink, Nucl. Phys. B198 (1982) 474.

17. J. Bagger, D. Nemeschansky, N. Seiberg, and S. Yankielowicz, Nucl. Phys. B289 (1987) 53; H. Neuberger, A.J. Niemi, and G.W. Semenoff, Phys. Lett. 181B (1986) 244.

18. K.S. Narain, Phys. Lett. 169B (1986) 41.

19. K.S. Narain, M.H. Sarmadi, and E. Witten, Nucl. Phys. B279 (1987) 369.

20. J.A. Harvey, in *Unified String Theories*, (World Science, 1986), p. 704; V.P. Nair, A. Shapere, A. Strominger, and F. Wilczek, Nucl. Phys. B287 (1987) 402.

21. M. Mueller and E. Witten, Phys. Lett. 182B (1986) 28.

22. Y. Watabiki, Phys. Lett. 180B (1986) 340.

23. G. Cristofano, G. Maiella, R. Musto, F. Nicodemi, R. Pettorini, Int. J. Mod. Phys. A2 (1987) 729; H. Itoyama and T.R. Taylor, Phys. Lett. 186B (1987) 129; P. Ginsparg and C. Vafa, Nucl. Phys. B289 (1987) 414.

24. W. Lerche and D. Lüst, Phys. Lett. 187B (1987) 45; W. Lerche, D. Lüst, and A.N. Schellekens, Nucl. Phys. B287 (1987) 477.

25. K.S. Narain, M.H. Sarmadi, and C. Vafa, Nucl. Phys. B288 (1987) 551.

26. P. Goddard, W. Nahm, and D. Olive, Phys. Lett. 160B (1985) 111; P. Goddard, A. Kent, and D. Olive, Commun. Math. Phys. 103 (1986) 105.

27. I. Antoniadis, C. Bachas, C. Kounnas, and P. Windey, Phys. Lett. 171B (1986) 51.

28. H. Kawai, D.C. Lewellen, S.-H.H. Tye, Phys. Lett. 191B (1987) 63.

29. I. Antoniadis, C.P. Bachas, and C. Kounnas, Nucl. Phys. B289 (1987) 87.

30. R. Bluhm, L. Dolan, and P. Goddard, Nucl. Phys. B289 (1987) 364.

31. L. Dixon, V. Kaplunovsky a. C. Vafa, "On four-dimensional gauge theories from type II superstrings," SLAC preprint SLAC-PUB-4282 (3/87); W. Lerche, B.E.W. Nilsson, and A.N. Schellekens, "Covariant lattices, superconformal invariance and strings," CERN preprint TH.4692/87 (3/87).

32. Y. Nambu, Lectures at the Copenhagen symposium, 1970.

33. L. Brink, P. DiVecchia, and P. Howe, Phys. Lett. 65B (1976) 471; S. Deser and B. Zumino, Phys. Lett. 65B (1976) 369.

34. A.M. Polyakov, Phys. Lett. 103B (1981) 207; Phys. Lett. 103B (1981) 211.

35. D. Friedan, in *Recent Advances in Field Theory and Statistical Mechanics,* eds. J.B. Zuber and R. Stora (Elsevier, 1984), p. 839.

36. O. Alvarez, Nucl. Phys. B216 (1983) 125.

37. A.A. Belavin, A.M. Polyakov, and A.B. Zamolodchikov, Nucl. Phys. B241 (1984) 333.

38. D. Friedan, E. Martinec, and S. Shenker, Nucl. Phys. B271 (1986) 93.

39. P. Goddard and D. Olive, Int. J. Mod. Phys. A1 (1986) 303.

40. C. Montonen, Nuovo Cim. 19A (1974) 69; D.B. Fairlie and D. Martin, Nuovo Cim. 18A (1973) 373.

41. L. Brink and J.O. Winnberg, Nucl. Phys. B103 (1976) 445; E. Martinec, Phys. Rev. D28 (1983) 2604.

42. L. Castellani, R. D'Auria, F. Gliozzi, and S. Sciuto, Phys. Lett. 168B (1986) 47; R. Bluhm and L. Dolan, Phys. Lett. 169B (1986) 347.

43. G. Waterson, Phys. Lett. 171B (1986) 77.

44. D. Friedan, Z. Qiu, and S. Shenker, Phys. Lett. 151B (1985) 37.

45. V. Knizhnik and A.B. Zamolodchikov, Nucl. Phys. B247 (1984) 83; D. Gepner and E. Witten, Nucl. Phys. B278 (1986) 483; J. Fuchs, Nucl. Phys. B286 (1987) 455.

46. T.L. Curtright and C.K. Zachos, Phys. Rev. Lett. 53 (1984) 1799; P. Di Vecchia, V.G. Knizhnik, J.L. Petersen, and P. Rossi, Nucl. Phys. B253 (1985)701; R. Rohm, Phys. Rev. D32 (1985) 2849; E. Abdalla and M.C.B. Abdalla, Phys. Lett. 152B (1985) 59.

47. E.B. Kiritsis and G. Siopsis, Phys. Lett. 184B (1987) 353.

48. H. Sugawara, Phys. Rev. 170 (1968) 1659; C. Sommerfield, Phys. Rev. 176 (1968) 2019.

49. V.G. Kac and I.T. Todorov, Commun. Math. Phys. 105 (1985) 337.

50. D.Z. Freedman and N.P. Warner, Phys. Lett. 176B (1986) 87; D.Z. Freedman, P. Ginsparg, C.M. Sommerfield, and N.P. Warner, "String ghost interactions and the trace anomaly," MIT preprint CTP#1444 (1/87).

51. G. Moore, "Atkin–Lehner symmetry," Harvard preprint HUTP-87/A013 (3/87).

# FOUR-DIMENSIONAL STRINGS

L. Dolan

Physics Department, Box 272
The Rockefeller University
New York, NY 10021

## INTRODUCTION

It has been a puzzle to explain why the world we live in has four space-time dimensions. In recent years, unified superstring models have had many successes in describing within one theory the fundamental interactions of the strong, weak and electromagnetic forces of elementary particle physics, as well as providing the most ambitious approach to date for formulating a quantum theory of gravity.[1]

This paper reviews the original work [2,3] on the compactification of the ten-dimensional type II superstring giving rise to an arbitrary dimension eighteen semi-simple Lie group as the gauge group of a closed oriented supersymmetric string model in four dimensions. There are just three such groups: the one of rank 6, $SU(2)^6$, and two others, namely SU(3) x SO(5) and SU(2) x SU(4) which both turn out to be of rank 4, exactly the rank of, and containing, the standard group of the fundamental interactions SU(3) x SU(2) x U(1). As we shall see, this mechanism, although it does not explain independently why we live in four dimensions, has with its radically new way of introducing symmetry exactly matched the fact that we live in four dimensions with the size of the symmetry group which governs four-dimensional interactions.

In the original spirit of Kaluza-Klein, this compactification technique reduces the string theory with no internal gauge group for the matter in ten dimensions and inserts the non-abelian gauge symmetry of the strong, weak and electromagnetic interactions now coupled to gravity in four dimensions. It is to be distinguished from the compactification of various other ten-dimensional strings in that the world sheet supersymmetry, which in light-cone gauge insures space-time Lorentz invariance and in covariant gauge the absence of ghosts, gives rise in the extended supergravity examples discussed here, to an N=1 super-Virasoro algebra with central charge equal to 15 rather than 26, and thus may lead to a more economical and direct derivation of the realistic four-dimensional spectrum. Further questions relating to the correct space-time fermion content, calculation of the Weinberg angle, Yukawa couplings and proton decay may have to be answered non-perturbatively or in a non-conventional manner, and are under current investigation. For further references see Reference 4.

## THE PARTITION FUNCTION

The cosmological constant $\Lambda$ in a string theory is sometimes called the partition function. It is related to the function which counts the number of states at each mass level. By conformal invariance, $\Lambda$ vanishes at the tree level for any string model. In superstring theory, $\Lambda$ vanishes at one-loop and it is conjectured this remains true to all orders.

To describe consistent interacting strings, one must in general check that all the scattering amplitudes are modular invariant, finite, and unitary. A guide to this program is the calculation of the one-loop partition function which can be checked for modular invariance, albeit a quantity equal to zero. In the original papers [2,3], the four-point scattering amplitude was computed for the four-dimensional strings coming from type II. Here, only the partition function for these models is discussed. This allows for the classification of the particle spectrum. And together with a set of vertex rules, or the direct calculation of the n-point amplitudes, it leads to a consistent theory.

For closed strings, the one-loop cosmological constant is defined by

$$\Lambda \equiv \tfrac{1}{2}\mathrm{tr}\ \ln\Delta^{-1} \tag{1}$$

where

$$\Delta^{-1} = \alpha'(p^2+m^2)$$

$$\frac{\alpha' m^2}{2} = \alpha' m_L^2 + \alpha' m_R^2$$

so in D space-time dimensions, for $\omega = e^{2\pi i\tau}$,

$$\Lambda = -\tfrac{1}{2}(2\pi)^{-1}(\alpha')^{-\frac{D}{2}}\int_F d^2\tau(\mathrm{Im}\tau)^{-2-\left(\frac{D-2}{2}\right)} \tag{2}$$

$$\cdot\sum_{\text{all sectors}} \mathrm{tr}[\bar{\omega}^{\alpha' m_L^2}\ \omega^{\alpha' m_R^2}\ \text{possible projections}].$$

F is a fundamental region of the modular group:

$-\tfrac{1}{2}\leq \mathrm{Re}\tau\leq \tfrac{1}{2}$; $|\tau|>1$.

In light-cone gauge, the transverse left-moving and right-moving degrees of freedom of the four-dimensional "compactified" type II superstring for $i = (1,\ldots 8) = \hat{i};I$; $\hat{i} = 1,2$; $I = 1,\ldots 6$; $n\epsilon Z$ are given by Eq.'s 3,4 in each sector. The different sectors have different ratios of Neveu-Schwarz (NS) to Ramond (R) fields given in Eq.'s 3c, 4c for $r\epsilon Z+\tfrac{1}{2}$ or $r\epsilon Z$ respectively. The three models with the eighteen-dimensional gauge groups and N = 4 supergravity have four sectors each. For smaller groups and N<4 there will be in general more sectors. In some cases, the internal world sheet fermions (i restricted to I) may be replaced by other degrees of freedom, which maintain the appropriate conformal anomaly.

$$\tilde{X}^{\hat{i}}(\bar{z}) = \frac{x^{\hat{i}}}{2} - \frac{i}{2}\alpha' p^{\hat{i}}\ln\bar{z} + \frac{i}{2}\sqrt{2\alpha'}\sum_{n\neq 0}\frac{1}{n}\tilde{A}_n^{\hat{i}}\bar{z}^{-n} \tag{3a}$$

$$\tilde{X}^I(\bar{z}) = \tilde{x}^I - i\alpha'\tilde{p}^I\ln\bar{z} + \frac{i}{2}\sqrt{2\alpha'}\sum_{n\neq 0}\frac{1}{n}\tilde{A}_n^I\bar{z}^{-n} \tag{3b}$$

$$\tilde{H}^i(\bar{z}) = \sum_r \tilde{h}_r^i\bar{z}^{-r} \tag{3c}$$

$$X^{\hat{i}}(z) = \frac{x^{\hat{i}}}{2} - \frac{i}{2}\alpha' p^{\hat{i}} \ln z + \frac{i}{2}\sqrt{2\alpha'} \sum_{n\neq 0} \frac{1}{n} A_n^{\hat{i}} z^{-n} \tag{4a}$$

$$X^{I}(z) = \bar{x}^{I} - i\alpha' \bar{p}^{I} \ln z + \frac{i}{2}\sqrt{2\alpha'} \sum_{n\neq 0} \frac{1}{n} A_n^{I} z^{-n} \tag{4b}$$

$$H^{i}(z) = \sum_r h_r^{i} z^{-r} \tag{4c}$$

The $SU(2)^6$ model corresponds to requiring the internal left and right momenta to be quantized on the hypercubic lattice $Z^6$ with basic vectors of unit length squared:

$$\sqrt{2\alpha'}\,\tilde{p}^{I} = \sum_{L=1}^{6} N^{L}\delta_{L}^{I}, \quad \sqrt{2\alpha'}\,\bar{p}^{I} = \sum_{L=1}^{6} N^{L}\delta_{L}^{I}, \quad (N^{L}\epsilon Z), \tag{5}$$

and additional sectors where $\sqrt{2\alpha'}\,\tilde{p}^{I}$ and $\sqrt{2\alpha'}\,\bar{p}^{I}$ take values on $(Z+\frac{1}{2})^6$. The other groups $SU(3) \times SO(5)$ and $SU(2) \times SU(4)$ can be shown to correspond to other internal lattices (after bosonization of the world sheet fermions), but were first discovered by fermionizing the internal bosons on $Z^6$ or $(Z+\frac{1}{2})^6$ by

$$\tilde{A}_n^{I} = -i\sum_r \tilde{h}_{(1)r}^{I}\, \tilde{h}_{(2)n-r}^{I}$$

$$A_n^{I} = -i\sum_r h_{(1)r}^{I}\, h_{(2)n-r}^{I}$$

(no sum on I), (6)

where for $Z^6$, $h_r$ are NS and for $(Z+\frac{1}{2})^6$, $h_r$ are R. The different gauge group algebras will be discussed in the next section. The four-dimensional mass formula in a given sector is

$$\alpha' m_L^2 = N_{\tilde{A}}^{\hat{i}} + N_{\tilde{h}}^{\hat{i}} + N_{\tilde{h}}^{a} - \tilde{\alpha}_0 \tag{7a}$$

$$\alpha' m_R^2 = N_{A}^{\hat{i}} + N_{h}^{\hat{i}} + N_{h}^{a} - \alpha_0 \tag{7b}$$

where

$$N_A^{\hat{i}} = \sum_{n=1}^{\infty} A_{-n}^{\hat{i}} A_n^{\hat{i}} \qquad \text{(sum on } \hat{i}\text{, etc.)}$$

$$N_h^{\hat{i}} = \sum_{r>0} r h_{-r}^{\hat{i}} h_r^{\hat{i}}$$

$$h_r^{a} = (h_r^{I}, h_{(1)r}^{I}, h_{(2)r}^{I}); \; a = 1,\ldots 18$$

$$N_h a = \sum_{r>0} r h^a_{-r} h^a_r$$

$$\alpha_0 = -\frac{1}{12} - \frac{b}{48} + \frac{d}{24} \quad . \tag{8}$$

(b,d) is the total number of N and R fields respectively. Eq. (8) holds separately for left- and right- movers.

The physical spectrum consists of direct product states ($|\rangle_L \times |\rangle_R$) of the Fock space of the left- and right-moving degrees of freedom. All physical states satisfy a constraint between left- and right- movers:

$$\alpha' m_L^2 = \alpha' m_R^2 \quad . \tag{9}$$

States which do not satisfy Eq. 9 can be shown to not be in the $\mathrm{Im}\tau \to \infty$ limit of the integral in Eq. 2 and therefore do not occur as intermediate states in the scattering amplitudes. Thus they do not occur in the spectrum of states on which these amplitudes are unitary.

Additional constraints, i.e. projections, usually also occur in consistent string theories. In the uncompactified ten-dimensional type II superstring there are four sectors, and in each sector, on left and right operators separately, these projections are

$$(-1)^{\sum_{s=\frac{1}{2}}^{\infty} b^i_{-s} b^i_s} = -1 \tag{10a}$$

and $$2^4 d_0^1 \ldots d_0^8 (-1)^{\sum_{n=1}^{\infty} d^i_{-n} d^i_n} = \pm 1. \tag{10b}$$

Eq's 10 are the standard GSO projections and correspond to keeping states with an odd number of NS fields $h_r = b_s$, and Ramond fields $h_r \equiv d_n$ which are restricted to be Majorana-Weyl. In the four-dimensional models, there are additional states labelled by $\tilde{p}^I$, $\bar{p}^I$ or $h^I_{(1)r}$ , $h^I_{(2)r}$. The projections relevant for the three eighteen-dimensional gauge groups models involve these internal operators as well. The ten-dimensional limit of these models is given by removing these extra states from the spectrum. In that limit, as we shall see, the four-dimensional projections become precisely

Eq. 10. In this sense, these models are "compactifications" of the ten-dimensional type II superstring.

The three four-dimensional string models with the eighteen-dimensional semi-simple gauge groups have the same set of four sectors, each sector being given by a specific world sheet fermion content and projections. The sectors, together with their contributions to the partition function are given in Eq. (13). ($\hat{i}$ = 1, 2; i = 1,...8; a = 1...18; λ = 1...12).

Define

$$P_1 \equiv \sum_{s=½}^{\infty} (\tilde{b}_{-s}^{\hat{i}}\tilde{b}_s^{\hat{i}} + \tilde{b}_{-s}^{a}\tilde{b}_s^{a} + b_{-s}^{\lambda}b_s^{\lambda}) \qquad (11a)$$

$$P_2 \equiv \sum_{s=½}^{\infty} b_{-s}^{i}b_s^{i} \qquad (11b)$$

$$(-1)^{P_3} \equiv 2^{16}\tilde{d}_0^1 \ldots \tilde{d}_0^{20} d_0^1 \ldots d_0^{12} (-1)^{\sum_{n=1}^{\infty}(\tilde{d}_{-n}^{\hat{i}}\tilde{d}_n^{\hat{i}} + \tilde{d}_{-n}^{a}\tilde{d}_n^{a} + d_{-n}^{\lambda}d_n^{\lambda})} \qquad (11c)$$

$$(-1)^{P_4} \equiv 2^4 d_0^1 \ldots d_0^8 (-1)^{\sum_{n=1}^{\infty} d_{-n}^{i}d_n^{i}} . \qquad (11d)$$

and, from Eq. 2

$$\Lambda \equiv -\frac{1}{4\Pi(\alpha')^2} \int_F d^2\tau (\mathrm{Im}\tau)^{-3} |f(\omega)|^{-24} |\omega|^{-1} \sum_{i=1}^{4} \Lambda_i . \qquad (12)$$

The four sectors are

I. $\alpha' m_L^2 = N_{\tilde{A}}\hat{i} + N_{\tilde{b}}\hat{i} + N_{\tilde{b}}a - ½$

$\alpha' m_R^2 = N_A\hat{i} + N_b i + N_b\lambda - ½$

$P_1$, $P_2$ = odd

$$\Lambda_1 = \frac{1}{4}(\theta_3^4 - \theta_4^4)(|\theta_3|^{12}\bar{\theta}_3^4 - |\theta_4|^{12}\bar{\theta}_4^4) \qquad (13a)$$

II. $\alpha' m_L^2 = N_{\tilde{A}}\hat{i} + N_{\tilde{b}}\hat{i} + N_{\tilde{b}}a - \frac{1}{2}$

$\alpha' m_R^2 = N_A\hat{i} + N_d i + N_b\lambda$

$$P_1 = \text{odd} \quad ; \quad (-1)^{P_4} = 1$$

$$\Lambda_2 = -\frac{1}{4}\theta_2^4(|\theta_3|^{12}\bar{\theta}_3^4 - |\theta_4|^{12}\bar{\theta}_4^4) \tag{13b}$$

III. $$\alpha' m_L^2 = \tilde{N}_A \hat{i} + \tilde{N}_d \hat{i} + \tilde{N}_d a + \frac{3}{4}$$

$$\alpha' m_R^2 = N_A \hat{i} + N_b i + N_d \lambda + \frac{1}{4}$$

$$P_2 = \text{odd}; \quad (-1)^{P_3} = 1$$

$$\Lambda_3 = \frac{1}{4}(\theta_3^4 - \theta_4^4)(-|\theta_2|^{12}\bar{\theta}_2^4) \tag{13c}$$

IV. $$\alpha' m_L^2 = \tilde{N}_A \hat{i} + \tilde{N}_d \hat{i} + \tilde{N}_d a + \frac{3}{4}$$

$$\alpha' m_R^2 = N_A \hat{i} + N_d i + N_d \lambda + \frac{3}{4}$$

$$(-1)^{P_3} = 1 \quad ; \quad (-1)^{P_4} = 1$$

$$\Lambda_4 = \frac{1}{4}|\theta_2|^{20} \tag{13d}$$

The world-sheet fermion content and their projections in these four sectors are pictorially represented in Fig. 1. Models with less gauge symmetry or supersymmetry can also be represented by such diagrams with more sectors and more boxes in each sector.

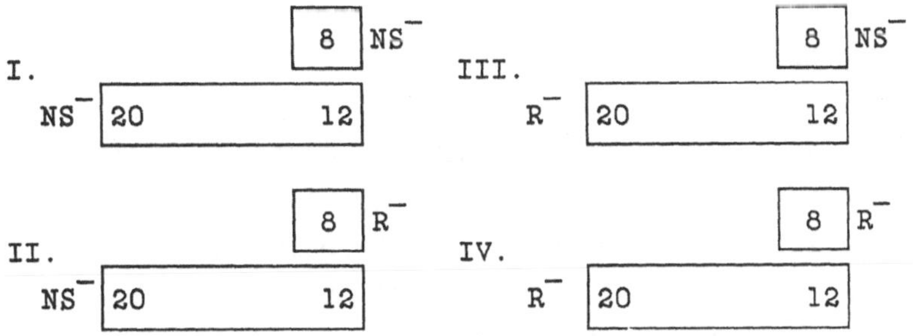


Fig. 1. The world sheet fermion content and projections for the four-dimensional N = 4, dimension eighteen gauge group model.

The one-loop partition function for the the three models discussed here is

$$\Lambda = -\frac{(\alpha')^{-2}}{4\pi}\int_F d^2\tau(\mathrm{Im}\tau)^{-3}|f(\omega)|^{-24}|\omega|^{-1}$$

$$\cdot\frac{1}{4}(\theta_3^4-\theta_4^4-\theta_2^4)(|\theta_3|^{12}\bar\theta_3^4-|\theta_4|^{12}\bar\theta_4^4-|\theta_2|^{12}\bar\theta_2^4). \qquad (14)$$

Eq. (14) vanishes by the identity $\theta_3^4-\theta_4^4-\theta_2^4=0$. Invariance of Eq. (14) under the group of modular transformations SL(2,I) generated by $\tau\to1$ and $\tau\to-\frac{1}{\tau}$ can be checked. The modular transformations on the Jacobi theta functions $\theta_i(0|\tau)$, and the forumlae for the number operators are given by

$$\tau\to\tau+1 \ : \quad \theta_3\to\theta_4;\ \theta_4\to\theta_3;\ \theta_2\to e^{\frac{i\pi}{4}}\theta_2 \qquad (15a)$$

$$\tau\to-\frac{1}{\tau} \ : \quad \theta_2\to(-i\tau)^{½}\theta_4;\ \theta_4\to(-i\tau)^{½}\theta_2;\ \theta_3\to(-i\tau)^{½}\theta_3;$$

$$\omega^{\frac{1}{24}}f(\omega)\to(-i\tau)^{½}\omega^{\frac{1}{24}}f(\omega) \qquad (15b)$$

$$\mathrm{tr}\,\omega^{N_{b^\alpha}} = \prod_{s=½}^{\infty}(1+\omega^s)^d = \left(\frac{\theta_3}{f}\right)^{\frac{d}{2}} \qquad (16a)$$

$$\mathrm{tr}\,\omega^{N_{A^\alpha}} = \prod_{n=1}^{\infty}(1-\omega^n)^{-d} = f^{-d} \qquad (16b)$$

$$\mathrm{tr}\,\omega^{N_{d^\alpha}} = \prod_{n=1}^{\infty}(1+\omega^n)^d = (2\omega^{\frac{1}{8}})^{-\frac{d}{2}}\left(\frac{\theta_2}{f}\right)^{\frac{d}{2}} \qquad (16c)$$

where $\alpha = 1,\ldots d$.

It follows from Eq. 14 that the number of states at each mass level is independent of which of the three eighteen-dimensional gauge groups is being discussed. These models are distinguished by the definition of the group structure constants, which appear in the gauge group generators and in the (light-cone-gauge) Lorentz generators (defined through the N = 1 super Virasoro generators). For Lorentz invariance, the Lorentz generators must commute with the gauge generators. Each sector and each model has a different definition of the super Virasoro generator $G_r$ in the super conformal algebra, and of the generators of the affine Kac-Moody Lie algebra.

Only sectors I. and II. contain massless states. From them one can show that the massless particles form a four-dimensional N = 4 supergravity multiplet (spin = $(\pm 2,\ 4(\frac{\pm 3}{2}),\ 6(\pm 1),\ 4(\pm\frac{1}{2}),\ 2(0))$ coupled to an N = 4 supersymmetric Yang-Mills multiplet (spin = $(\pm 1,\ 4(\pm\frac{1}{2}),\ 6(0))$ in the adjoint of the gauge group $SU(2)^6$ or $SU(3) \times SO(5)$ or $SU(2) \times SU(4)$.

## THE ALGEBRAIC STRUCTURE

In light-cone-gauge, for the models discussed here, the generators of the four-dimensional Lorentz algebra (SO(3,1) are defined in each sector through the generators of the N = 1 super Virasoro algebra with c = 15. On each side (left - and right - movers) the super conformal algebra is

$$[L_n, L_m] = (n-m)L_{n+m} + \frac{c}{12}(n^3-n)\delta_{n,-m} \tag{17a}$$

$$[L_n, G_r] = (\frac{n}{2}-r)G_{n+r} \tag{17b}$$

$$\{G_r, G_{r'}\} = 2L_{r+r'} + \frac{c}{3}(r^2-\frac{1}{4})\delta_{r,-r'} \tag{17c}$$

$L_n$ and $G_r$ may be split into a space-time part and an internal part:

$$L_n^{(ST)} = \tfrac{1}{2}(\sum_m :A_m^{\hat{i}} A_{n-m}^{\hat{i}}: + \sum_r (r-\frac{n}{2}):h_r^{\hat{i}} h_{n-r}^{\hat{i}}:) + \delta_{n,0}\,\frac{d}{16} \tag{18a}$$

$$\bar{L}_n = \tfrac{1}{2}\sum_m (r-\frac{n}{2}):h_r^a h_{n-r}^a: + \delta_{n,0}\,\frac{d}{16} \tag{18b}$$

$$G_r^{(ST)} = \sum_n A_n^{\hat{i}} h_{r-n}^{\hat{i}} \tag{18c}$$

$$\bar{G}_r = \frac{-i}{6} f_{abc} \sum_{r',r''} :h_{r''}^a h_{r''-r'}^b h_{r-r'}^c: \tag{18d}$$

The internal generators satisfy Eq. (17) with $c = 9$; d is the number of world sheet Ramond fields appearing in Eq. (18a,b). The structure constants fabc in Eq. (18d) are the structure constants of any semi-simple dimension eighteen Lie algebra g; i.e. $SU(2)^6$, $SU(3) \times SO(5)$, or $SU(2) \times SU(4)$. We see here that the generalization of the gauge group from $SU(2)^6$ corresponds to a redefinition of $G_r$.

In these models, the super Virasoro algebra in Eq. 17 forms a semi-direct product[5] with an affine Kac-Moody Lie algebra of level equal to the dual Coxeter number of the gauge group.

$$[T^a_n, T^b_m] = if_{abc}T^c_{n+m} + kn\delta_{n,-m}0^{ab} \tag{19a}$$

$$[T^a_n, h^b_r] = if_{abc}h^c_{n+r} \tag{19b}$$

$$\{h^a_r, h^b_r\} = \delta_{r,-r'}\delta^{ab} \tag{19c}$$

$$[L_n, T^a_m] = -mT^a_{n+m} \tag{20a}$$

$$[L_n, h^a_r] = -(\frac{n}{2}+r)h^a_{n+r} \tag{20b}$$

$$[G_r, T^a_n] = -n\sqrt{k}\ h^a_{n+r} \tag{20c}$$

$$\{G_r, h^a_{r'}\} = \frac{1}{\sqrt{k}}\ T^a_{r+r'} \tag{20d}$$

where $k = 1$, and $c_\psi$ the value of the quadratic Casimir in adjoint representation, has been normalized to 2:

$$fabcfabe = c_\psi\delta_{ce} = 2\delta_{ce}. \tag{21}$$

The affine Kac-Moody generators are constructed by

$$T^a_n = \frac{-i}{2}fabc \sum_r h^b_r h^c_{n-r}. \tag{22}$$

For example, in Sector II., (Eq. 13b), on the left,

$$\bar{L}^L_n \equiv \tfrac{1}{4}\sum_s (s-\frac{n}{2}) : \tilde{b}^a_s \tilde{b}^a_{n-s}: \tag{23a}$$

$$\bar{G}^L_s \equiv \frac{-i}{6} \text{ fabc} \sum_{s',s''} : \tilde{b}^a_{s''}\tilde{b}^b_{s'-s''}\tilde{b}^c_{s-s'} \tag{23b}$$

$$T^a_n \equiv \frac{-i}{2}\text{fabc} \sum \tilde{b}^b_s \tilde{b}^c_{n-s}. \tag{23c}$$

On the right,

$$\bar{L}^R_n = \tfrac{1}{2}\sum_m (m-\tfrac{n}{2}) : d^I_m d^I_{n-m} : +\tfrac{1}{2}\sum_s (s-\tfrac{n}{2}) : b^\lambda_s b^\lambda_{n-s} : \tag{24a}$$

$$F_n \equiv \frac{-i}{2}\bar{f}_{I\lambda\rho} \sum_{s,m} : d^I_m b^\lambda_{s-m} b^\rho_{n-s} : \tag{24b}$$

$$T^I_n \equiv \frac{-i}{6}\bar{f}_{I\lambda\rho} \sum_s b^\lambda_s b^\rho_{n-s}. \tag{24c}$$

In all four sectors, the dimension eighteen non-abelian gauge group $g$ has structure constants $f_{abc}$ appearing on the left as in Eq. (23c). From the right side, only a $U(1)^6$ gauge group survives, given in Eq. (24c). The structure constants in Eq. (24b,c) are the $U(1)^6$ embedding as a symmetric subgroup into $SU(2)^6$, i.e. for $\bar{f}abc$ the structure constants of $SU(2)^6$, then $\bar{f}_{abc} \equiv \{\bar{f}_{I\lambda p}, \text{ otherwise } = 0\}$.

In each sector, the SO(3,1) Lorentz algebra in light-cone-gauge is then given by

$$M^{+-} = l^{+-} \tag{25a}$$

$$M^{\hat{i}+} = l^{\hat{i}+} \quad ; \quad (l^{\mu\nu} \equiv x^\mu p^\nu - x^\nu p^\mu) \tag{25b}$$

$$M^{\hat{i}\hat{j}} = l^{\hat{i}\hat{j}} - i\sum_{n=1}^{\infty} \frac{1}{n}(\tilde{A}^{\hat{i}}_{-n}\tilde{A}^{\hat{j}}_n - \tilde{A}^{\hat{i}}_{-n}\tilde{A}^{\hat{j}}_n) - i\sum_{r>0}(\tilde{h}^{\hat{i}}_{-r}\tilde{h}^{\hat{j}}_r - \tilde{h}^{\hat{j}}_{-r}\tilde{h}^{\hat{i}}_r)$$
$$-i\sum_{n=1}^{\infty}\frac{1}{n}(A^{\hat{i}}_{-n}A^{\hat{j}}_n - A^{\hat{j}}_{-n}A^{\hat{i}}_n) - i\sum_{r>0}(h^{\hat{i}}_{-r}h^{\hat{j}}_r - h^{\hat{j}}_{-r}h^{\hat{i}}_r) \tag{25c}$$

and
$$M^{\hat{i}-} = l^{\hat{i}-} - 2i\sum_{n=1}^{\infty}\frac{1}{n}(\tilde{A}^{\hat{i}}_{-n}L^L_n - L^L_{-n}\tilde{A}^{\hat{i}}_n) - 2i\sum_{r>0}(\tilde{h}^{\hat{i}}_{-r}G^L_r - G^L_{-r}\tilde{h}^{\hat{i}}_r)$$
$$-2i\sum_{n=1}^{\infty}\frac{1}{n}(A^{\hat{i}}_{-n}L^R_n - L^R_{-n}A^{\hat{i}}_n) - 2i\sum_{r>0}(h^{\hat{i}}_{-r}G^R_r - G^R_{-r}h^{\hat{i}}_r). \tag{25d}$$

From Eq.'s (20a,c) and (25d), we see that the gauge generators $T_0^a$ commute with the Lorentz generators $M^{+-}$, $M^{\hat{i}\hat{j}}$, $M^{\hat{i}-}$.

The further symmetry breaking to models with realistic space-time fermion content is under current investigation.

REFERENCES


1. For a recent review see M. B. Green, J. H. Schwarz, and E. Witten, Superstring Theory, vol. 1,2. Cambridge: Cambridge University Press, 1987.

2. R. Bluhm and L. Dolan, Phys. Lett. B169, 347 (1986).

3. R. Bluhm, L. Dolan, and P.Goddard, Nucl. Phys. B289, 364 (1987).

4. The realistic eighteen-dimensional groups have been subsequently discussed by H. Kawai, D. C. Lewellen, S. H. H. Tye, CLNS 87/760; L. Dixon, V. Kaplunovsky, C. Vafa, HUTP-87/A034; W. Lerche, B. Nilsson, A. Schellekens, CERN-TH. 4692/87; and I. Antoniadis and C. Bachas, CERN-TH.4767/87.

5. P. Goddard, W. Nahm, and D. Olive, Phys, Lett. B160, 111 (1985).

# STRING THERMODYNAMICS AND COSMOLOGY [+]

F. Englert [++]

Université Libre de Bruxelles

and the Hebrew University of Jerusalem

## INTRODUCTION

We shall show that theories involving only closed strings with at least four non-compact space-time dimensions exhibit a phase transition. The high temperature phase is characterized by a condensate of arbitrarily long strings with Hausdorff dimension two (area filling curves). We suggest that this stringy phase is the ancestor of the adiabatic era. Fundamental strings could then both drive the inflation and seed, in a way reminiscent of the cosmic string mechanism, the large structures in the universe. These results may lead to testing string theory itself and eventually provide some hints for constructing a more elaborate unification of gravity and matter.

## THE PHASE TRANSITION

One might have expected that in string theories, very high mass elementary excitations would necessarily hide beyond the Planck scale. This is not the case. We shall show that there are string theories which experience a transition from a low energy density phase describable in terms of ordinary quantum fields to a phase where a finite fraction of the energy density condenses into "infinite" strings of Hausdorff dimension two.

---

\+ The work presented in this talk is due to Y. Aharonov, F. Englert and J. Orloff. [1)]

++ Postal address : Université Libre de Bruxelles, Service de Physique Théorique, Campus Plaine C.P. 225, Boulevard du Triomphe, 1050 Bruxelles, BELGIUM.

These are theories involving only closed strings with at least four non-compact space-time dimensions. It is suggested that this stringy phase is the ancestor of the adiabatic era. A scenario is proposed in which the macroscopic fondamental strings drive both the inflation and the formation of large scale structures in the universe.

To understand in simple terms the origin of the phase transition, let us first consider the following toy model. A box of arbitrarily large volume $V$ in $D-1$ spatial dimensions contains a total energy $E$ shared between two constituants in thermal equilibrium : a gas of massless particles and a macroscopic "string" characterized only by a density of states growing with its energy $E_s$ as $\exp(\beta_0 E_s)$ .
Denoting by $S_g$ , $S_s$ , $S_{g+s}$ respectively the entropies of the gas, the string, and the string-gas system we have

$$S_g = \frac{D}{D-1} V'^{1/D} E^{(D-1)/D} , \qquad (1)$$

$$S_s = \beta_0 E , \qquad (2)$$

$$S_{g+s} = \frac{D}{D-1} V'^{1/D} (E-E_s)^{(D-1)/D} + \beta_0 E_s . \qquad (3)$$

Here $V' = \xi V$ where $\xi$ is a number and the temperature of the gas is $\beta^{-1} = (E_g/V')^{1/D}$ where $E_g = E$ in (1) and $E - E_s$ in (3).

Equilibrium of the two-phase system implies $\partial S_{g+s}/\partial E_s = \beta_0 - \beta = 0$ and it is a stable one. Thus we can rewrite (3) as

$$S_{g+s} = \beta_0 E + \frac{1}{D-1} \beta_0^{1-D} V' . \qquad (4)$$

In terms of the total energy density $\sigma \equiv E/V'$ we then have for the corresponding specific entropies $s_g$ , $s_s$ and $s_{g+s}$

$$s_g = \frac{D}{D-1} \sigma^{(D-1)/D} , \qquad (5)$$

$$s_s = \beta_0 \sigma \qquad , \qquad (6)$$

$$s_{g+s} = \beta_0 \sigma + \frac{1}{D-1} \beta_0^{1-D} \qquad . \qquad (7)$$

Clearly $s_{g+s} > s_s$ and, for $E_s/V > 0$, $s_{g+s} > s_g$ if $\sigma > \sigma_c \equiv \beta_0^{-D}$; when $\sigma = \sigma_c$, $ds_{g+s}/d\sigma = ds_g/d\sigma = \beta_0$. Therefore when the gas reaches the critical energy density $\sigma_c$, any increase of density in the gas would "condense" into the "string" at the temperature $T = \beta_0^{-1}$.

For weak coupling we now consider in a flat background genuine strings of total energy $E$ enclosed in an arbitrarily large volume $V$. The main qualitative new feature, ignored in the toy model, is the existence of a mass spectrum which extends from the zero mass states to macroscopic masses. The asymptotic density of states of mass $M$ is

$$\rho(M) = C \alpha'^{(-a+1)/2} M^{-a} e^{\beta_0 M} \qquad (8)$$

where $\alpha'$ measures the inverse string tension and $a$ depends on the model; the constant $C$ will not play a significant role in what follows. Thus in addition of the zero mass homogeneous background gas we have to take into account a background of strings of any size.
Clearly the prefactor coefficient $a$ plays a determinantal role in controling the behaviour of the string background. This was analysed in references 2) and 3) and later generalized to the case of relativistic strings [4). The main results can be summarized as follows. First examine, in thermal equilibrium, the set of fields defined by the quantized string states. After summing over field quanta, one gets for the cananical partition function

$$\log Z = \frac{V}{(2\pi)^{D-1}} \int d^{D-1}k \int_0^\infty \rho(M)\, dM \log\left(1 \pm e^{-\beta\sqrt{k^2+M^2}}\right)^{\pm 1} \qquad (9)$$

where $M$ and $k$ are the mass and momenta of the fields; fermionic and bosonic fields in (9) are characterized respectively by + or - signs. Separating out the zero mass fields, using $\beta M \gg 1$ for large masses, we get by expanding the logarithm to first order and using the non relativistic limit $(k^2 + M^2)^{1/2} \simeq M + k^2/2M$ ,

$$\log Z \simeq \eta \beta^{1-D} V + C \alpha'^{(-a+1)/2} (2\pi\beta)^{(1-D)/2} V \cdot \int_{M_0}^{\infty} M^{-a+\frac{D-1}{2}} e^{(\beta_0-\beta)M} dM, \tag{10}$$

where $\eta$ is a number and $M_0$ a typical low mass of the spectrum. Thus if

$$-a + \frac{D-1}{2} < -1 \tag{11}$$

the free energy density tends to a finite limit when $T \to \beta_0^{-1}$ from below. If in addition

$$-a + \frac{D-1}{2} < -2 \tag{12}$$

the energy density tends to a finite value $\sigma_c$. These results are confirmed by an analysis of the grand microcanonical ensemble which permits to study the case $\sigma > \sigma_c$. Clearly, if $\sigma$ is increased above the critical density $\sigma_c$, entropy will favour, as in the toy model, condensation of very long strings whose structure will be analysed later on. However, one finds that the instability of the canonical background can occur already if (11) is satisfied.[2] This is because the integrated distribution of string masses $n(M)$ for $T \to \beta_0^{-1}$, which from (10) has the asymptotic behaviour (up to a constant)

$$n(M) \to V M^{-a+\frac{D-1}{2}}, \tag{13}$$

is then finite. Thus, when energy is poured into the system at a temperature close to $\beta_0^{-1}$, the entropy increase in the string distribution is quenched by the homogeneity of the canonical background and very long string will condense; but as long as $-2 \leq -a + (D-1)/2 < -1$ the critical energy density for condensation increases with the volume so that in the thermodynamic limit of large volume the canonical distribution remains stable as the temperature approches $\beta_0^{-1}$[3]. Thus, as expected from the toy model, the inequality (12) indeed governs, in the infinite volume limit, the transition from a canonical phase characterized by an homogeneous zero mass gas in equilibrium with strings distributed according to (13) to a stringy phase containing infinite strings. While the canonical phase is describable in terms of an infinite numbers of free fields in thermal equilibrium, the stringy phase would reveal directly the existence of extended fondamental objects in physics.

Having reviewed in this way the known string thermodynamics, we now examine when and how "realistic" fondamental string theories lead to a phase transition in the thermodynamical limit of large $V$. We shall first evaluate the prefactor "$a$", assuming that the relevant compactification can be described on a torus by a Frenkel-Kac mechanism, with or without twisted boundary conditions in 2-dimensional fermion language. We shall then obtain a complete description of the stringy phase.

## THE CRITICAL SPACE-TIME DIMENSION

Let us first consider a bosonic open string theory in $D+\delta$ dimensions with $\delta$ dimensions compactified on a torus of radius $R \simeq \sqrt{\alpha'}$. We write the asymptotic estimate of the $N^{th}$ level degeneracy $P(N)$[5], up to a multiplicative factor of order 1 (in Planck units), in the form

$$P(N) \longrightarrow N^{-\frac{3}{4}-\frac{D-2+\delta}{4}} \quad e^{2\lambda N} \tag{14}$$

with the saddle parameter $\lambda$ given by

$$N = (D-2+\delta)\sum_{m=1}^{\infty}\frac{m}{e^{\lambda m}-1} \longrightarrow \frac{D-2+\delta}{\lambda^2}\;\frac{\pi^2}{6}\,. \tag{15}$$

To get the asymptotic degeneracy $P(M)$ of the mass $M$ we use $N = \alpha'(M^2 - \vec{p}^{\,2})$, where $\vec{p}$ is the compactified momentum, and sum $P(N)$ over all lattice momenta. The leading term is obtained by using the "non relativistic limit" $N^{1/2} \simeq \sqrt{\alpha'}(M - \vec{p}^{\,2}/2M)$ in the exponent $2\lambda N \sim N^{1/2}$, replacing $N$ by $M^2$ in the prefactor and approximating the sum by an integral. Performing this Gaussian integral we thus get

$$P(M) \to M^{-(D+1)/2}\, e^{\beta_0 M}\;;\quad \beta_0 = \left[\frac{2\pi^2}{3}(D-2+\delta)\,\alpha'\right]^{1/2}. \tag{16}$$

As the $M^2$ level separation is of order one, in Planck units, we obtain for $\rho(M)$

$$\rho(M) \longrightarrow M\, P(M)\,. \tag{17}$$

Any dependence of the prefactor on $\delta$ has dropped out. This is not unexpected : compactified bosons are equivalent to two Majorana fermions. Performing the saddle point evaluation of $\mathcal{P}(\mathcal{N}')$ where $\mathcal{N}'$ is the sum of fermionic and bosonic transverse oscillators, with $\delta = 0$ but with $2\delta$ fermionic degrees of freedom one recovers (16) using $\mathcal{N}' = \alpha' M^2$. This is because fermionic degrees of freedom do not contribute to the prefactor and add one half to $D$ in the exponent for each fermion. Thus we may for instance use (16) for type I open superstrings by putting $D = 10$ and $\delta = 4$. One gets in this way the known prefactor [4] and critical temperature [6] of type I open superstrings. More generally we can use (16), (17) for any theory of open strings containing fermions and bosons with stringy torus compactification to $D$ non compact space-time dimensions.

Similarly we use the bosonized form for any closed string theory with or without stringy compactification satisfying modular invariance [7]. From the factorization property of left and right mass spectra we can use (14), (15), (16) and (17) separately for left and right movers, replacing $\alpha'$ by $\alpha'/4$. Introducing $L$ $(R)$-subscripts we thus get in terms of $\mathcal{P}_L(M/2)$ and $\mathcal{P}_R(M/2)$

$$\rho(M) \rightarrow M^{-1}\mathcal{P}_L(M/2)\,\mathcal{P}_R(M/2) \rightarrow M^{-D} e^{\beta_0 M};$$

$$\beta_0 = \beta_{0,L} + \beta_{0,R}\,; \quad \beta_{0,L(R)} = \left[\frac{\pi^2}{6}(D - 2 + \delta_{L(R)})\alpha'\right]^{1/2}, \qquad (18)$$

where $\delta_L$ and $\delta_R$ are the number of left and right compactified bosons. We can immediatly check that for the particular case of uncompactified type II and heterotic superstrings, the known results are recovered for the exponent and for the prefactor [4], [6]. Comparing (16), (17) and (8) with (12) we see that string theories with or without stringy compactification, containing open strings cannot lead to a stringy phase. Theories with only closed strings have a stringy phase if $D \geq 4$ !

## THE STRINGY PHASE

To understand the nature of this phase we study the shape of high mass strings in a Euclidean space at a time t. To this effect we choose a Weyl gauge such that the world sheet metric is equal to the metric induced from the host space. We parametrize the world sheet coordinates $\sigma$ and $\tau$ so that $g_{\sigma\tau} = 0$, $g_{\sigma\sigma} = -g_{\tau\tau}$, and we choose $\tau = t$. In this way a point of the string is labeled by $D-1$ cartesian components

$x^i(\sigma, t)$ . Boosting to the infinite moment frame we are left with $D-2$ independent $x^i_\infty(\sigma, t_\infty)$ where $t_\infty$ is a light-like variable. They satisfy the usual light-cone expansion

$$x^i_\infty(\sigma, t_\infty) = \left(\frac{\alpha'}{2}\right)^{1/2} \sum_{m\neq 0} \frac{1}{m} \left[\alpha^i_{m,L}\, e^{-2im(t_\infty-\sigma)} + \alpha^i_{m,R}\, e^{-2im(t_\infty+\sigma)}\right] \tag{19}$$

where we omitted the zero modes. The average transverse size $L_{tr}$ of the string in the rest frame, which measures the correlation lenght, can now be obtained from (19) using isotropy :

$$L^2_{tr} = \frac{D-1}{D-2} \sum_{i=1}^{D-2} \langle x^{i\,2}_\infty \rangle = \alpha'(D-1) \sum_{m=1}^{\infty} \frac{1}{m} (n_{m,L} + n_{m,R}) \tag{20}$$

where from (15) $n_{m,L(R)} = (\exp \lambda_{L(R)} m - 1)^{-1}$.
We have

$$\sum_{m=1}^{\infty} m^{-1}\, n_{m,L(R)} \rightarrow \frac{1}{\lambda_{L(R)}} \sum_{m=1}^{\infty} m^{-2} = \frac{1}{\lambda_{L(R)}} \frac{\pi^2}{6}. \tag{21}$$

From (20), (21) and (15), we get using $N_{L(R)} = (\alpha'/4)(M^2 - 4\vec{p}^{\,2}_{L(R)})$ and neglecting the compactified momenta which are of order $\sqrt{M}$

$$L^2_{tr} = \alpha'^{3/2} M (D-1) \frac{\pi}{2\sqrt{6}} \left[(D-2+\delta_L)^{-1/2} + (D-2+\delta_R)^{-1/2}\right]. \tag{22}$$

For $\delta_L = \delta_R = 0$ , one recovers the bosonic string result [8].
Similarly we evaluate the total length of the string

$$L = \int_0^\pi d\sigma \left[\frac{D-1}{D-2} \sum_{i=1}^{D-2} \left\langle \left(\frac{dx^i_\infty}{d\sigma}\right)^2 \right\rangle\right]^{1/2} ; \tag{23}$$

the result is

$$L = \alpha' M (D-1)^{1/2} \pi \left[(D-2+\delta_L)^{-1} + (D-2+\delta_R)^{-1}\right]^{1/2}. \tag{24}$$

Eq (24) expresses the fact that the mass per unit lentgh is constant. Thus $L$ grows like $L^2_{tr}$ : high mass strings always have a Hausdorff dimension two and are area filling.

In a volume $V$ of space, local density fluctuations at $\sigma > \sigma_c$ will condense the longest strings fitting into this volume which have thus a mass $M \sim V^{2/(D-1)}$ and a macroscopic number of long strings will be produced in this way. In the limit $V \to \infty$, the condensate appears thus in the form of an infinite number of infinite strings spanning random walks. From (21) and (13), the background strings span asymptotically, a scale invariant distribution of random walks [8]. Thus in $D \geq 4$ the fundamental closed strings theories yield, above the critical density, the same distribution as the one obtained, either from simulation [9] as from theoretical considerations [8] similar to ours, for the formation of hypothetical cosmic strings generated by a symmetry breaking mechanism.

## COSMOLOGICAL IMPLICATIONS

We now examine the cosmological implications of the stringy phase. From now on we restrict ourselves to the critical dimension $D = 4$. We must take gravity, which is in fact contained in the string theory, into account. For scales large compared to the Planck size we may assume that the classical background metric obeys Einstein's equations. If the space curvature is negligeable and if thermodynamical equilibrium is realized in the early universe at any time, we are led to a natural hypothesis which will eventually justify these inputs. Namely the adiabatic era stems from a stringy phase when the temperature drops below $\beta_c^{-1}$ As $\beta_c^{-1}$ can be much smaller than the Planck temperature if $\alpha'$ is large compared to one in Planck units, a natural assumption in the weak coupling limit, we see that the critical density $\sigma_c$ can be sizeably smaller than the Planck density. Therefore we can investigate our tentative hypothesis by applying general relativity to the stringy phase itself.

The main question is what would cause the large scale homogeneity of the stringy phase as we cannot rely on the homogeneity of the background radiation which has precisely to be explained. There seem to be a natural answer which would justify at the same time thermal equilibrium and the assumed flatness of space : the stringy phase produces inflation which in turn garantees its homogeneity and its stability. Indeed de Sitter space is the only cosmological solution of Einstein's equations coupled to quantized matter for which the event horizon produces the constant temperature required for the stringy phase.

This suggests that thermodynamical consistency requires

$$\frac{H}{2\pi} = \beta_c^{-1} \tag{25}$$

where $H$ is the Hubble constant. In this way the total energy density $\sigma = 3H^2/8\pi = 3\pi/2\beta_0^2$. For $\alpha'$ larger than one this can easily exceed the critical density $\sigma_c$ while remaining small compared to the Planck density. Thus strings would be continuously created as a quantum effect described by the temperature of the event horizon. This process appears as a possible realization of the original picture of inflation as produced by massive objects created quantum mechanically out of the classical expansion energy in a self-consistent way [10], when the massive objects have a large internal entropy [11]. In this way, as discussed in references (10) and (11), the de Sitter equation of state $\sigma = -p$ would arise automatically.

The continuous creation will eventually terminate. Indeed, strings of size comparable to the horizon size would be unstable and their decay would bring the inflation to an end and mark the onset of the adiabatic era. However a network of macroscopic out of thermal equilibrium should persist in the radiation dominated era and their evolution could lead, as in the cosmic string hypothesis, to seeding the large scale structure of the universe.

We were thus led to suggest that macroscopic fundamental strings could both drive inflation and then large scale structures in a way similar to cosmic strings. However the difference in the nature of the seeds may lead to different predictions for correlations at large scales. On the other hand, the very existence of a stringy phase may indicate the need for a more elaborate theory of unified gravity and matter : this phase suggests the existence of local fields, and an underlying local description of what we presently call strings may perhaps reveal a more fundamental structure than the strings themselves. A more detailed study of the stringy phase in the cosmological context is thus desirable both for confronting the astrophysical datas and for understanding the nature of the theory at a fundamental level.

The above scenario rests on the hypothesis of weak coupling and this may possibly conflict with present attempts to understand low energy phenomenology. If the weak coupling constraint turns out ultimately to be inconsistent, a more elaborate description of an early stringy phase may be required. Namely the emergence of macroscopic strings would induce large deficiency angles which would prevent the birth of space-time. The adiabatic era and space-time itself would then appear at once as the remmant of a correlated quantum state. Clearly, one should first investigate the possibility of a less drastic departure from conventional physics.

Finally we wish to emphasize that the above construction of a macroscopic world can only be envisaged if $D \geq 4$. Hopefully an analysis of infrared effect would select the critical dimension $D = 4$ as the only consistent one at large scales.

ACKNOWLEDGEMENT

We are very grateful to R. Brout, M. Henneaux, S. Nussimov, P. Spindel and L. Susskind for many helpful and most enjoyable discussions.

REFERENCES

1. Y. Aharonov, F. Englert and J. Orloff, "Macroscopic Fundamental Strings in Cosmology" U.L.B. Preprint TH 87/08, to be published in Physics Letters.
2. S. Frautschi, Phys. Rev. D3 (1971) 2821.
3. R.D. Carlitz, Phys. Rev. D5 (1972) 3231.
4. B. Sundborg, Nucl. Phys. B254 (1985) 583.
   M.J. Bowick and L.C.R. Wijewardhana, Phys. Rev. Lett. 54 (1985) 2485.
   M. Gleiser and J.G. Taylor, Phys. Lett. 164B (1985) 36.
   S.-H. Tye, Phys. Lett. 158B (1985) 388.
   E. Alvarez, Nucl. Phys. B269 (1986) 596.
5. K. Huang and S. Weinberg, Phys. Rev. Lett. 25 (1970) 895.
6. D. Gross, J.A. Harvey, E. Martinec and R. Rohm, Nucl. Phys. B256 (1985) 253.
7. See for instance W. Lerche, D. Lüst and A.N. Schellekens, Nucl. Phys. B287 (1987) 477.
8. D. Mitchell and N. Turok, Phys. Rev. Lett. 58 (1987) 1577.
9. T. Vachaspati and A. Vilenkin, Phys. Rev. D30 (1984) 2036.
10. R. Brout, F. Englert and E. Gunzig, Ann. Phys. 115 (1978) 78; Gen. Rel. & Grav. 10 (1979) p 1; for an explicit check of see R. Brout, F. Englert and P. Spindel, Phys. Rev. Lett. 43 (1979) 417.

11. This idea was suggested by A. Casher and F. Englert (Phys. Lett. 104B (1981) 117) who tentatively interpreted the massive objects as black holes. The suggestion of using strings as entropy reservoirs in the early universe was made by Y. Aharanov and A. Casher (Phys. Lett. 166B (1986) 289) and F. Englert ("String Theory and Quantum Cosmology", Proceedings of the XVII G.I.F.T. International Seminar on Cosmology and Particle Physics (1986), World Scientific Publ. Ed. by Alvarez, Dominguez, Ibanez and Quiros).

## ORBIFOLD COMPACTIFICATION TOWARD STANDARD MODEL

Jihn E. Kim

Department of Physics
Seoul National University
Seoul, Korea

## INTRODUCTION

Superstrings are supposed to be a candidate theory for unifying all known interactions. It will fail to be a physical theory if it turns out to be successful only for gravity. Therefore, the ultimate hope is to find a standard model from the string approach. One route to this objective is to go through D=10 N=1 supergravity compactified on a Calabi -Yau space.[1] Another more promising approach is to obtain 4 dimensional string models through orbifold compactifications.[2,3] In this talk, I will concentrate on compactifying the heterotic $E_8 x E_8$ superstring in an attempt to obtain a standard model superstring. We have not succeeded in obtaining a standard model superstring yet, but some models are very close to the standard model.[4]

The standard model superstring must satisfy a lot of observed phenomena,

(i) a correct gauge group $SU(3)xSU(2)xU(1)x(U(1))^{n-1}$
(ii) chiral matter fields
(iii) 3 quark-lepton families
(iv) (N=1 D=4 supersymmetry)
(v) existence of Weinberg-Salam Higgs doublets
(vi) stability of proton
(vii) strong CP invariance
(viii) weak CP violation
(ix) $\sin^2\theta_W \cong 0.23$, $\rho = 1$
(x) no exotic particles lighter than 40 GeV
(xi) (almost) massless neutrinos, etc.

We will see that some of the above constraints are satisfied by four dimensional superstring models constructed by the orbifold method.

## ORBIFOLD COMPACTIFICATION

Orbifold are twisted tori.[2,3] The toroidal compactification of 22 left-moving coordinates and 6 right moving coordinates gives 4 dimensional string models[5] as the compactification of just 16 left moving coordinates gives the ten dimensional heterotic string.[6] For the modular invariance, the lattice must be even, Lorentzian and self-dual.[5] In this way, rank 22 groups and N=4 supergravity (nonchiral) have been obtained. Phenomenologically preferred ones are N=1. So one needs to kill three massless gravitinos. Moding of strings can achieve this need, and can produce chirality.

We use the bosonic construction since it is particularly useful for our approach. The toroidally compactified string is "moded" by the discrete isometries $M_R$ and $M_L$ of the tori $T_R^6$ and $T_L^{22}$

$$T_L^{22} \otimes T_R^6 \xrightarrow{\text{moding}} T_L^{22}/M_L \otimes T_R^6/M_R \tag{1}$$

Since the space-time supersymmetry comes from the right movers and we prefer to have one unbroken supersymmetry the isometry $M_R$ must be a discrete subgroup of the $SU(3) \subset SO(6)$ where SO(6) is the tangent group of the extra 6 dimensions. Furthermore, moding the left extra 6 dimensions in the same way as the right extra 6 dimensions, we construct a "symmetric orbifold"

$$O_s = T_L^{16}/G \otimes T_L^6/P \otimes T_R^6/P \tag{2}$$

If they are different, we obtain an asymmetric orbifold. For phenomenological purposes, we concentrate on the symmetric orbifolds.[7] We restrict ourselves furthermore that the 6 extra left and right coordinates are compactified on the same torus, and $T_L^{16}$ is taken as $T_{8x8}^L$. Namely, the models we obtain can be interpreted as four dimensional superstrings, but these can also be obtained from compactifying the ten dimensional $E_8 \times E_8$ heterotic string. We will see that even in this restricted class of 4 dimensional superstrings we can go very closely to the standard model. Then the symmetric orbifold can be defined as

$$O = T_{8x8}^L/G \otimes R_{LxR}^6/S \tag{3}$$

where S is the space group consisting of the point group P plus the shifts in $R^6$,

$$S = (\theta, v^i);\ \theta \in P,\ i=1,2,\ldots 6 \tag{4}$$

An element $(\theta, v^i)$ of the space group acts on the 6 space-time coordinates as

$$(S\,X)^i = (\theta\,X)^i + v^i,\ i=1,2,\ldots 6 \tag{5}$$

where $v^i$ is constructed from the defining vectors of the 6-torus $e_a^i$ $(a = 1,2,\ldots 6)$,

$$v^i = m_a e_a^i \tag{6}$$

In the bosonic construction, $\theta$ is the automorphism of the 6 dimensional lattice. The action of the space group can be accompanied by the action on the 16 extra gauge coordinates, I=1,2,....16. This is denoted as the gauge twisting group G,

$$G = (\Theta, v^I), \quad I=1,2,\ldots.16 \tag{7}$$

The action G on the gauge coordinates is

$$(G\, X)^I = (\Theta\, X)^I + v^I, \quad I=1,2,\ldots.16 \tag{8}$$

Because the gauge coordinate lattice is the $E_8 \times E_8$ lattice, $\Theta$ represents some automorphism of the $E_8 \times E_8$ lattice, i.e. $p^I \longrightarrow p^I + v^I$. Rewriting the above equations,

$$X^i(\sigma=2\pi) = (\theta\, X(\sigma=0))^i + v^i, \quad i=1,\ldots 6$$
$$X^I(\sigma=2\pi) = (\Theta\, X(\sigma=0))^I + v^I, \quad I=1,\ldots 16 \tag{9}$$

Let us represent the twisting as

$$m = (\theta, v^i; \Theta, v^I) \tag{10}$$

There are two types of closed strings, untwisted and twisted. Twisted strings are closed on orbifolds but not on tori. Untwisted strings are closed on tori. For $Z_2$ orbifold with PX=-X, the fixed points and different kinds of string are shown in Fig. 1

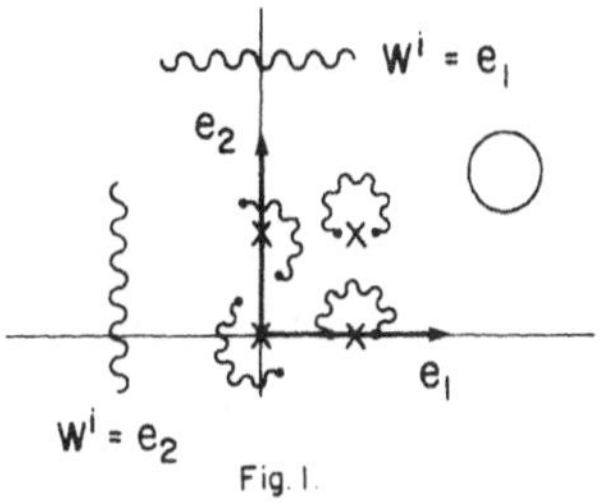

Fig. 1.

Untwist strings satisfy $X^i(\sigma=2\pi) = X^i(\sigma=0) + w^i$. Twisted strings propagate around fixed points; thus the number of twisted strings is the number of fixed points.

Phenomenologically interesting orbifold is the $Z_3$ orbifold where the point group is $(1, P, P^2)$. Here P is a $120^0$ rotation of $SU(3)^3$ root lattice defining the 6-torus as shown in Fig. 2

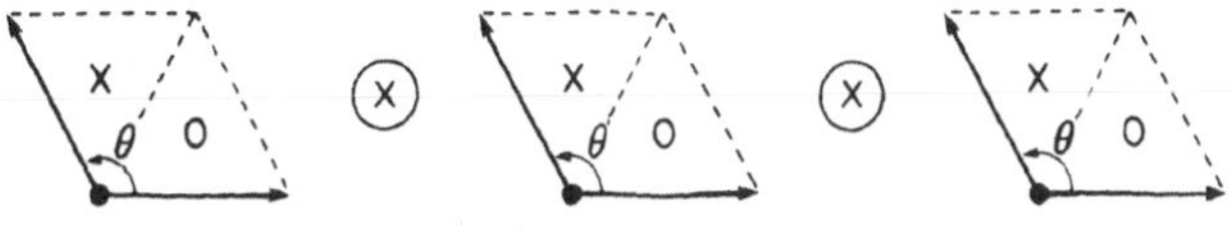
Fig. 2

There are $3^3=27$ fixed points. For simple case of $G=(1, v^I)$, three consecutive rotations brings back to the original point, i.e. $(\theta,$

$0;\ 1,\ v^I)^3=1$. Thus $3v^I$ must belong to the $E_8$ lattice. The modular invariance conditi.n restricts to four possible nontrivial shifts which have been studied by Dixon et al,[2]

$$v^I = (\tfrac{1}{3}\ \tfrac{1}{3}\ \tfrac{2}{3}\ 0\ 0\ 0\ 0\ 0)\ (0\ 0\ 0\ 0\ 0\ 0\ 0\ 0)$$
$$v^I = (\tfrac{1}{3}\ \tfrac{1}{3}\ 0\ 0\ 0\ 0\ 0\ 0)\ (\tfrac{2}{3}\ 0\ 0\ 0\ 0\ 0\ 0\ 0)$$
$$v^I = (\tfrac{1}{3}\ \tfrac{1}{3}\ \tfrac{2}{3}\ 0\ 0\ 0\ 0\ 0)\ (\tfrac{1}{3}\ \tfrac{1}{3}\ \tfrac{2}{3}\ 0\ 0\ 0\ 0\ 0) \tag{11}$$
$$v^I = (\tfrac{2}{3}\ \tfrac{1}{3}\ \tfrac{1}{3}\ \tfrac{1}{3}\ \tfrac{1}{3}\ 0\ 0\ 0)\ (\tfrac{2}{3}\ 0\ 0\ 0\ 0\ 0\ 0\ 0)$$

Each case breaks the gauge group down to $E_6 x SU(3) x E_8'$, $E_7 x U(1) x SO(14)' x U(1)'$, $[E_6 x SU(3)]^2$ and $SU(9) x SO(14)' x U(1)'$, respectively. These models have too many families, typically 36 families.

INQ MECHANISM

Ibanez, Nilles and Quevedo[8] observed that there are an additional class of 4 dimensional strings obtained from the orbifold compactification. These are the ones in which tori are wrapped with U(1) gauge background fields

$$\int_{(a)} A_i^I\, dx^i = A_i^I\, e_a^i \equiv a_a^I \tag{12}$$

which is called as Wilson lines. Now strings propagate on the orbifolds in the presence of background gauge fields. Then the Wilson lines can be understood as the lattice translation on the $E_8 x E_8$ lattice, $e_a^i \longrightarrow \gamma_a \in E_8 x E_8$. Since the pair of wrapping with gauge fields commute $[(1,e_a^i),(1,e_b^j)]=0$, the associated Wilson lines must commute $[\gamma_a, \gamma_b]=0$. However, in general it may not commute with the twist $g \in E_8 x E_8$, $[g, \gamma_a] \neq 0$. Nevertheless, we restrict to the simpler case $[g, \gamma_a]=0$. Then g must be the shift on the gauge lattice. Thus our study is

$$\begin{aligned} &(\theta,\ 0;\ 1,\ v^I) && \text{twist} \\ &(1,\ e_a^i;\ 1,\ a_a^I) && \text{Wilson lines} \end{aligned} \tag{13}$$

where $i=1,\ldots 6$, $a=1,\ldots .6$ and $I=1,2,\ldots .16$. There are some constraints on the Wilson lines coming from the group law properties of the space group. Firstly, suppose that the twist is of order n. Since $\theta^n=1$ and all $e_a^j$ are rotated by $\theta$, we obtain $n\, a_a^I \in E_8 x E_8$ lattice as $nv^I \in E_8 x E_8$. Secondly, all 6 Wilson lines are not independent. For $Z_3$ orbifold $e_a^i$ is rotated to $(\theta e_a)^i$. With $e_1^i$, $e_2^j$, etc as bases, $\theta e_1 = e_2$, etc. Therefore,

$$a_1^I = a_2^I,\ a_3^I = a_4^I,\ a_5^I = a_6^I \tag{14}$$

Thirdly, again there are untwisted and twisted strings. The untwisted strings are the same as before but with additional constraints. Now the twisted strings are further distinguished. When we twist, there is a possibility of wrapping the torus with gauge fields, i.e. adding Wilson lines. The Wilson lines can be added or subtracted. Previously, we considered only a twisted sector $v^I$ ( In addition, $-v^I$ can be considered but the resulting spectrum is the CPT conjugates). For one Wilson line $a_1^I=a_2^I$, the 27 twisted sector splits into 3 classes, $v^I$ (without wrapping the torus with gauge fields), $v^I+a_1^I$ (wrapping in the positive direction), and $v^I-a_1^I$(wrapping in the negative direction). These are called as twisted trivially and twisted nontrivially, respectively. Each sector has 9 identical copies. With two Wilson lines, we obtain 9 classes each of which has 3 identical copies.

## ON THE ROAD TOWARD STANDARD MODEL SUPERSTRING

Strings come in untwisted and twisted sectors.

Since the untwisted strings are closed before moding, their mass formula is the same as the one obtained from torus compactification

$$\frac{M^2}{8} = \tilde{N} + \frac{p^2}{2} - 1 \tag{15}$$

where $\tilde{N}$ is the left-handed oscillator number, $p^I$ is the winding states in the $E_8xE_8$ lattice and -1 is the zero point energy. Twisting, however, picks up invariant states on

$$O = R^6_{L+R}/S \otimes T^{16}_{8x8}/G \tag{16}$$

Thus SxG invariant states are picked up among massless states satisfying (15). There appears the usual supergravity multiplet. The gauge multiplet must be chosen in such a way that they are SxG invariant. The particle spectrum is obtained from $|>_R x|>_L$. $|>_R$ contains both bosonic and fermionic states, $|i>_R$ and $|a>_R$ with i, a=1,...8. These can be split into $3+\bar{3}+1+\bar{1}$ under $SU(3)_R$. Thus we assign phases $\alpha=e^{2\pi i/3}$ and $\alpha^2$ for 3 and $\bar{3}$ under the operation of the point group P. On the other hand, gauge nonsinglets from $|>_L$ can be split into

$$248 = (1,78) + (8,1) + (3,27) + (\bar{3},\overline{27}) \tag{17}$$

under $SU(3)xE_6$. This decomposition is useful for the point group P. The symmetric orbifold identifies $SU(3)_R$ and $SU(3)_L$. Therefore, gauge multiplet is of the form $(1)_R x(1,78)_L$. The P x G invariant states satisfy P x G = 1 x $e^{2\pi i p.v}$=1. Thus,

gauge multiplet :

$$p^I v^I = n = \text{integer} \tag{18}$$

$$p^I a^I_a = \text{integer, for all a} \tag{19}$$

For the matter, the spectrum is obtained from $(\bar{3})_R x(3,27)_L$ plus antiparticles. Thus, the projection is PxG = $\alpha^2 \exp(2\pi i p.v)$=1. Thus, matter from untwisted sector

$$p^I v^I = \frac{1}{3} \quad \text{mod integer} \tag{20}$$

$$p^I a^I_a = \quad \text{integer, for all a} \tag{19}$$

The second condition comes from the masslessness condition

$$M^2/4 = (p_R^{i2}/2 + N) + (P^{A2}/2 + \tilde{N} - 1)$$

where $i=1,...6$ and $A=1,...22$. Therefore, the massless mode must require $p_R^i = \frac{1}{2}(P^i - P^I A_I^i) = 0$. Since $A_I^i = a_I^a e_a^i$, we obtain $p^I.a_a^I$=integer. These particles come in 3 copies.

For twisted sector, the vacuum energy for each moded oscillator is $(-1/24 + 1/4\,\eta\,(1-\eta))$, $0 \leq \eta = r/n \leq 1$, where n is the order of twisting. In our case $\eta=1/3$ or $2/3$. Since there are 6 left-handed moded oscillators we obtain

$$\frac{M_L^2}{8} = \frac{(p^I+v^I+...)^2}{2} + \tilde{N} - \frac{2}{3} = 0 \tag{21}$$

where dots denote appropriate combinations of $a_a^I$ depending on Wilson lines. The massless modes in twisted sectors give

matter from twisted sector

$$\begin{aligned}(p^I+v^I+...)^2 &= \frac{4}{3} \text{ for } \tilde{N} = 0 \\ (p^I+v^I+...)^2 &= \frac{2}{3} \text{ for } \tilde{N} = \frac{1}{3} \text{ (multiplicity 3)}\end{aligned} \tag{22}$$

Having obtained all the formulae for the orbifold compactification, let us present one 4 dimensional superstring which has many desirable features.

The model has the shift and Wilson lines,

$$\begin{aligned}v^I &= (\tfrac{1}{3}\,\tfrac{1}{3}\,\tfrac{1}{3}\,\tfrac{1}{3}\,\tfrac{2}{3}\,0\,0\,0)\ (\tfrac{1}{3}\,\tfrac{1}{3}\,0\,0\,0\,\tfrac{1}{3}\,\tfrac{1}{3}\,\tfrac{2}{3}) \\ a_1^I &= (\tfrac{1}{3}\,\tfrac{1}{3}\,\tfrac{1}{3}\,\tfrac{2}{3}\,\tfrac{1}{3}\,0\,0\,0)\ (0\,0\,0\,0\,0\,\tfrac{2}{3}\,0\,0) \\ a_3^I &= (0\,0\,0\,0\,0\,0\,0\,\tfrac{2}{3})\ (\tfrac{1}{3}\,\tfrac{1}{3}\,\tfrac{1}{3}\,\tfrac{2}{3}\,0\,0\,0\,\tfrac{1}{3})\end{aligned} \tag{23}$$

Untwisted Sector $(p^2 = 2)$

gauge bosons : p.v = integer, p.a = integer

$p = \pm(\underline{1\ {-1}\ 0};\ 0\ 0;\ 0\ 0\ 0)\ (0\ 0;\ 0\ 0;\ 0\ 0\ 0\ 0)$ SU(3)

$p = \pm(0\ 0\ 0;\ 1\ 1;\ 0\ 0\ 0)\ (0\ 0;\ 0\ 0;\ 0\ 0\ 0\ 0)$ SU(2)

$p = \pm(0\ 0\ 0;\ 0\ 0;\ 0\ 0\ 0)\ (1 {-1};\ 0\ 0;\ 0\ 0\ 0\ 0)$ $SU(2)'_1$

$p = \pm(0\ 0\ 0;\ 0\ 0;\ 0\ 0\ 0)\ (0\ 0;\ 1\ 1\ ;\ 0\ 0\ 0\ 0)$ $SU(2)'_2$

where underline means permutation. Thus gauge group is

$$SU(3)xSU(2)xU(1)^5xSU(2)'_1xSU(2)'_2xU(1)'^6$$

matter fields: p.v = 2/3, p.a = integer

$$\left.\begin{array}{l} p = (\ \underline{1\ 0\ 0};\ 1\ 0;\ 0\ 0\ 0\ )\ (\ 0\ 0\ 0\ 0\ 0\ 0\ 0\ 0\ ) \\ p = (\ \underline{1\ 0\ 0};\ 1\ {-1};\ 0\ 0\ 0\ )\ (\ 0\ 0\ 0\ 0\ 0\ 0\ 0\ 0\ ) \end{array}\right\}\ 3(3,2)$$

$$\left.\begin{array}{l} p = (\ 0\ 0\ 0\ 0\ 0\ 0\ 0\ 0\ )\ (\ \underline{-1\ 0};\ 1\ 0;\ 0\ 0\ 0\ 0\ ) \\ p = (\ 0\ 0\ 0\ 0\ 0\ 0\ 0\ 0\ )\ (\ \underline{-1\ 0};\ 0\ {-1};\ 0\ 0\ 0\ 0\ ) \end{array}\right\}\ 3(2,2)'$$

$$p = (\ 0\ 0\ 0\ 0\ 0\ 0\ 0\ 0\ )\ (\ \underline{1\ 0};\ 0\ 0;\ 0\ 0\ 0\ {-1}\ )\qquad 3(2,1)'$$

plus singlets

Twisted Sector

Twenty seven fixed points are divided into 9 groups. Each repeats 3 times. As an exampe, consider

$$\text{TNT7: } v+a_1-a_3 = (\ \tfrac{2}{3}\ \tfrac{2}{3}\ \tfrac{2}{3}\ 1\ 1\ 0\ \tfrac{1}{3}\ {-\tfrac{1}{3}}\ )\ (\ 0\ 0\ {-\tfrac{1}{3}}\ {-\tfrac{2}{3}}\ 0\ 1\ \tfrac{1}{3}\ \tfrac{1}{3}\ )$$

There are some p's satisfying $(p+v+a_1-a_3)^2 = 4/3$,

$$p = (\ \underline{-1\ {-1}\ 0};\ {-1}\ {-1};\ 0\ 0\ 0\ )\ (\ 0\ 0\ 0\ 1\ 0\ {-1}\ 0\ 0\ )$$

$$p = \begin{cases} (\ -\tfrac{1}{2}\ -\tfrac{1}{2}\ -\tfrac{1}{2}\ -\tfrac{1}{2}\ -\tfrac{1}{2}\ +\tfrac{1}{2}\ -\tfrac{1}{2}\ +\tfrac{1}{2}\ )\ (\ 0\ 0\ 0\ 1\ 0\ {-1}\ 0\ 0\ ) \\ (\ -\tfrac{1}{2}\ -\tfrac{1}{2}\ -\tfrac{1}{2}\ -\tfrac{3}{2}\ -\tfrac{3}{2}\ +\tfrac{1}{2}\ -\tfrac{1}{2}\ +\tfrac{1}{2}\ )\ (\ 0\ 0\ 0\ 1\ 0\ {-1}\ 0\ 0\ ) \end{cases}$$

p = 12 singlets

which gives after consideration of opposite chiralities of untwisted and twisted sectors

$$3(3^*,1) + 3(1,2) + 12.\underline{1}$$

Because the modular invariance is satisfied, massless particles from twisted sectors make the spectrum anomaly free. Indeed this happens in our case also,

| Sector | Matter fields |
|---|---|
| UT | 3(3,2) |
| TNT1 $(v+a_1)$ | $3(3^*,1)+3(1,2)$ |
| TNT7 $(v+a_1-a_3)$ | $3(3^*,1)+3(1,2)$ |
| TNT8 $(v-a_1+a_3)$ | 9(1.2) |

Thus we obtain

$$3\,(3,2) + (1,2) + 2(3^*,1) + (1,1)$$

$$+12(1,2) + 3(2,2)' + 18(2,1)'$$

$$+18(1,2)' + \text{singlets} \qquad (24)$$

CONCLUSION

We have shown that we are able to obtain four dimensional models which are very close to a standard model superstring. The example we found satisfies (i), (ii), (iii), (iv), (v) and (vi) given in INTRODUCTION. In particular, the longevity of proton is obtained from the simple reason that there is no color triplets dangerous for proton decay. On the other hand, there are too many Higgs doublets. We are now trying to reduce the number of Higgs doublets, and possibly the rank of the gauge group also.[9]

ACKNOWLEDGMENTS

This research has been supported by the Korean Science and Engineering Foundation.

REFERENCES

1. P. Candelas, G. Horowitz, A. Strominger and E. Witten, Nucl. Phys. B258:46 (1985).

2. L. Dixon, J.A. Harvey, C. Vafa and E. Witten, Nucl. Phys. B261:678 (1985); Nucl. Phys. B274:285 (1986).

3. L.E. Ibanez, talk presented at 6th Symposium on Theoretical Physics, Korea, July (1987)

4. L.E. Ibanez, J.E. Kim, H.P. Nilles and F. Quevedo, Phys. Lett. B191:282 (1987).

5. K.S. Narain, Phys. Lett. B169:41(1986); K.S. Narain, M.H. Sarmadi and E. Witten, Nucl. Phys. B279:369(1987); P. Ginsparg, Harvard preprint HUTP-86/A053 (1986).

6. D.J. Gross, J.A. Harvey, E. Martinec and R. Rohm, Phys. Rev. Lett. 54:502 (1985); Nucl. Phys. B256:253 (1985).

7. Asymmetric orbifords so far considered seem to lead to vectorlike models (Private communications with L.E. Ibanez and H.P. Nilles).

8. L.E. Ibanez, H.P. Nilles and F. Quevedo, Phys. Lett. B187:25 (1987).

9. L.E. Ibanez, H.P. Nilles and F. Quevedo, Phys. Lett. B192:332 (1987).

# Chapter V
# Polyakov Strings

# LIOUVILLE STRINGS

Jean-Loup Gervais

Physique Théorique Ecole Normale Supérieure
Paris France

## I INTRODUCTION

In 1981 Polyakov [1,2] showed that the integration measure over surfaces embedded into flat space-times of dimension D not equal to 26 ( or 10), involves an additional scalar field with an exponential potential. With this motivation I, together with A. Neveu [3-8] and A. Bilal [9-12], extensively studied this so-called Liouville dynamics and developed the associated string theories, which are of a novel type. As we shall see, two new critical values of the space-time dimension D appea,r both for the purely bosonic and for the Neveu-Schwarz-Ramond (NSR) case. They are equal to 7 and 13, and to 3 and 5 respectively. Our results possess an interesting structure which I intend to summarize, as much as posssible, in the present lecture notes.

A preliminary remark is in order, concerning the numerology, of these values. As we shall argue, the new string theories in D dimensions have D-1 physical degrees of freedom. Altogether the number of such physical degrees of freedom is thus equal to three times two to the power n (resp two to the power n) for the bosonic (resp NSR) models, with n=1,2, for the new models, and n=3 for the standard ones. In each case, the new values are a natural extension of the old one. Moreover, for the NSR model, these numbers coincide with the real dimensions of the algebras of complex numbers, quaternions and octonions, respectively. Besides the real numbers, these are the only known division algebras. The existence of a division algebra is crucial in the building of superstrings field as we shall see.

The quantum Liouville solution, discussed here has noticeable implications for critical systems in two dimensions [6,12]. For lack of space, I shall leave them aside here and concentrate on the string theory aspects.

## II THE BASIC FEATURES OF THE BOSONIC LIOUVLLE STRING DYNAMICS

### II-1 The relevance of the Liouville dynamics for string theories

In this section we briefly review the arguments of Polyakov [1]. Following Brink, Di Vecchia, and Howe [13], one takes the string position X and the world sheet metric tensor $g_{ab}$ as independant degrees of freedom. The action is

$$\mathcal{S} = \frac{1}{2\pi} \int d\sigma d\tau \sqrt{-\det g}\; g^{ab}\, \partial_a X^\mu\, \partial_b X^\mu \quad ; \quad g^{ab} = (g^{-1})_{ab} \tag{2.1}$$

$\sigma$ and $\tau$ are the parameters of the surface generated by the string motion. This action is invariant under reparametrization of this world sheet and under the Weyl transformations

$$X \Rightarrow X \quad ; \quad g_{ab} \Rightarrow f(\sigma,\tau)\, g_{ab} \tag{2.2}$$

where f is an arbtrary function of $\sigma$ and $\tau$. In the partition function

$$Z = \int \mathcal{D}X\, \mathcal{D}g\; e^{i\mathcal{S}} \tag{2.3}$$

we first integrate over X. Consider

$$Z_0(g) = \int \mathcal{D}X\; e^{i\mathcal{S}} \tag{2.4}$$

The well-known formula [14]

$$g^{ab} \frac{\delta Z_0}{\delta g^{ab}} = g^{ab} < T_{ab} > = \frac{D}{24\pi} (R + cste) \tag{2.5}$$

relates the trace anomaly of the energy-momentum tensor T to the world sheet scalar curvature R. This trace anomaly breaks the Weyl invariance but preserves the reparametrization invariance. We shall fix the parametrization by imposing

$$g_{ab} = e^{\phi} \begin{vmatrix} -1 & 0 \\ 0 & 1 \end{vmatrix} \tag{2.6}$$

In this case, equation (2.5) leads to

$$Z_0(\phi) = \exp\Big(- \frac{iD}{48\pi} \int d\sigma d\tau [(\partial_\tau \phi)^2/2 - (\partial_\sigma \phi)^2/2 \; + e^{\phi} \,]\Big) \tag{2.7}$$

On the other hand [1], the measure of integration over $\phi$ involves the Jacobian (Faddeev-Popov determinant) which is equal to

$$\Delta(\phi) = \exp\Big( \frac{i26}{48\pi} \int d\sigma\, d\tau\, [(\partial_\tau \phi)^2/2 - (\partial_\sigma \phi)^2/2 \; + e^{\phi} \,]\Big) \tag{2.8}$$

The coefficient of $\exp(\phi)$ may be freely changed by adding an irrelevant constant to $\phi$. We took advantage of this in order to get rid of the arbitrary constant in formula (2.5) and set this coefficient equal to one.

Remarkably, equations (2.7) and (2.8) both involve the same Liouville action and the coefficients in front are opposite at D=26. If one assumes that the integral which is common to eqs. (2.7) and (2.8) remains finite at D=26, it follows that the Liouville dynamics disappears at the critical dimension. This common wisdom, which leads to the so-called Polyakov measure, assumes, however, that $\phi$ is a background field. Indeed, in the limit D going to 26, the Planck constant of the Liouville dynamics goes to infinity.

In this limit, the functional integral over $\phi$ will be dominated by very irregular configurations, unless it is a priori restricted to continuous functions. If $\phi$ is considered as a full-fledged quantum field, one cannot draw any definite conclusion about the behavior at D=26, from the above discussion, since the kinetic term of (2.7,8) is unbounded in the limit. One must first solve the quantum Louville theory. This is the program which I started to follow with A. Neveu in 1982 [3-8]. We studied the quantum dynamics by operator methods in order to avoid any ambiguity. A brief account of the key features of our work follows next.

II-2 Highlights of the Liouville dynamics

As a starting point, we shall take the action

$$S = \frac{1}{16\pi\hbar} \int d\sigma\, d\tau\, [(\partial_\tau\phi)^2/2 - (\partial_\sigma\phi)^2/2 + e^{\phi}] \tag{2.9}$$

This particular rewriting of the factor in front is convenient since the parameter $\hbar$ so defined will indeeed play the role of Planck's constant. In view of formulae (2.7), (2.8), Polyakov's argument which, we just outlined, gives

$$\hbar_P = 3 / (26 - D) \tag{2.10}$$

As we shall see, however, this formula will be modified by finite renormalization effects, and it is better to leave $\hbar$ as a free parameter to begin with.

Consider, first the classical Louville dynamics [3]. The field equation

$$\partial^2_{uv}\phi = 4\, e^{\phi} \quad ; \quad u = \tau + \sigma \quad ; \quad v = \tau - \sigma \tag{2.11}$$

can be easily used to derive the relations

$$\partial_u \left(e^{\phi/2}\, \partial^2_{vv}\, e^{-\phi/2}\right) = \partial_v \left(e^{\phi/2}\, \partial^2_{uu}\, e^{-\phi/2}\right) = 0 \tag{2.12}$$

It follows that we can write equations of the form

$$[\, -\partial^2_{uu} + U^{(L)}(u)]\, e^{-\phi/2} = [\, -\partial^2_{vv} + U^{(R)}(v)]\, e^{-\phi/2} = 0 \tag{2.13}$$

where $U^{(L)}$ and $U^{(R)}$ are functions of a single variable. This allows to separate the right movers and the left movers in a natural way, since the general solution of the last partial differential equations is

$$e^{-\phi/2} = [\Psi^{(R)}_1(v)\, \Psi^{(L)}_2(u) - \Psi^{(R)}_2(v)\, \Psi^{(L)}_1(u)]/(2\sqrt{2}) \tag{2.14}$$

where the psi's are functions of one variable which obey the Schrödinger equations

$$[\, - \frac{d^2}{dx^2} + U^{(L)}(x)]\, \Psi^{(L)}_{1,2}(x) = [\, - \frac{d^2}{dx^2} + U^{(R)}(x)]\, \Psi^{(R)}_{1,2}(x) = 0 \tag{2.15}$$

Let us next display the basic features of the canonical formalism and of the quantization following ref. [4]. Since this part deals with the left and right movers separately and in exactly the same way we drop all R and L superscripts. At the classcal level, it is convenient to define

$$\lambda(x) = \frac{1}{\Psi_1(x)} \frac{d\Psi_1(x)}{dx} \quad ; \quad \tilde{\lambda}(x) = \frac{1}{\Psi_2(x)} \frac{d\Psi_2(x)}{dx} \tag{2.16}$$

The boundary counditions are always such that U is periodic with a period which we choose to be $2\pi$. Generically, the two solutions of (2.15) may be selected so that $\lambda$ and $\tilde{\lambda}$ are periodic. We then let

$$\lambda(x) = \sum_n \lambda_n e^{-inx} \quad ; \quad \tilde{\lambda}(x) = \sum_n \tilde{\lambda}_n e^{-inx} \tag{2.17}$$

Standard Wronskian arguments show that $\lambda_o + \tilde{\lambda}_o = 0$. As is well-known, these logarithmic derivatives satisfy the Riccati equations

$$U = (\lambda)^2 + d\lambda/dx = (\tilde{\lambda})^2 + d\tilde{\lambda}/dx \tag{2.18}$$

One can show [4], that the canonical Poisson brackets associated with (2.9) are equivalent to

$$i\left\{\lambda(x_1)\,,\,\lambda(x_2)\right\} = 2\pi\hbar\ \delta(x_1 - x_2) \tag{2.19}$$

or to the same relation with $\lambda$ replaced by $\tilde{\lambda}$. This Poisson bracket structure is such that if we let

$$U(x) = 2\hbar \sum_n \Lambda_n e^{-inx} \tag{2.20}$$

the operators $\Lambda_m$ satisfy the classical Poisson bracket Virasoro algebra

$$i\left\{\Lambda_m\,,\,\Lambda_n\right\} = (m-n)\,\Lambda_{m+n} + \frac{n^3}{4\hbar}\,\delta_{m+n,0} \tag{2.21}$$

One may verify that the psi fields transform as conformal fields ([15], [16]) through Poisson brackets. Their conformal weights is 1/2, and equation (2.14) shows that $\exp(-\phi/2)$ has conformal weights $\delta = \bar{\delta} = 1$. This is consistent since it follows that the product $\exp(\phi)$dudv which appears in the action (2.9) is invariant under conformal transformations.

Let us now move on to the quantum case. As shown in refs [4,5], the structure summarized above extends with surprisingly small modifications. In equation (2.19), the Poisson bracket is replaced by a commutator. Eq (2.18) becomes

$$U = N[(\lambda)^2] + d\lambda/dx = \tilde{N}[(\tilde{\lambda})^2] + d\tilde{\lambda}/dx \tag{2.22}$$

where N and $\tilde{N}$ denote normal ordering with respect to the modes of $\lambda$ and $\tilde{\lambda}$ respectively. The key property, derived in ref.[4] and further discussed in ref. [8], is that, generically, equation (2.22), together with the condition $\lambda_o + \tilde{\lambda}_o = 0$, establishes a quantum correspondance between $\lambda$ and $\tilde{\lambda}$ which is such that, if $\lambda_m$ satisfies the quantum analog of (2.19), the same is true for $\tilde{\lambda}_m$. One thus has

$$[\lambda_m\,,\,\lambda_n] = [\tilde{\lambda}_m\,,\,\tilde{\lambda}_n] = \hbar\ m\ \delta_{m+n,0} \tag{2.23}$$

The quantum version of (2.16) is

$$\Psi_1(x) = N[\exp(i\eta \sum_n \frac{\lambda_n}{n} e^{-inx})]; \quad \Psi_2(x) = \tilde{N}[\exp(i\eta \sum_n \frac{\tilde{\lambda}_n}{n} e^{-inx})] \tag{2.24}$$

Equation (2.23) shows that the two sets of operators $\lambda_m$ and $\tilde{\lambda}_m$ separatly satisfy standard free field commutation relations, and $\psi_1$ and $\psi_2$ are given by expressions which are both identical to a standard string vertex operator for a single string component. The above expression is slightly abreviated. The zero modes are to be handled in the usual way (see [5] for details). The factor $\eta$, which is a pure number, exhibits a quantum modification with respect to (2.16). Its value is determined as follows. $\psi_1$ and $\psi_2$ are conformal fields ([15], [16]) with the same conformal weight. This can be verified from the formulae we just recalled, together with the quantum version of (2.20) which gives

$$\Lambda_n=\frac{1}{2\hbar}N[\sum_r\lambda_r\lambda_{n-r}-in\lambda_n+\tfrac{1}{4}\delta_{n,0}]=\frac{1}{2\hbar}\tilde{N}[\sum_r\tilde{\lambda}_r\tilde{\lambda}_{n-r}-in\tilde{\lambda}_n+\tfrac{1}{4}\delta_{n,0}] \tag{2.25}$$

An easy computation shows that the Virasoro algebra is satisfied with central charge

$$c = 1 + \frac{3}{\hbar} \tag{2.26}$$

From the short distance expansion of products of $\psi_1$ and $\psi_2$ one next generates a conformal family following [4], [6], [8]. In the same way as in the classical case, some of its members must have a conformal weight equal to one since one should be able to build a conformally invariant action with an interaction term that is a marginal operator. Classically $\psi_1$ and $\psi_2$ raised to the power minus two do have unit conformal weight. Imposing the same condition at the quantum level leads to the equation

$$2\hbar\,\eta^2 - \eta + 1 = 0 \tag{2.27}$$

Another motivation for this choice of $\eta$ is that the psi's so defined satisfy a second order differential equations that is the quantum analogue of (2.15). (see [5], [6],and [8]). It plays a key role, but we shall not consider it explicitely. This differential equation was independantly derived in ref. [16] and applied to two-dimensional critical systems.

Equation (2.27) has the two solutions

$$\eta_\pm = \frac{1}{4\hbar}(1 \pm \sqrt{1 - 8\hbar}) \tag{2.28}$$

Substituting them into (2.24) we get four operator psi's instead of the two we had at the quantum level. This is not surprising since for $\hbar \longrightarrow 0$, $\eta_+$ blows up. Using formula (2.28) with $\eta_+$ leads to quantum operators which have no smooth classical limit. Nevertheless, the conformal family generated by performing operator products with both solutions of (2.27) has a remarkable property [6], [8]. Indeed, the corresponding set of conformal weights is given by

$$\epsilon(m,n)= \frac{1}{48}[(13-c)(m^2+n^2)-24mn-(1-c)+(n^2-m^2)\sqrt{(c-1)(c-25)}] \tag{2.29}$$

where m and n are arbitrary positive or negative integers. This generalizes Kac's formula [17]. Indeed, if m and n are restricted to have the same sign, it is precisely equal to the highest weight which is such that a zero of a Kac determinant occurs at the level m times n. It is crucial that the set of conformal weights derived in the discussion summarized above is more general that Kac's formula. Indeed, for c>1, the null states only exist for negative highest weights and are unphysical (the corresponding representation may never be unitary). In general, the conformal weights (2.31) are not all positive and only a subset may appear in the spectrum of the Liouville theory.

Since the string has a finite length, the Liouville field theory is considered with a finite extension in $\sigma$ and boundary conditions must be specified. We only consider the free string case where the world sheet parameters vary over a strip parallel to the $\tau$ axis, and begin with the boundary conditions relevant for open strings. By convention, $\sigma$ is taken to run from $0$ to $\pi$. Let us go back, for a short while to the classical case. The conditions are [3]

$$\partial_\sigma e^{-\phi/2} = \frac{1}{\sqrt{2}} \text{ at } \sigma=0 \;;\; \partial_\sigma e^{-\phi/2} = -\frac{1}{\sqrt{2}} \text{ at } \sigma=\pi \tag{2.30}$$

This may seem complicated, but one easily sees from (2.14) that these conditions only involve the Wronskian of the solutions of (2.15) and are easily implemented. One finds that $U^{(R)}$ and $U^{(L)}$ are equal and periodic with period $2\pi$. Equations (2.15) then show that $\psi^{(R)}_{1,2}$ and $\psi^{(L)}_{1,2}$ are linearly related and one ends up with a formula of the form [3]

$$e^{-\phi/2} = \sum_{m,n\,=\,1,2} \alpha_{mn}\, \Psi_m(u)\Psi_n(v) \tag{2.31}$$

where the coefficients are determined up to one unknown constant that reflects the possibility of an arbitrary time translation. For closed string one would expect that $\phi$ should be simply taken as periodic. Although it is conceptually simpler to begin with, this case turned out to be much more delicate than the boundary condition relevant for open string; and we shall not discuss it. The closed strings will come out later from the one-loop open-string diagrams

As shown in [6] and [7], the spectrum of the quantum Liouville theory may be very simply summarized from formula (2.29), since the physical conformal weights are obtained by choosing all values of m and n that give positive highest weights. We next discuss the Liouville spectrum by applying this simple rule. In this connection, the square root of formula (2.29) plays a key role. Clearly it is real only if c<1 or if c>25. The first case is the standard region of two-dimensionnal critical systems. We shall not discuss it since it is excluded by condition (2.26) assuming of course that $\hbar$ is positive. In the region c>25, one has $\hbar$ <1/8, and the solutions of (2.27) are both real. This is the weak coupling regime of the Liouville theory, that includes its semi classical limit $\hbar \longrightarrow 0$. The quantum solution is thus constructed only using $\eta_-$ . This allows to extend (2.31) to the quantum case [5] in such a way that locality in $\sigma$ is satisfied. The requirement that the operator so defined can be restricted to the subset of positive highest weights of (2.29) completely determines the quantum solution [5]. Moreover this is really completely consistent [5] when c takes the discrete set of values

$$c = 1 + \frac{6(N+1)^2}{N} \quad ; \quad N \text{ positive integer} \tag{2.32}$$

The set of physical conformal weights is then given by ($\nu$ is a positive integer)

$$\delta(m) = \epsilon(1\ ,\ 2m-N) \quad ; \quad 0 \leq m \leq \nu \quad ; \quad N = 2\nu+1 \tag{2.33}$$

$$\delta(m) = \epsilon(1\ ,\ 2m-1-N) \quad ; \quad 1 \leq m \leq \nu \quad ; \quad N = 2\nu \tag{2.34}$$

Remarkably one finds that c must be such that the square root of formula (2.29) is rational. This is similar to the series put forward by Belavin Polyakov and Zamolodchikov [16] in the region c<1. Such a rationality requirement is natural since if it is not satisfied the set of values taken by (2.29) for m and n arbitrary positive or negative is dense over the whole real line.

There remains to discuss the region 1<c<25. There, the square root of (2.29) is pure imaginary. One gets real conformal weights only if m and n are equal or opposite. The first possibility is excluded since it only gives negative or vanishing values. One finds [7] the spectrum

$$\delta(m) = \epsilon(m\ ,\ -m) \equiv \frac{1}{24}\ [(m^2(25 - c) + (c - 1)] \tag{2.35}$$

The quantum Liouville theory only exist [7] if c takes the values 7 , 13 , or 19. This discrete set comes out from the condition that it must be possible to build an algebra of local operators restricted to the subset of values (2.35) that are real and positive. These local fields involve both solutions of (2.27), which are on the same footing in this case since they are conplex conjugate.

In general, the infinite conformal family built up from products of the psi operators play the role of building blocks which are to be used in order to construct the physical conformal operators. These should generate a conformal family that is closed over conformal fields with positive conformal weights. This conformal bootstrap requirement is so stringent that it can be fulfilled only for the special values of c indicated above. As we next discuss this fixes the values of the space-time dimension where the string may be embedded.

## II-3 The Liouville string theories and their modular properties

The Virasoro generators $L_m$ of the X modes commute with the $\Lambda_m$ operators. The consistency of the string theory requires two conditions. First the total central charge must be 26. The Liouville coupling constant is thus determined by

$$c = 26 - D \quad ; \quad \hbar = \frac{3}{25 - D} \tag{2.36}$$

this value of $\hbar$ differs from (2.10). This is due to fact that formula (2.26) involves a quantum contribution besides the classical term $3/\hbar$ . Second, the open string spectrum is to be determined from the mass shell condition

$$(L_0 + \Lambda_0 - 1)|\,\text{phys.}> = 0 \tag{2.37}$$

These two conditions ensure, in particular that there exist a BRST nilpotent operator, so that two degrees of freedom are unphysical.

From condition (2.36), one sees that the weak coupling regime of Liouville only corresponds to D<1. The associated string theories may be given a meaning only after continuation in D. One obtains

$$D = 1 - \frac{6(N-1)^2}{N} \quad ; \text{ N positive integer} \tag{2.38}$$

Although they are unphysical these cases exhibit an interesting structure [5].

For physical string building, we are however only interested in the values of D such that 1<D<25. This precisely selects the region 1<c<25 where the square root of (2.29) is pure imaginary. From the allowed values c=7, 13, 19, we conclude that Liouville string theories may exist only if D=7 ;13 ; 19. Condition (2.37) leads to the mass spectrum

$$M_{m,\mathcal{N}}^{2} = (m^2 - 1)\,\frac{D-1}{12} + 2\,\mathcal{N} \tag{2.39}$$

where m is the integer that determines the highest weight of the Liouville theory ( see formula (2.35), and where the integer $\mathcal{N}$ is determined by the occupation numbers describing the harmonic excitations both of the Liouville mode and of the X modes.

So far we have only imposed the conditions that are necessary for the consistency of the tree diagrams.The loop diagrams are consistent only if additional modular properties are satisfied. For open strings, which we we discuss at present, the point to consider is whether the closed string singularities that appear in the orientable non-planar one loop diagrams, are poles as is required by unitarity (For the standard bosonic string, this gave the first derivation of the D=26 condition ). Details are given in ref.[9]. Consider the partition function of the Liouville conformal weights:

$$Z_L(q_1) \equiv \sum_m q_1^{\,2\delta(m)} = q_1^{\,(D-1)/12} \sum_m q_1^{\,m^2(D-1)/12} \tag{2.40}$$

The nature of the closed string singularities is exhibited by performing a Jacobi transformation on the integrand of the open string amplitude. Define the variable q through the equation

$$\ln(q)\,\ln(q_1) = \pi^2 \tag{2.41}$$

One may see that one will get poles if the last factor of the above formula is such that the product

$$\Big(\sum_m q_1^{\,m^2(D-1)/12}\Big)\theta'_1(q_1) \tag{2.42}$$

is a meromorphic function of q. We adopt the standard notation [18] for the theta functions with vanishing first argument. This condition is indeed fulfilled for D=7 and D=13, since the last term of (2.40) is given by theta functions of $q_1$ or $q_1^2$ . For instance, at D=13, equation (2.42) may be rewritten as

$$\Big(\sum_m q_1^{\,m^2(D-1)/12}\Big)\theta'_1(q_1) \equiv \theta_3(q_1)\;\theta'_1(q_1)\;\alpha\,\theta_3(q)\;\theta'_1(q) \tag{2.43}$$

The last equality is a consequence of the modular properties of the theta functions [18].

It is remarkable that, the values of D that are selected by the present requirement of modular properties coincide with two of the three values for which the

Liouville dynamics exist. The dimensions D=7 and 13 thus appear as leading to string theories that are completely consistent. They are new critical dimensions much on the same footing as D=26. On the other hand, this is not the case for the third possible value D=19. It does not lead to a consistent string theory and we shall not discuss it any further.

Finally, one may discuss the closed string spectrum by assuming that it is, as usual, given by two copies of the open string one discussed above. The partition function is indeed modular invariant at D=7 and 13 [7].

## III THE LIOUVILLE SUPERSTRING THEORIES

In this section we study the Neveu-Schwarz (NS) Ramond (R) Liouville string theories that incorporate the 2D supersymmetric Liouville (Sliouville) theory, following the original sugestion of Polyakov [2]. The general structure of these theories is very similar to the one of the purely bosonic theories which we just reviewed. There again appear two new critical dimensions. Their values are D=3 and D=5. The properties of the purely bosonic case which we just summarized may be straightforwardly extended to the present case, and we shall not elabarate upon them.

### III-1 The equality between Neveu-Schwarz and Ramond partition functions

The distinctive feature of the standard superstrings, as compared with the purely bosonic ones, is that they are space-time supersymmetric in 10 dimensions. The first hint in this direction was obtained by Gliozzi Scherk and Olive [19] who proved that the Neveu-Schwarz (NS) and Ramond (R) partition functions are equal. In this section we study the NSR Sliouville strings from this viewpoint, and show that they satisfy similar equalities. The present section is a summary of ref. [9]

The weak coupling solution of refs.[3-5] has been extended to the Sliouville theory by Arvis [20] and Babelon [21]. The associated representation of the conformal algebra reads

$$[\Lambda_m , \Lambda_n] = (m-n)\, \Lambda_{m+n} + \frac{c}{8}\, (m^3-\text{"}n\text{"}) \quad ; \quad c = 1 + \frac{2}{\hbar} \tag{3.1}$$

In this formula, and in the following, quantities between quotation marks are only present in the NS sector. In the R sector they are to be replaced by zero. The weak coupling regime is now for c>9. As shown in ref. [20] and [21], the Sliouville field theory can be solved through a complicated transformation to free fields similar to $\lambda$ and $\tilde{\lambda}$ . These fields have the same structure as one component of the standard space-time NSR fields. In particular, there is a NS Sliouville field and a R Sliouville field; and the Sliouville theory has its own NS and R sectors. Formula (2.31) becomes

$$\epsilon(m,n)=\frac{1}{32}[(5-c)(m^2+n^2)-8mn-\text{"}(1-c)\text{"}+(n^2-m^2)\sqrt{(c-1)(c-8)}] \tag{3.2}$$

The string theory will be consistent if c + D = 10, and if the spectrum is determined from the mass shell condition

$$[L_0 + \Lambda_0 - "1/2"]| \text{phys.} > = 0 \tag{3.3}$$

These conditions are equivalent to the existence of a nilpotent BRST operator in each sector. From (3.2), one sees that, in the same way as in the bosonic case, the interesting region for string theory building coincides with the interval where the square root of formula (3.2) is pure imaginary. The physical spectrum correspond to m+n=0 as before. The mass spectrum is thus given by:

$$M^2(m, N_1, N_2) = \frac{(D-1)}{8}(m^2-"1") + 2N_1 + 2N_2 \tag{3.4}$$

$N_1$ ($N_2$) are the contributions of the non zero 2D bosonic (fermionic modes) to $\Lambda_0$ and $L_0$. It is easy to derive the partition functions of the physical string states:

$$Z = \sum_{m, N_1, N_2} q^{M^2(m, N_1, N_2)/2} Z_X \tag{3.5}$$

Using formula (3.4), one obtains

$$Z^{\pm}_{NS} = (\sum_m q^{(m^2-1)(D-1)/8})\frac{1}{2}[\prod_{n>0}(1+q^{n-1/2})^{D-1} \pm \prod_{n>0}(1-q^{n-1/2})^{D-1}] \tag{3.6}$$

$$Z_R = d_s \sum_m q^{m^2(D-1)/8} \prod_{n>0}(1+q^n)^{D-1} Z_X \tag{3.7}$$

where the common factor $Z_X$ comes from the transverse components of X and from the Liouville harmonic bosonic modes:

$$Z_X = f(q)^{D-1} \quad ; \quad f(q) = \prod_{n>0}(1-q^n)^{-1} \tag{3.8}$$

The string theory is built up in such a way that the ghost killing mechanism removes two degree of freedom. In D dimensions the total number of degrees of freedom is D+1, and hence the number of physical degrees of freedom is D-1. This is why the partition functions of the non zero modes are standard harmonic oscillator partition functions raised to the power D-1. The factor $d_s$ represents the number of Dirac physical components of the zero-mode spinor of Ramond. It will be determined below. In the NS sector, we already separated the two G-parity sectors, since this was a key point in the work of G.O.S. In practice, the equality between $Z_{NS}$ and $Z_R$ is only possible if they can be expressed as Jacobi theta functions of q or $q^2$. This is true for any values of D for all factors in (3.6) and (3.7) apart from the contribution of the zero-modes of the Sliouville field theory. For 1<D<9, this latter factor takes a suitable form only for the new critical dimensions D=3 and 5.

For D=5, the basic identities are

$$2Z^{(0)}\prod_{n>0}\left(1+q^n\right)^4 = q^{-1/4}Z^{(1/2)}\frac{1}{2}\left[\prod_{n>0}(1+q^{n-1/2})^4 + \prod_{n>0}(1-q^{n-1/2})^4\right] \tag{3.9}$$

$$2Z^{(1/2)}\prod_{n>0}\left(1+q^n\right)^4 = q^{-1/4}Z^{(0)}\frac{1}{2}\left[\prod_{n>0}(1+q^{n-1/2})^4 - \prod_{n>0}(1-q^{n-1/2})^4\right] \tag{3.10}$$

We have introduced the Sliouville zero-mode partition functions

$$Z^{(0)}=\sum_{p=-\infty}^{+\infty} q^{p^2} = \theta_3(q);\ Z^{(1/2)}=\sum_{p=-\infty}^{+\infty} q^{(p+1/2)^2} = \theta_2(q) \qquad (3.11)$$

which are equal to standard theta functions [18] with vanishing first argument. This particular rewriting will be convenient below. The proof of equalities (3.9) and (3.10) follows from standard properties of the theta functions. Ultimately they are equivalent to

$$\sqrt{\theta_3^4(q) - \theta_4^4(q)} = \pm\theta_2^2(q) \qquad (3.12)$$

These last relations are the square roots of the one which gives the GSO [19] formula for the standard superstring. Indeed, by equating the products of the left and of the right members of equations (3.9) and (3.10), one obtains the famous GSO relation

$$8\prod_{n>0}\left(1+q^n\right)^8=q^{-1/2}\ \frac{1}{2}\Big[\prod_{n>0}(1+q^{n-1/2})^8 - \prod_{n>0}(1-q^{n-1/2})^8\Big] \qquad (3.13)$$

It is now our task to connect (3.9) and (3.10) with the string state partition functions. This is done by selecting appropriate vamues of m in (3.6,7). Since $Z_\chi$ is a common factor, we need not discuss it. Apart from an overall factor, the left hand side of eq. (3.9) (resp. (3.10)) coincides with (3.7) if the summation over m is restricted to the even (resp odd) values. Similarly, the right hand side of (3.9) (resp (3.10)) corresponds to formula (3.6) with the plus (resp minus ) sign if the summation is restricted to the odd (resp even ) values of m. Consider, now, the factor $d_S$. The zero mode Sliouville R field anticommute with the D zero-mode Dirac matrices of the space-time R field. It follows that altogether we have a realization of the Dirac algebra with D+1=6 Dirac matrices. The Dirac spinors thus have 8 complex components. Imposing the Weyl condition and the Dirac equation leaves 4 real degrees of freedom and we conclude that $d_S$ is equal to this number. Both sides of equations (3.9) and (3.10) must be multiplied by 2. For the R sector this gives the right factor $d_S$ , but, in the NS sector, we are led to assume that each value of m is doubly degenerate.

One sees that, contrary to the ten-dimensional superstring, both G-parities lead to supersymmetric theories. For the even G-parity NS sector, the lightest particles are massless scalars and spin one half particles. For the odd case, the lightest bosonic particle has spin one as in standard superstrings. It is massive, however. Expanding the right hand side of (3.10), on finds that its mass is equal to 1/2.

Finally, the case D=3 also leads to equalities between NS and R partition functions for both G parity sectors. The interested reader will find the details in ref.[9].

III-2 Group triality for the D=5 model

Another key feature of the Green-Schwarz superstring is the remarkable triality property of O(8) that appears as the group of transverse rotation in the light-cone formalism. Summarizing ref.[11], we shall next show that for the D=5 Sliouville string theory, a different group emerges, in a dynamical way, with similar triality properties as O(8). It is the basis of the superstring formalism which will be the subject of the following section. (The D=3 model is discussed along the same line in ref. [12], but we shall not consider this case here).

We shall hereafter make use of a six-dimensional notation that combines the space-time fields $X_\mu$, and $\Gamma_\mu$ whose indices run from 1 to 5, together with the Sliouville free fields denoted $X_6$ and $\Gamma_6$. One may visualise the present D=5 Sliouville string theory as a six-dimensional one where the Lorentz invariance is broken down to the D=5 Lorentz group. This breaking is rather subtle, and cannot be reduced to a trivial compactification of one space dimension. It is important to stress that the theory is strictly five-dimensional even if we use a six-dimensional language. In particular, the equality between partition functions which we discussed above is for the five-dimensional masses. From now on, we make use of the light-cone formulation of string theories. A space-time vector has components

$$V_\pm = (V_1 \pm V_2)/\sqrt{2} \quad , \text{ and } V_i \quad \text{with} \quad i=3,4,5 \tag{3.14}$$

By convention, lower case latin indices run from 3 to 5 and refer to the true transverse coordinates. When the Liouville fields are included, it is convenient to use a O(4) notation and to introduce two complex components:

$$V^3 = \frac{(V_3 + iV_4)}{\sqrt{2}} \; ; \quad V^{\bar{3}} = (V^3)^+ \quad ; \quad V^5 = \frac{(V_5 + iV_6)}{\sqrt{2}} \; ; \quad V^{\bar{5}} = (V^5)^+ \tag{3.15}$$

Lower and upper indices describe real and complex components respectively. The lower index 6 refers to the Liouville degrees of freedom.

In the light-cone formulation, the plus and minus components of X and $\Gamma$ are eliminated. Let us next bosonize the four remaining NSR fields. They are equivalent to two bosonic modes $\alpha^1$ and $\alpha^2$. Define

$$\Omega^{\pm\pm}_{1,2}(z) = \sqrt{z}\, K_1^{\pm 1} K_2^{\pm 1}\, z^{(\pm\alpha_0^1 \pm \alpha_0^2)/\sqrt{2}}\, N\Big(\exp\Big[-\sum_{n\neq 0} \frac{z^{-n}}{n} \frac{(\pm\alpha_n^1 \pm \alpha_n^2)}{\sqrt{2}}\Big]\Big) \tag{3.16}$$

The indices 1 and 2 of $\Omega$ are for later convenience. The operators K are the usual zero-mode shift operators:

$$[\alpha_0^1 \, , K_1] = K_1/\sqrt{2} \quad ; \quad [\alpha_0^2 \, , K_2] = K_2/\sqrt{2} \tag{3.17}$$

The four NSR fields are given by

$$\Gamma^3(z) \sim \Omega^{++}_{1,2}(z); \; \Gamma^{\bar{3}}(z) \sim \Omega^{--}_{1,2}(z); \; \Gamma^5(z) \sim \Omega^{+-}_{1,2}(z); \; \Gamma^{\bar{5}}(z) \sim \Omega^{-+}_{1,2}(z) \tag{3.18}$$

For compactness, we leave aside the well known cocycle factors. By standard computations one may verify that the fields so defined satisfy the anticommutation relations

$$[\Gamma_I(e^{i\sigma_1}) \, , \Gamma_J(e^{i\sigma_2})]_+ = \delta(\sigma_1 - \sigma_2)\, \delta_{I,J} \; ; \quad I,J = 3,4,5,6 \tag{3.19}$$

The two bosonic modes we just introduced generate together the space-time and Slouville Fermi Fock spaces. By the standard quark current construction [22], one derives the generators of the O(4) Kac-Moody algebra, which is of course the direct product of two SU(2) Kac-Moody algebras. The first SU(2) generators are given by:

$$I_{+}^{(1)} = :\Gamma^{5} \Gamma^{3}:; \quad I_{-}^{(1)} = :\Gamma^{\bar{3}} \Gamma^{\bar{5}}:; \quad I_{3}^{(1)} = ( :\Gamma^{3} \Gamma^{\bar{3}}: + :\Gamma^{5} \Gamma^{\bar{5}}: )/2 \tag{3.20}$$

The other SU(2) generators are deduced from these relations by exchanging 5 and $\bar{5}$. In bosonic terms, both operators take the Frenkel-Kac form [22] (s=1, 2):

$$I_{\pm}^{(s)}(z) \sim z (K_s)^{\pm 2} z^{\pm \alpha_0^s \sqrt{2}} N\left( \exp\left[ \mp \sum_{n \neq 0} \frac{z^{-n}}{n} \sqrt{2} \alpha_n^s \right] \right); \quad I_3^{(s)}(z) = \sum_n \alpha_n^s z^{-n} / \sqrt{2} \tag{3.21}$$

Eqs (3.17) show that the eigenvalues of the zero modes $\alpha_0^1$ and $\alpha_0^2$ are discrete. The corresponding two-dimensional lattice is of course the weight lattice of O(4), which is the direct sum of two SU(2) weight lattices. Its points may be separated into four cosets. One is the set of points with coordinates integer mutiples of $\sqrt{2}$. Denoted (0,0) it corresponds to product of two integer spin representations of SU(2). Two more cosets are obtained by shifting the preceding one by $1/\sqrt{2}$ along the two SU(2) axes successively. Denoted (0,1/2) and (1/2,0) they describe representations which are products of one integer and one halph integer spin representations. Finally, the set (1/2,1/2) which is the product of two halph integer spin representations is obtained after shifting the first coset in both directions simultaneously by $1/\sqrt{2}$. The group generators do not mix the cosets. The NSR fields mix ( 0 , 0 ) and ( 1/2 , 1/2 ) on the one hand (this is the NS sector), and ( 0 , 1/2 ) and ( 1/2 , 0 ) on the other hand (this is the R sector). ( 0 , 0 ) and ( 1/2 , 1/2 ) are the odd and even G-parity sectors of NS respectively. The other two cosets correspond to the two chirality projections of the R sector.

The construction of the NSR fields and of the group generators which we just carried out is not specific to O(4). The O(8) NSR fields are constructed in a similar way [22]. The distinctive feature of this last case is that there are additional lattice vectors of unit length, besides the ones that give the NSR fields, that can be used to construct the light-cone superfields of Green and Schwarz. This of course reflects the triality of O(8) [22]. In the present case the only unit vectors have coordinates $(\pm 1/\sqrt{2}, \pm 1/\sqrt{2})$. They all appear in formula (3.18), and there is no way to build additional spinor fields besides the NSR ones. This is not surprising, since the Dynkin diagram of O(4) has no triality. There is thus no way to achieve triality from the NSR fields alone even including the Sliouville NSR field as we did above. We shall next show, however that triality is recovered if one further includes the Sliouville bosonic mode . This agrees with its key role in the equality between the partition functions of the space-time bosons and fermions which we derived above.

At D=5, the spectrum of highest weights may be rewritten as

$$\delta(r) = \frac{(r\sqrt{2})^2}{2} + \text{"}\frac{1}{4}\text{"} \; ; \text{ or } \delta(r) = \frac{(1/\sqrt{2}+r\sqrt{2})^2}{2} + \text{"}\frac{1}{4}\text{"}; \quad r \text{ integer} \tag{3.22}$$

where we separated the even and odd values. From the way we wrote these expression it is clear that, apart from the last constant term, the bosonic Sliouville zero mode spectrum precisely corresponds to the one-dimensional weight lattice of SU(2) that just appeared above in the O(4) bosonization. Since the bosonic Sliouville spectrum is made up with the harmonic excitations of these ground states, one may carry out the Frenkel-Kac construction in this Hilbert space, thereby generating the level one representation of SU(2). We thus have another SU(2) at our disposal which comes entirely from the Sliouville bosonic modes, and is purely dynamical. It describes internal quantum numbers in the present physical five-dimensional world. The trick will be to treat it on the same footing as the generators we built up --see formula (3.21)-- from the NSR fields. We shall denote all quantities relevant to this third SU(2) by the same symbols as for the first two, apart from the value of the upper indices which is set equal to 3. In particular, $\alpha^3$ denotes the harmonic oscillators of the Sliouville bosonic modes and the third SU(2) generators are given by formula (3.21) with s=3. Altogether with have an SU(2)$\otimes$SU(2)$\otimes$SU(2) lattice and triality is due to the exact mathematical equivalence between these three SU(2)'s. The complete weight lattice is now tridimensional and cubic. There are eight cosets, corresponding to the two choices of integer or halph integer spin representations for the three axes. The NS and R sectors are each made up with two disconnected pieces corresponding to the two cosets of the third SU(2). Concerning each SU(2) Kac-Moody algebra, the two cosets made up with the integer and halph integer representations respectively are the irreducible components of the level one representations [22]. The corresponding zero-mode partition functions already appeared in formula (3.11).

Let us now apply this analysis to rederive the equality between partition functions which we discussed in the preceding section. The mass shell conditions (3.3) becomes, after bosonization,

$$\left(\frac{M^2}{2} - \Xi_0\right) |\;\; \rangle = 0; \quad \Xi_0 = \frac{1}{2}\sum_{s=1}^{3}\sum_{n} :(\alpha_n^s\, \alpha_{-n}^s): - \frac{1}{4} + \frac{1}{2}\sum_{n=1}^{\infty} \Pi_{i,-n}\Pi_{i,n} \tag{3.23}$$

where the last term is the contribution of the transverse X modes. The partition functions (3.5) is thus given by the partition function of the operator $\Xi_0$ which is completely symmetric in the three $\alpha$ modes.

It now becomes very easy to understand the equalities between partition functions. First it is clear that, for each of the two cosets (1/2,0,0) and (0,0,1/2) the partition function is equal to

$$q^{1/4}Z^{[1/2]}\,[Z^{(0)}]^2\,f^3(q) \tag{3.24}$$

The first coset describes a O(4) chiral R sector, while the second is the odd G-parity NS sector. The equality between space-time boson and fermion partition functions is a direct consequence of the equivalence between the three SU(2), which therefore plays the same role as theSO(8) triality of the Green-Schwarz superstring. The identity(3.10) is now immediately rederived by writing the partition functions of both cosets of formula (3.18) in the NSR language. The second case of equality between partition functions, is treated in exactly the same way. The two cosets are (1/2,1/2,0), and (0,1/2,1/2). Both partition functions are equal to

$$q^{-1/4}\,Z^{(0)}\,[Z^{(1/2)}]^2\,f^3(q) \tag{3.25}$$

The first coset describes the even G-parity sector of NS, while the second is a O(4) chiral sector of R.

One sees that the O(8) triality has been replaced by the $SU(2)^3$ triality. Remarkably the physics of the D=5 Sliouville string theory is precisely such that this is possible. It is worth noting that the $SU(2)^3$ triality was investigated, prior to ref [11], in order to perform the Frenkel-Kac construction for non-sinply-laced groups [23].

The O(8) superstring makes an essential use of the special properties of the O(8) gamma matrices which are themselves a consequence of the properties of octonions. In the present case, there are four NSR fields and quaternions play a key role. Let us define

$$\sigma_I\ ;\ I=3,4,5,6\quad ;\quad \sigma_3=\begin{vmatrix}0&1\\1&0\end{vmatrix}\ ;\ \sigma_4=\begin{vmatrix}0&-i\\i&0\end{vmatrix}\ ;\ \sigma_5=\begin{vmatrix}1&0\\0&-1\end{vmatrix}\ ;\ \sigma_6=\begin{vmatrix}i&0\\0&i\end{vmatrix} \tag{3.26}$$

In the quaternionic language, the NSR fields are represented by

$$\Gamma = \Gamma_I\,\sigma_I = \vec{\Gamma}\,\vec{\sigma} + i\,\Gamma_6 \tag{3.27}$$

where we introduced a convenient vector notation in the true transverse coordinate space. The transformation laws of under O(4) are

$$[\,\Gamma\,,\,I^{(1)}_r] = -\Gamma\,\frac{\sigma_r}{2}\quad ;\quad [\,\Gamma\,,\,I^{(2)}_r] = \frac{\sigma_r}{2}\,\Gamma \tag{3.28}$$

It is convenient, next to discuss the graded algebra satisfied by the fermionic operators: (primes denote derivatives with respect to ln(z))

$$\mathcal{G}^{(1)}_{\alpha\beta} = (\,\Pi^{+}\Gamma + \Gamma'\sqrt{2})/2\quad ;\quad \mathcal{G}^{(2)}_{\alpha\beta} = (\,\Pi\,\Gamma^{+} - \Gamma^{+\prime}\sqrt{2})/2\,. \tag{3.29}$$

They are such that their traces are equal to the total superconformal generators:

$$\mathcal{G}_n = \oint\frac{dz}{2\pi iz}\,z^n\,(\Pi_I\Gamma_I + i\sqrt{2}\,\Gamma_6{}')\quad ;\quad \Pi_I = iX_I' \tag{3.30}$$

It is straightforward to compute the commutators of the operators without multiplying the 2x2 matrices. Such is a common practice in 2-d integrable systems, and we use the same notation. One obtains, in particular

$$[G^{(s)}_m \overset{\otimes}{,} G^{(s)}_n]_+ = -\sigma_I\tau_I \Big( \mathcal{L}_{m+n} + 2\delta_{m,-n}(m^2-\text{"}1/4\text{"})\Big)/2$$

$$+ (m-n)\Big(\sigma_I\tau_I \ \vec{\tau}\vec{I}^{(s)}_{m+n} - \sqrt{2}\ \vec{\sigma}\vec{\Pi}_{m+n}\ \sigma_I\tau_I\Big)/2 \qquad (3.31)$$

where the $\mathcal{L}$'s are the total Virasoro generators

$$\mathcal{L}_n = \oint \frac{dz}{2\pi iz} z^n \Big( N(\Pi_I\Pi_I)+:\Gamma_I'\Gamma_I:+i\sqrt{2}\ \Pi_6'\Big)/2 + \text{"}1/4\text{"}\delta_{n,0} \qquad (3.32)$$

The quaternions which appear in the first and second operators of the left-hand side are noted and respectively and are to be considered as acting in different spaces. The algebra of the zero-mode components is simply:

$$[G^{(s)}_0 \overset{\otimes}{,} G^{(s)}_0]_+ = -\sigma_I\tau_I\ L_0/2 \qquad (3.33)$$

The linear terms of eqs. (3.29) are needed in order to recover the full Virasoro operator (3.32) in (3.31). They are responsible for the contribution to the commutation relation that is proportional to the $\Pi$ field. Apart from this, formula (3.31) coincides with the N=4 superconformal algebra commutation relation [24]. This is a novel application of this mathematical structure, where the index of the fermionic generators partly corresponds to the transverse components of the $X_i$ and $\Gamma_i$ space-time fields.

Next, one can show that these two sets of N=4 generators satisfy

$$[\tilde{\mathcal{G}}^{(1)} \overset{\otimes}{,} \tilde{\mathcal{G}}^{(2)}]_+ = -\vec{\sigma}\vec{I}^{(1)}\ \vec{\tau}\vec{I}^{(2)} + \tfrac{1}{4}\vec{\sigma}\vec{\tau}(\Pi_6^2-\vec{\Pi}^2) + \tfrac{1}{2}(\vec{\Pi}\vec{\sigma})(\vec{\Pi}\vec{\tau}) - \tfrac{1}{2}\Pi_6\epsilon_{ijk}\Pi_i\sigma_j\tau_k$$

where, in order to simplify, we removed all linear terms and subtracted the traces from the generators (this is why we put wiggles above them). This algebra is intriguing. There appear new operators of conformal weight 2 on the right hand side. Commuting them leads to enlarge the algebra further, and this process does not seem to stop. This structure remains to be investigated.

### III-3 The light-cone superstring formalism

We sall again restrict ourselves to the D=5 case. The superstring fields are obtained by exchanging the role of the three SU(2). They are given by formulae similar to (3.18). One ontains two new sets of fermionic fields

$$S^3(z)\sim \Omega^{+\,+}_{1,3}(z);\ S^{\bar{3}}(z)\sim \Omega^{-\,-}_{1,3}(z);\ S^5(z)\sim \Omega^{+\,-}_{1,3}(z);\ S^{\bar{5}}(z)\sim \Omega^{-\,+}_{1,3}(z)$$

$$T^3(z)\sim \Omega^{+\,+}_{2,3}(z);\ T^{\bar{3}}(z)\sim \Omega^{-\,-}_{2,3}(z);\ T^5(z)\sim \Omega^{+\,-}_{2,3}(z);\ T^{\bar{5}}(z)\sim \Omega^{-\,+}_{2,3}(z) \qquad (3.34)$$

where the new $\Omega$ fields are defined by formulae similar to (3.16). The fiels S and T commute with the G parity and the chirality projections. They are indeed the equivalent of the Green-Schwarz fields and are to be used in order to construct the supersymmetry generators. We shall for,instance, discuss the S fields. There are two type of light-cone supersymmetry generators. As usual, the first type is essentially proportional to the zero mode of the superstring field. Indeed, on has

$$Q_+ \sim \sqrt{2p_+}\, \underset{\sim}{S}_0 \; ; \quad \underset{\sim}{S} = \begin{pmatrix} S^{\bar{5}} \\ S^3 \end{pmatrix} \tag{3.35}$$

The other supersymmetry generator is proportional to the zero mode component of the fermionic generator $\mathcal{S}$ obtained from $\mathcal{G}$ by exchanging the axes 2 and 3, in the lattice. Let us define (the linear term of eq. (3.30) is irrelevant here)

$$\underset{\sim}{\hat{\mathcal{S}}} = (\vec{\pi}\vec{\sigma} + i\,\alpha^2)\, \underset{\sim}{\hat{S}} \tag{3.36}$$

The other supersymmetry generator is given by

$$Q_- \sim \sqrt{\frac{2}{p^+}}\, \underset{\sim}{\hat{\mathcal{S}}}_0 \; ; \quad \underset{\sim}{\hat{\mathcal{S}}} = \begin{pmatrix} \underset{\sim}{\hat{\mathcal{S}}}^{\bar{5}} \\ \underset{\sim}{\hat{\mathcal{S}}}^3 \end{pmatrix} \tag{3.37}$$

Of course, both $Q_+$ and $Q_-$ are two component spinors. They satisfy the "light-cone supersymmetry algebra":

$$[Q^a_+ \,,\, \bar{Q}^b_+]_+ = 2\, p^+\, \delta^{ab}; \quad [Q^{\dot{a}}_- \,,\, \bar{Q}^{\dot{b}}_-]_+ = \frac{2}{p^+}\, \mathcal{L}_0\, \delta^{\dot{a}\dot{b}}$$

$$[Q^{\dot{a}}_- \,,\, \bar{Q}^b_+]_+ = \sqrt{2}\, p_I\, \sigma_I^{\dot{a}b}; \; [Q^a_+ \,,\, \bar{Q}^{\dot{b}}_-]_+ = \sqrt{2} p_I (\sigma_I^\dagger)^{a\dot{b}} \tag{3.38}$$

The reason for the above quotation marks is that, in the second line, there appears a quantity which we called $P_6$ since it is contracted with $\sigma_6$. It is, however given by

$$P_6 = \sqrt{2}\, (I_3^{(2)})_0 \tag{3.39}$$

The Lorentz properties are thus very unusual, since this "sixth" component is not a five-dimensional Lorentz scalar.

The theory is Lorentz invariant in five dimensions, since one can verify [11] the closure of the Lorentz algebra. This is basically due to the fact that we carefully arranged everything in such a way that the central charge of the superVirasoro generators (3.30) is equal to 8. The interplay between Lorentz invariance and supersymmetry is rather novel, however, and remains to be investigated.

For completeness let us mention that the three-dimensonal model may be discussed along the same line [13], but its properties are even more surprising.

REFERENCES


[1] A.M.Polyakov, Phys.Lett. 103B(1981)207.
[2] A.M.Pokyakov, Phys.Lett. 103B(1981)211.
[3] J.-L.Gervais, A.Neveu, Nucl.Phys. B199(1982)59; B209 (1982) 125.
[4] J.-L.Gervais, A.Neveu, Nucl.Phys. B209(1982)125.
[5] J.-L.Gervais, A.Neveu, Nucl.Phys. B238(1984)125; 396
[6] J.-L.Gervais, A.Neveu, Nucl.Phys. B257[FS14](1985)59.
[7] J.-L.Gervais, A.Neveu, Phys.Lett. 151B(1985)271.
[8] J.-L.Gervais, A.Neveu, Com.Math.Phys. 100 (1985) 15; Nucl. Phys. B264 (1986) 557
[9] A.Bilal, J.-L.Gervais, Nucl.Phys. B284(1987)397.
[10] A.Bilal, J.-L.Gervais, Phys.Lett. 187B(1987)39.
[11] A.Bilal, J.-L.Gervais, Nucl.Phys. B293(1987)1.
[12] A.Bilal, J.-L.Gervais, Liouville superstring and Ising model in three dimensions; preprint LPTENS 87/26.
[13] L.Brink, P.di Vecchia, P.Howe, Phys.Lett. 65B(1976)471
[14] see, e.g., A.S.Schwartz Com.Math.Phys. 64(1979)233.
[15] J.-L.Gervais, B.Sakita, Nucl.Phys. B34(1971)477.
[16] A.Belavin, A.Polyakov, A.Zamolodchikov, Nucl.Phy.B241 (1980)333.
[17] V.G.Kac, Proc. Int. Congress of Mathematicians, 1978 Helsinki; Lecture Notes in Physics, vol. 94 (Springer, New York, 1979) p. 441.
[18] Higher Transcendental Functions, Erdélyi, Magnus, Oberhetinger, Tricomi, Bateman Project, Vol.2, McGraw--Hill 955.
[19] F.Gliozzi, D.Olive, J.Scherk, Nucl.Phys.B122(1977) 253
[20] J.F.Arvis, Nucl.Phys.B212(1983)151; B218(1983)309.
[21] O.Babelon, Phys.Lett.141B(1984)353; Nucl.Phys. B258 (1985) 680.
[22] For a pedagogical review, see P. Goddard, D.Olive, Int. J.Mod.Phys.A1 (1986) 303.
[23] P.Goddard, W.Nahm, D.Olive, A.Schwimmer, Comm.Math. Phys.107(1986)179; A.Schwimmer in Proceedings of the 1985 Bonn Firenze Johns Hopkins meeting, World Scientific; P.Goddard, D.Olive, A.Schwimmer, Phys.Lett. 157B(1985)393
[24] M.Ademollo, L.Brink, A.D'Adda, R.D'Auria, E.Napolitano, S.Sciuto, E.Del Giudice, P.Di Vecchia, S.Ferrara, F.Gliozzi, R.Musto, R.Pettorino, Nucl.Phys. B114 (1976) 297.

# A POLYAKOV PATH INTEGRAL WITH GHOSTS

Carlos R. Ordonez* and Mark A. Rubin*

Theory Group
Rockefeller University
New York, NY 10021

Roberto Zucchini†

Physics Department
New York University
New York, NY 10003

Presented by M. A. R.

## ABSTRACT

We present a method for explicit evaluation of the Polyakov path integral with ghosts. The free propagator for the bosonic string, both with and without ghosts, is computed.

## INTRODUCTION

The Polyakov path integral[1] and covariant string field theory[2] are the two primary methods for addressing questions in string theory in a covariant manner. The former technique has the advantage of manifest duality, while the latter should be applicable to nonperturbative problems. On very general grounds, it is of interest to understand as precisely as possible the relationship between the two formalisms.

---

* Work supported in part by U.S.Department of Energy contract DEAC 02-87-ER40325 - Task B1

† Work supported in part by National Science Foundation grant PHY - 8413569

Specifically, the path integral may be of use in deriving vertices in covariant string field theory, and in obtaining "sewing rules" for constructing string-field-theoretic Feynman diagrams.[3] However, the field operators, and hence the n-point functions (more correctly, "n-curve functions"!) of covariant string field theory have as arguments not only the coordinates $X^{\mu}(\sigma)$ describing the string's embedding in spacetime, but also anticommuting ghosts coordinates $b(\sigma)$, $c(\sigma)$. Therefore, applicability of the path integral to covariant string field theory requires that we perform path integrals over worldsheets with boundaries specified by fixed values of spacetime coordinates $X^{\mu}(\sigma)$ and ghost coordinates $b(\sigma)$, $c(\sigma)$. We present here a technique for performing such path integrals, and apply it to the tree-level two-point amplitude for the open bosonic string.*

## THE PROPAGATOR WITHOUT GHOSTS

We first compute the propagator in the Polyakov formalism without ghosts. The relevant techniques, especially that of gauge-fixing to the conformal gauge, have been dealt with extensively in the literature[6,7]; in particular, these techniques have been applied to the closed string propagator without ghosts by Cohen et al.[8] Since this part of our work parallels that reference, we will omit most computational details.

The amplitude for a string to propagate from an initial curve in spacetime $X_i{}^{\mu}(\sigma)$ to a final curve $X_f{}^{\mu}(\sigma)$ is given by

$$A\,(X_i\to X_f)=\int d\Sigma_i d\Sigma_f\,\frac{[Dg][DX]}{V_{W-GC}}\exp(-S) \tag{1}$$

where S is the Polyakov action

$$S = \frac{1}{2}\int d^2\sigma\sqrt{g}\; g^{ab}\partial_a X^{\mu}\partial_b X^{\mu} \tag{2}$$

The path integral is to be performed over worldsheets (two-dimensional surfaces -- throughout, we take these surfaces

---

* Subsequent to the completion of this work we have become aware of related computations by Blau et. al.[4] and by Lee.[5]

and the 26-dimensional "spacetime" in which they are embedded to be of Euclidean signature) with all possible two-dimensional metrics $g_{ab}(\vec{\sigma})$ and spacetime embeddings $X^\mu(\vec{\sigma})$, provided the embedding $X^\mu(\vec{\sigma})$ matches, at part of it's boundary, the initial and final curves $X_i{}^\mu(\sigma)$ and $X_f{}^\mu(\sigma)$. We also sum over all reparametrization $\Sigma_i$, $\Sigma_f$ of the initial and final boundaries. Since we are performing a tree-level computation, the worldsheets in the path integral have the topology of a square, with the "bottom" and "top" edges matching $X_i{}^\mu(\sigma)$, $X_f{}^\mu(\sigma)$, respectively, while the "left" and "right" edges satisfy the usual open string boundary conditions.

Upon fixing the conformal gauge and dividing out by $V_{W-GC}$, the infinite gauge volume of two dimensional Weyl transformations and diffeomorphisms (general coordinate transformations), eq. (1) becomes, for the problem we are considering,

$$A(X_i \to X_f) = \int d\Sigma_i d\Sigma_f \int_0^\infty d\lambda \, \exp(-S[\hat{g}, X_{cl}])(\det\Delta)^{-13} \frac{(\psi^T|d\hat{g}/d\lambda)}{(\psi^T|\psi^T)^{1/2}} \, (\det P^\dagger P)^{1/2}. \tag{3}$$

The Teichmuller metric $\hat{g}_{ab}$ depends on a single Teichmuller parameter $\lambda$. (Latin indices indicate worldsheet tensors.) The operator P maps vectors $\eta^a$ to traceless symmetric rank-two tensors:

$$(P\eta)_a = \nabla_a \eta_b + \nabla_b \eta_a - \hat{g}_{ab} \nabla_c \eta^c, \tag{4}$$

where $\nabla_a$ is the torsion-free covariant derivative compatible with $\hat{g}_{ab}$. The formal adjoint of P maps traceless symmetric rank-two tensors $\zeta_{ab}$ to vectors (indices are raised and lowered with $\hat{g}_{ab}$):

$$(P^\dagger \zeta)^a = -2\nabla_b \zeta^{ba} \tag{5}$$

(The boundary conditions sufficient for $P^\dagger$ to be, in fact, adjoint to P will be discussed below.) The Teichmuller deformation $\psi^T{}_{ab}$ satisfies

$$(P^\dagger \psi^T)^a = 0. \qquad (6)$$

The inner product between rank-two symmetric traceless tensors, which appears in (3), is

$$(\xi^{(1)} | \xi^{(2)}) = \int d^2\sigma \sqrt{\hat{g}}\ \xi^{(1)}_{ab} \xi^{(2)ab}. \qquad (7)$$

$\Delta$ is the Laplacian operator for scalars on the worldsheet with metric $\hat{g}_{ab}$. $S[\hat{g}, X_{cl}]$ is the action (2) evaluated at $g_{ab}=\hat{g}_{ab}$ and $X^\mu = X_{cl}{}^\mu$, where $X_{cl}{}^\mu$ is the solution to the classical equations of motion for $X^\mu$ following from (2), with metric $\hat{g}_{ab}$, satisfying $X_{cl}{}^\mu$ at the initial boundary, $X_{cl}{}^\mu = X_f{}^\mu$ at the final boundary, and open-string boundary conditions at the other two edges.

To evaluate explicity the amplitude (3), we must assign boundary conditions to the respective vector fields $\eta^a$ and tensor fields $\zeta_{ab}$ on which P and $P^\dagger$ act. We choose worldsheet coordinates such that $\sigma^2=0$ ($\sigma^2=1$) is the initial (final) boundary, while $\sigma^1=0$ ($\sigma^1=1$) is the left (right) free boundary. Then the physical requirement that diffeomorphisms generated by $\eta^a$ not move the initial or final boundary in spacetime tells us that

$$\eta^2=0, \ \sigma^2=0,1. \qquad (8)$$

In addition, we impose two mathematical requirements: 1) The square is to be viewed as a "flattened cylinder": That is it is obtained from a cylinder with longitudinal coordinate $0 \leq \tilde{\sigma}^2 \leq 1$, axial coordinate $-1 \leq \tilde{\sigma}^1 \leq 1$, and "flattened" embedding $X^\mu(-\tilde{\sigma}^1, \tilde{\sigma}^2) = X^\mu(\tilde{\sigma}^1, \tilde{\sigma}^2)$, by the mapping $(\pm\tilde{\sigma}^1, \tilde{\sigma}^2) \rightarrow (\sigma^1, \sigma^2)$. 2) The surface term in the expression relating P and $P^\dagger$, and in the first and second Green's identities for $P^\dagger P$ and $PP^\dagger$, must vanish. The following conditions, in addition to (8), are sufficient to meet these requirements:

$$\partial_1 X^\mu = \eta^1 = \partial_1 \eta^2 = \zeta_{12} = \partial_1 \zeta_{11} = 0 \quad \text{at } \sigma^1 = 0,1 \tag{9a}$$

$$\partial_2 \eta^1 = \zeta_{12} = \partial_2 \zeta_{22} = 0 \quad \text{at } \sigma^2 = 0,1. \tag{9b}$$

The flattening procedure applied to the Teichmuller metric on the cylinder yields the Teichmuller metric on the square

$$\hat{g}_{ab} = \begin{pmatrix} 1 & 0 \\ 0 & \lambda^2 \end{pmatrix}. \tag{10}$$

The range of $\lambda$ must be taken from zero to infinity, rather than zero to one, because the different boundary conditions at the edges $\sigma^2=0,1$, as opposed to those at $\sigma^1=0,1$, do not allow us to interchange them. (For a more extensive discussion of boundary conditions, see Zucchini[9]).

The factors in (3) are thus found to be:

$$\exp(-S[\hat{g}, X_{cl}]) = \exp(-\tfrac{1}{2\lambda}(\vec{X}_{f,0} - \vec{X}_{i,0})^2) \tag{11}$$

$$\cdot \exp\left(-\sum_{m=1}^{\infty} \frac{\pi m}{4 \mathrm{sh}(\pi m \lambda)} \left[ (\vec{X}^2_{f,m} + \vec{X}^2_{i,m}) \,\mathrm{ch}(\pi m \lambda) - 2\, \vec{X}_{f,m} \cdot \vec{X}_{i,m} \right]\right)$$

$$\det P^\dagger P = \lambda \exp\left(\frac{-\pi\lambda}{6}\right) \prod_{m=1}^{\infty} (1 - \exp(-2\pi m \lambda))^2 \tag{12}$$

$$\det \Delta = \lambda^{1/2} (\det P^\dagger P)^{1/2} \tag{13}$$

$$\frac{(\psi^T | d\hat{g}/d\lambda)}{(\psi^T|\psi^T)^{1/2}} = \lambda^{-1/2} \tag{14}$$

In (11), $X_i{}^\mu{}_m$ ($X_f{}^\mu{}_m$) is the $m^{th}$ expansion coefficient of $X_i{}^\mu(\sigma)(X_f{}^\mu(\sigma))$ in a Fourier cosine series. Eq. (3) therefore becomes

$$A(X_i \to X_f) = \int d\Sigma_i \, d\Sigma_f \tag{15}$$

$$\cdot \int_0^\infty d\lambda \, \lambda^{-13} \prod_{n=1}^{\infty} (1 - \exp(-2\pi n \lambda))^{-12} \exp(-S[\hat{g}, X_{cl}] + \pi\lambda)$$

Define the harmonic-oscillator Hamiltonian for the $m^{th}$ Fourier coefficient:

$$\hat{H}_m = \frac{\hat{p}^\mu{}_m \hat{p}^\mu{}_m}{2\pi} + \frac{\pi m^2}{2} \hat{X}_m{}^\mu \hat{X}^\mu_m \tag{16}$$

Hats here denote operators. $\hat{p}^{\mu}_{m}$ is the momentum conjugate to $\hat{X}^{\mu}_{m}$. Defining eigenstates

$$\hat{X}^{\mu}_{m}|X^{\mu}_{m}\rangle = X^{\mu}_{m}|X^{\mu}_{m}\rangle \tag{17}$$

we find that

$$A\ (X_i \to X_f) \tag{18}$$

$$= \int d\Sigma_i\ d\Sigma_f \sum_{m=-\infty}^{\infty} (-1)^{m} \langle\, X_f|(\hat{L}_0 - 1 + 3m^2 + m)^{-1}|X_i\rangle$$

where $|X_i\rangle = \prod_{m=0}^{\infty} \prod_{\mu=1}^{26} |X_i{}^{\mu}{}_{m}\rangle$, $|X_f\rangle = \prod_{m=0}^{\infty} \prod_{\mu=1}^{26} |X_f{}^{\mu}{}_{m}\rangle$, and the zero-mode Virasoro operator $\hat{L}_0$ is

$$\hat{L}_0 = \sum_{m=0}^{\infty} \hat{H}_m \tag{19}$$

In the limit of pointlike initial and final strings,

$$X_i{}^{\mu}{}_{m} = X_f{}^{\mu}{}_{m} = 0, \quad m \neq 0 \tag{20}$$

the amplitude (18) becomes

$$A(X_i \to X_f,\ \text{pointlike}) = \sum_{m=0}^{\infty} a_m \langle\, X_f|\ \left(\frac{\hat{p}^{\mu}{}_{0}\hat{p}^{\mu}{}_{0}}{2\pi} + 2n - 1\right)^{-1}|\ X_i\rangle \tag{21}$$

where the nonnegative coefficients $a_m$ are defined by

$$\sum_{m=0}^{\infty} a_m z^m = \prod_{m=1}^{\infty} (1 - z^m)^{-12} \tag{22}$$

The amplitude (21) is identical to that obtained in the lightcone gauge for pointlike open strings

## THE PROPAGATOR WITH GHOSTS

We now wish to construct the amplitude $A(X_i, b_i, c_i \to X_f, b_f, c_f)$ corresponding to $A(X_i \to X_f)$.

Define the ghost action

$$S_{gh}= -\int d^2\sigma\sqrt{\hat{g}}\ \hat{g}_{ab}c^a(P^\dagger b)_a \qquad (23)$$

The ghost field $c^a(\vec{\sigma})$ is an anticommuting Grassmann-imaginary vector; the antighost field $b_{ab}(\vec{\sigma})$ is an anticommuting Grassman-real traceless symmetric two-tensor. We demand that $c^a$ and $b_{ab}$ satisfy the respective boundary conditions for vectors and tensors in eqs. (9). Then we can expand $c^a$ and $b_{ab}$ as

$$c^a(\vec{\sigma})=\sum_\alpha C_\alpha \Phi_\alpha{}^a(\vec{\sigma}),\ b_{ab}(\vec{\sigma})=\sum_\mu B_\mu \Psi_{\mu ab}(\vec{\sigma}) \qquad (24a,b)$$

where

$$(P^\dagger P\Phi_\alpha(\vec{\sigma}))^a=E_\alpha\Phi_\alpha{}^a(\vec{\sigma}),\ (PP^\dagger\Psi_\mu(\vec{\sigma}))_{ab}=\tilde{E}_\mu\Psi_{\mu ab}(\vec{\sigma}) \qquad (25a,b)$$

The abstract indices $\alpha$ and $\mu$ label the distinct eigenvectors of $P^\dagger P$ and $PP^\dagger$, respectively, $E_\alpha$ and $\tilde{E}_\mu$ being the corresponding eigenvalues. The nonzero eigenvalues of $PP^\dagger$ are equal to those of $P^\dagger P$, with equal degeneracy, and the corresponding (normalized) eigenvectors of $PP^\dagger$ may be expressed as[6]

$$\Psi_{\alpha ab}=E_\alpha{}^{-1/2}(P\Phi_\alpha)_{ab} \qquad (25c)$$

if $\Phi_\alpha$ is equivalently normalized. So (24b) may be written as

$$b_{ab}(\vec{\sigma})=\sum_\alpha B_\alpha E_\alpha{}^{-1/2}(P\Phi_\alpha)_{ab}+\mathring{B}\Psi^T \qquad (26)$$

where $\Psi^T$ is the Teichmuller deformation. (Recall that in the case we are considering there is but a single one.) Using the expansions (24a) and (26), the ghost action (23) becomes

$$S_{gh}= \sum_\alpha E_\alpha^{1/2}C_\alpha B_\alpha \qquad (27)$$

The coefficient of the Teichmuller mode $\mathring{B}$ is absent from (27). We therefore define a path integral measure

$$[Dc\tilde{D}b] = \prod_\alpha dC_\alpha dB_\alpha \qquad (28)$$

Then

$$\int[Dc D\tilde{b}] \exp(-S_{gh}) = \prod_\alpha E_\alpha^{1/2} = (\det{}' P^\dagger P)^{1/2} \qquad (29)$$

Equivalently,

$$\int[Dc Db]\mathring{B} \exp(-S_{gh}) \equiv \int[Dc D\tilde{b}] d\mathring{B}\mathring{B} \exp(-S_{gh}) = (\det{}' P^\dagger P)^{1/2} \qquad (30)$$

where we have used the fact that, for Grassman integration, the Dirac delta-function of a variable is the variable itself,

$$\int d\theta\theta f(\theta) = f(0). \qquad (31)$$

The representations (29) and (30) for the determinant suggest that we construct the amplitude $A(X_i, b_i, c_i \rightarrow X_f, b_f, c_f)$ by replacing, in eq. (3), $(\det P^\dagger P)^{1/2}$, the "path integral of $\exp(-S_{gh})$ over all ghost and antighost configurations", by a suitably defined "path integral of $\exp(-S_{gh})$ over ghost and antighost configurations with certain values at the initial and final boundaries." The boundary conditions (9), which are implicit in the path integral measure (38), (30), force the components $c^2$ and $b_{12}$ to zero at the initial and final boundaries, and $b_{22}$ is proportional to $b_{11}$ everywhere due to the tracelessness of $b_{ab}$. So we are free to independently assign the values of $c^1$ and $b_{11}$ at $\sigma^2=0$ and 1. We set

$$c^1(\sigma^1,0) = c_i(\sigma^1) = \sum_{m>0} c^i{}_m \sin(m\pi\sigma^1)$$

$$b_{11}(\sigma^1,0) = b_i(\sigma^1) = \sum_{m\geq 0} b^i{}_m \cos(m\pi\sigma^1)$$

$$c^1(\sigma^1,1) = c_f(\sigma^1) = \sum_{m>0} c^f{}_m \sin(m\pi\sigma^1)$$

$$b_{11}(\sigma^1,1) = b_f(\sigma^1) = \sum_{m\geq 0} b^f{}_m \cos(m\pi\sigma^1) \qquad (32)$$

Taking into account (31), we define

$$A_{gh}(X_i, b_i, c_i \rightarrow X_f, b_f, c_f)$$

$$= \int[Dc D\tilde{b}] \prod_{\sigma^1} (b_{11}(\sigma^1,0) - b_i(\sigma^1))(c^1(\sigma^1,0) - c_i(\sigma^1))$$

$$(b_{11}(\sigma^1,1) - b_f(\sigma^1))(c^1(\sigma^1,1) - c_f(\sigma^i))\exp(-S_{gh}) \qquad (33)$$

Fourier expanding each of the four Grassmann delta functions in (33), and using (24a), (26) and (30), we see that, ignoring an irrelevant λ-independent determinant, we can rewrite (33) as

$$A(X_i,b_i,c_i \to X_f,b_f,c_f)$$

$$=\int[Dc\tilde{D}b]\prod_{m>0}\Gamma^+\Gamma^-\prod_{n\geq 0}\Delta^+_n\Delta^-_n\exp(-S_{gh}) \qquad (34)$$

where

$$\Gamma^{\pm}_m \equiv \Sigma_{n\binom{even}{odd}\geq 0}\Sigma_{i=1,2}\,\gamma^{\pm}_{(m,n,i)}\,C_{(m,n,i)}-c_{m,\pm},\quad m>0 \qquad (35a)$$

$$\Delta^{\pm}_m \equiv \Sigma_{n\binom{even}{odd}\geq 0}\Sigma_{i=1,2}\,\beta^{\pm}_{(m,n,i)}\,B_{(m,n,i)}-b_{m,\pm},\quad m\geq 0 \qquad (35b)$$

In (35) we have used the explicit eigenspectrum of $P^\dagger P$ on our worldsheet, and defined $c_{m,\pm}=\frac{1}{2}(c^i{}_m \pm c^f{}_m)$, $b_{m,\pm}=\frac{1}{2}(b^i{}_m \pm b^f{}_m)$. $\gamma^{\pm}{}_{(m,n,i,)}$ and $\beta^{\pm}_{(m,n,i)}$ are certain λ-dependent constants. Upon performing the Grassmann integration we find

$$A_{gh}=(\det P^\dagger P)^{\frac{1}{2}}b^i{}_0 b^f{}_0$$

$$\cdot\prod_{m=1}^{\infty}\exp\left[\frac{1}{sh(\pi\lambda m)}(ch(\pi\lambda m)(c^i{}_m b^i{}_m+c^f{}_m b^f{}_m)-c^f{}_m b^i{}_m-c^i{}_m b^f{}_m)\right] \qquad (36)$$

As in the bosonic case, this can be rewritten as the matrix element of an operator. Define ghost and antighost operators $\hat{c}_m,\hat{b}_m,-\infty<m<\infty$, according to the usual conventions,[10] and a ghost vacuum satisfying

$$\hat{c}_m|0\rangle_{gh}=0,\ m>0,\quad \hat{b}_m|0\rangle_{gn}=0,\ m\geq 0 \qquad (37a,b)$$

Define also, following Gross and Jevicki,[11] the generalized coherent state $|w,p\rangle$, a function of the Grassmann-imaginary numbers $p_m$ and Grassmann-real numbers $w_m$, $n>0$:

$$|w,p\rangle=\prod_{m=1}^{\infty}\exp(-w_m p_m-2^{\frac{1}{2}}w_m\hat{c}_{-m}-2^{\frac{1}{2}}p_m\hat{b}_{-m}-\hat{c}_{-m}\hat{b}_{-m})\ |0\rangle_{gh} \qquad (38)$$

Then

$$A_{gh}=\lambda^{\frac{1}{2}}\ e^{\frac{-\pi\lambda}{12}}\ b^i{}_0 b^f{}_0\ \langle b^f,c^f|\ \exp(-\lambda\pi\hat{H}_{gh})|b^i,c^i\rangle$$

where $\hat{H}_{gh}$ is the normal-ordered ghost Hamiltonian.

Substituting $A_{gh}$ for $(\det P^{\dagger}P)^{1/2}$ in (3) we obtain

$$A(X_i,b^i,c^i\rightarrow X_f,b^f,c^f)$$
$$=\int d\Sigma_i d\Sigma_f b^f{}_0 b^i{}_f \langle X_f,b^f,c^f|(\hat{L}_0{}^{BRST})^{-1}|X_i,b^i,c^i\rangle \quad (40)$$

where

$$\hat{L}_0{}^{BRST}=\hat{L}_0+\hat{H}_{gh}-1$$

is the total zero-mode BRST Virasoro operator. To recover $A(X_i\rightarrow X_f)$ we integrate (40) with respect to all initial and final ghost and antighost coordinates:

$$\int \prod_{m\geq 0} db^i{}_0 db^f{}_0 dc^i{}_0 dc^f{}_0 A(X_i,b^i,c^i\rightarrow X_f,b^f,c^f) \quad (41)$$
$$=A(X_i\rightarrow X_f)$$

We note that, in covariant string field theory,[12] we may expand the open-string field in a Grassmann Taylor series:

$$\Phi[X,b,c]=\psi[X,\tilde{b},c]+b_0\phi[X,\tilde{b},c] \quad (42)$$

where $\tilde{b}$ denotes all antighosts modes <u>except</u> $b_0$. In the Seigel gauge $b_0\Phi=0$, the Green's function for the string field is

$$\langle\langle\Phi[X_f,b^f,c^f]\Phi[X_i,b^i,c^i]\rangle\rangle$$
$$=b^f{}_0 b^i{}_0 \langle\langle\phi[X_f,\tilde{b}^f,c^f]\phi[X_i,\tilde{b}^i,c^i]\rangle\rangle \quad (43)$$

The form of the dependence of (43) on its arguments is the same as that of the integrand in (40). Furthermore, the quadratic part of the open-string field theory action in the Siegel gauge is of the form $\int\Phi\hat{c}_0\hat{L}_0{}^{BRST}\Phi$. This suggests that the path integral we have performed, <u>prior to integration over boundary reparametrizations</u>, yields the string-field-theoretic Green's function in the Siegel gauge. Use of the completeness relation for the states (38) shows that this is, in fact, the case; specifically, the path integral is the propagator for the field $\phi$ in (42).

ACKNOWLEDGEMENTS

One of us (M.A.R.) would like to thank P.G.O. Freund, K.T. Mahanthappa, the University of Colorado and NATO for a productive workshop; and S. Carlip, M. Clements, and S. Della Pietra for interesting conversations.

## REFERENCES


1. A. M. Polyakov, Quantum geometry of bosonic strings. Phys. Lett. 103B: 207 (1981); Quantum geometry of fermionic strings, Phys. Lett. 103B: 211 (1981).
2. W. Siegel and B. Zweibach, Gauge string fields, Nucl. Phys. B 263: 105 (1986).
   H. Hata, K. Itoh, T. Kugo, H. Kunitomo and K. Ogawa, Manifestly covariant field theory of interacting string I, Phys. Lett. 172B: 186 (1986); Manifestly covariant field theory of interacting string II, Phys. Lett. 172B: 195 (1986).
   A. Neveu, J. H. Schwarz and P. C. West, Gauge symmetrics of the free bosonic string field theory, Phys. Lett. 164B; 63 (1985).
   M. Kaku, Gauge field theory of covariant strings, Nucl. Phys. B267: 125 (1986).
   T. Banks, D. Friedan, E. Martinec, M. Peskin and C. Preitschopf, All free string theories are theories of forms, Nucl. Phys. B274: 71 (1986).
   E. Witten, Non-commutative geometry and string field theory, Nucl. Phys. B268: 253 (1986); Interacting field theory of open superstrings, Nucl. Phys. B276: 291 (1986).
3. S. K. Blau, S. Carlip, M. Clements, S. Della Pietra, V. Della Pietra, The string amplitude on surfaces with boundaries and crosscaps, University of Texas preprint UTGH-14-87.
4. ____________________, unpublished.
5. T. Lee, Closed string in the proper-time gauge, University of Washington preprint 40048-06 P7; Open bosonic string in the proper-time gauge, University of Washington preprint 40048-07 P7.
6. O. Alverez, Theory of strings with boundaries: Fluctuations, topology, and quantum geometry, Nucl. Phys. B216: 125 (1983).
7. J. Polchinski, Evaluation of the one loop string path integral, Comm. Math. Phys. 104: 37 (1986).
   S. Chandhuri, H. Kawai, and S.-H. H. Tye, Path integral formulation of closed strings, Cornell University preprint CLNS - 86/723.
8. A. Cohen, G. Moore, P. Nelson and J. Polchinski, An Off-shell propagator for string theory, Nucl. Phys. B267: 143 (1986).
9. R. Zucchini, On the Polyakov theory of open string off-shell Green functions, New York University preprint NYU/TR7/87.
10. M. B. Green, J. H. Schwarz and E. Witten, "Superstring Theory," Vol. 1, Cambridge University Press, Cambridge (1987).
11. D. J. Gross and A. Jevicki, Operator formulation of interacting string field theory, Nucl. Phys. B283: 1 (1987).
12. B. Zweibach, Gauge invariant string actions, in: "Unified String Theories," M. Green and D. Gross, eds., World Scientific, Singapore (1986).

# STRING PATH INTEGRALS FOR OPEN OR UNORIENTED WORLD SHEETS

Steven Carlip

Institute for Advanced Study
Princeton, NJ 08540

Minot Clements

University of Washington
Seattle, WA 98195

Stephen Della Pietra

University of Texas
Austin, TX 78712

## ABSTRACT

The Polyakov path integral is evaluated for an arbitrary open or unoriented string world sheet. The integrand is expressed in terms of known functions on the double of the world sheet, and the region of integration is described as a region of a real slice of the Teichmüller space of the double.

## INTRODUCTION

The field theory of closed strings appears to be much more complicated than that of open strings. For the Polyakov path integral, on the other hand, it is closed string theory that is well understood. In particular, the path integral for a closed oriented world sheet can be written exactly as a finite dimensional integral over moduli space of an expression made up of known functions (theta functions, Abelian differentials, etc.).[1,2] The result is not pretty, but in principle it contains all there is to be known about perturbative closed string theory.

Until now, no such general expression has been known for open or unoriented world sheets. This is a serious deficiency, not only because of the importance of open and unoriented string theories, but because path integrals for surfaces with boundaries give off-shell closed string Greens functions.[3] Such functions may provide valuable insight into closed string field theory: the three string Greens function, for example, should contain information about the field theory vertex.

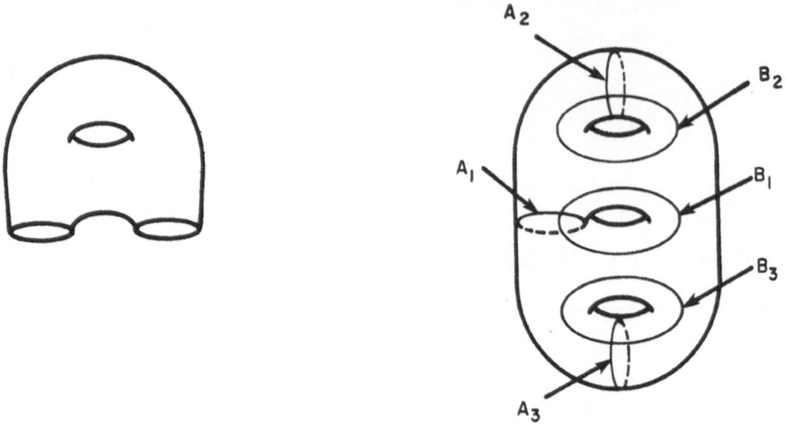


*Figure 1. A two-holed torus and its double, a genus three surface. A homology basis for the double is shown.*

Recently, Steven Blau, Vincent Della Pietra and the authors have studied the Polyakov path integral for a general open and/or unoriented world sheet.[4] We have succeeded in expressing the resulting amplitudes in terms of known quantities on an associated closed oriented surface. The basic idea – the use of the double of a surface – is not new,[5] and expressions for a few low genus surfaces have been known for some time.[6] Our goal has been to generalize these results, using the modern post-Polyakov technology of Riemann surfaces and moduli space, in order to write down an expression for the path integral for a world sheet of arbitrary topology.

## DOUBLES

Figure 1 illustrates the double of a surface with boundary, the torus with two boundary curves. The double $\overline{\Sigma}$ of an arbitrary surface with boundary $\Sigma$ can be formed similarly, by glueing together two copies of $\Sigma$ along their boundaries. Clearly, the extension of a metric on $\Sigma$ to $\overline{\Sigma}$ will not generally be smooth; but any metric on $\Sigma$ is conformal to one which extends smoothly to $\overline{\Sigma}$. Equivalently, any complex structure on $\Sigma$ extends to the double.

Similarly, any unoriented surface has a double, its oriented double covering. This is illustrated in Figure 2 for the Klein bottle. Again, any metric on the unoriented surface is conformal to one which extends smoothly to the double. The case of an unoriented surface with boundary is somewhat more complicated, and will not be discussed further in this paper, but it can be treated by the same techniques.[4]

In each case, the double $\overline{\Sigma}$ consists of two copies of $\Sigma$, along with an involution $I$ which sends points on one copy to the corresponding points on the other. Fixed points of $I$ correspond to boundary curves of $\Sigma$, and one can write $\Sigma = \overline{\Sigma}/I$. The key feature of $I$ is that it is anticonformal, that is, it anticommutes with the complex structure:[7]

$$I_* J = -J I_* \tag{2.1}$$

As a consequence, $I$ behaves nicely with respect to differential operators on $\overline{\Sigma}$ ; for example,

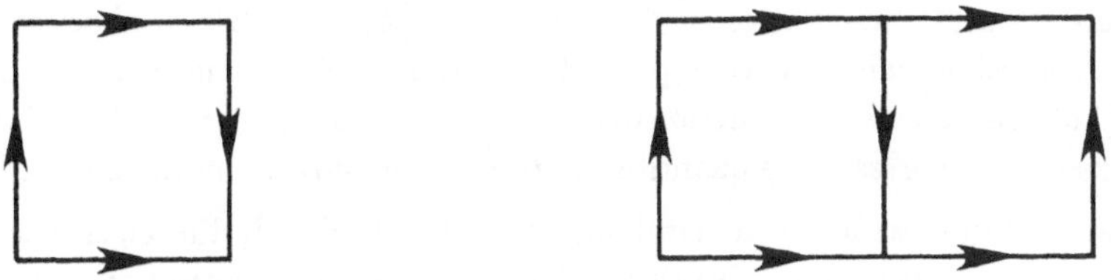

*Figure 2. The Klein bottle and its double, the torus.*

$$I_* \partial = \bar{\partial} I_* \tag{2.2}$$

Objects on $\overline{\Sigma}$ such as eigenfunctions of Laplacians thus break up naturally into pieces even and odd under $I$. These, in turn, can be interpreted in terms of objects on the original surface $\Sigma$ . For instance, if $\Sigma$ is oriented with boundary, an eigenfunction of the scalar Laplacian $\Delta$ on $\overline{\Sigma}$ which is odd under $I$ corresponds to an eigenfunction of $\Delta$ on $\Sigma$ which vanishes on the boundary; an eigenfunction on $\overline{\Sigma}$ which is even under $I$ corresponds to one on $\Sigma$ with vanishing normal derivative.

The action of $I$ on homology and on the period matrix $\tau$ will be important later. Figure 1b shows a homology basis on $\overline{\Sigma}$ ; observe that the involution $I$ takes A cycles to A cycles and B cycles to B cycles. A basis with this property can be found for an arbitrary double, and one can write

$$I^* \omega_i = \Gamma_{ij} \bar{\omega}_j \tag{2.3}$$

for some matrix $\Gamma$ of integers. If one defines

$$\tau^{\pm} = (1 \pm \Gamma)\tau(1 \pm \Gamma) \tag{2.4}$$

it can be shown that the ratio

$$R = det' Im \tau^+ / det' Im \tau^- \tag{2.5}$$

is independent of the choice of homology basis.[8]

## THE INTEGRAND

The Polyakov path integral for a surface $\Sigma$ can be written[9]

$$A_\Sigma[X_i] = \int_{T^\Sigma / G^\Sigma} \sigma_\Sigma \int (\prod dC_i) e^{-S_{cl}[X_i \circ C_i]} \tag{3.1}$$

with

$$\sigma_\Sigma = \left(\frac{det' \Delta_\Sigma}{\int_\Sigma \sqrt{g}}\right)^{-13} (det(P^\dagger P)_\Sigma)^{\frac{1}{2}} d\mu_{WP}(\Sigma) \tag{3.2}$$

Here $T^\Sigma$ is the Teichmüller space of $\Sigma$, and $G^\Sigma$ is the mapping class group; their quotient is moduli space. The measure $d\mu_{WP}$ is the Weil-Petersson measure; $\Delta$ and

$P^\dagger P$ are the scalar and vector Laplacians; and $S_{cl}[X_i]$ is the classical action for those parameterized boundary curves representing initial and final string states. The integral over $\prod dC_i$ represents a summation over reparameterizations of these boundary curves. Our goal is to express these quantities in terms of quantities on the double $\overline{\Sigma}$.

Let us begin with the determinant $detP^\dagger P$. Under $I$, the eigenspace of $P^\dagger P$ splits into even and odd subspaces. The even subspace corresponds to eigenvectors whose normal components vanish and whose tangent components have vanishing normal derivatives at the boundaries of $\Sigma$. These are precisely the boundary conditions discussed by Alvarez[9] for $P^\dagger P$ for a surface with boundary; writing $(P^\dagger P)^\pm = \frac{1}{4}(1 \pm I_*)P^\dagger P(1 \pm I_*)$, we have

$$det(P^\dagger P)_\Sigma = det(P^\dagger P)^+_{\overline{\Sigma}} \tag{3.2}$$

On the other hand, the complex structure J maps even vectors into odd vectors, providing an isomorphism between the even and odd eigenspaces of $(P^\dagger P)_{\overline{\Sigma}}$. So using the fact that $det(P^\dagger P)_{\overline{\Sigma}} = det(P^\dagger P)^+_{\overline{\Sigma}} det(P^\dagger P)^-_{\overline{\Sigma}}$, we find that

$$det(P^\dagger P)_\Sigma = (det(P^\dagger P)_{\overline{\Sigma}})^{\frac{1}{2}} \tag{3.3}$$

The evaluation of $det\Delta_\Sigma$ is somewhat harder. We have $det\Delta_\Sigma = det\Delta^+_{\overline{\Sigma}}$ for a closed unoriented surface or an open surface with Neumann boundary conditions, and $det\Delta_\Sigma = det\Delta^-_{\overline{\Sigma}}$ for an open surface with Dirichlet boundary conditions. But in contrast with the vector case, the complex structure J has no natural action on scalars, and the even and odd eigenspaces of $\Delta$ are not isomorphic. Instead, one must look at the variation of $det\Delta^+/det\Delta^-$ with the moduli, or, equivalently, at the relationship of the stress energy tensors of $\Sigma$ and its double. We show in [4] that for Dirichlet boundary conditions,

$$det\Delta_\Sigma = \left(\frac{det'\Delta_{\overline{\Sigma}}}{\int_{\overline{\Sigma}}\sqrt{g}}\right)^{\frac{1}{2}} R^{-\frac{1}{2}} \tag{3.4}$$

where R is defined in equation (2.5); there is a similar expression for Neumann boundary conditions.

The classical action $S_{cl}$ can be written in terms of Dirichlet Greens functions on $\Sigma$, which in turn can be expressed in terms of the Greens function (or the prime form and the holomorphic differentials) on $\overline{\Sigma}$. As for the Weil-Petersson measure, it is essentially a wedge product of quadratic differentials. But if the metric on $\overline{\Sigma}$ is symmetric under $I$, it is not hard to show that the space of quadratic differentials on $\overline{\Sigma}$ splits into even and odd subspaces which are interchanged by J. Thus for symmetric metrics we have

$$d\mu_{WP}(\overline{\Sigma}) = d\mu_{WP}(\Sigma) \wedge Jd\mu_{WP}(\Sigma) \tag{3.5}$$

## THE REGION OF INTEGRATION

Combining the results of the last section, we can express the integrand (3.2) completely in terms of quantities on $\overline{\Sigma}$; in fact, it is very nearly the square root of the integrand for $\overline{\Sigma}$. We have yet to consider the region of integration, however. To see the potential difficulties, consider the simplest case of the cylinder and its double, the torus.

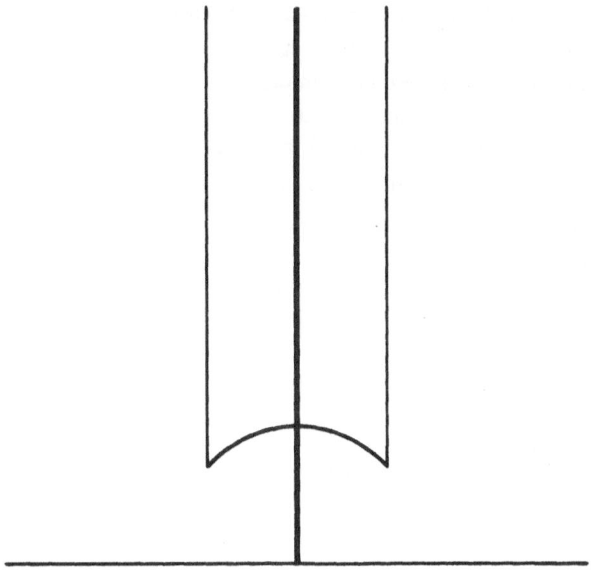

*Figure 3. The moduli space of the cylinder (vertical line) and its double, the torus.*

The moduli space of the torus is the standard "keyhole" region shown in Figure 3. The moduli space of the cylinder is the entire positive imaginary axis, and is thus not a subspace of the moduli space of the torus. What has happened is that the mapping class group of the torus is too large – it includes elements which are not in the mapping class group of the cylinder – so the moduli space is too small.

On the other hand, the Teichmüller space of the torus is the whole upper half plane, and the Teichmüller space of the cylinder – again, the positive imaginary axis – is a subspace. In fact, it is a very nice subspace. The Teichmüller space of the torus inherits an involution, $\tau \mapsto -\bar{\tau}$, from the involution $I$, and the Teichmüller space of the cylinder is precisely the fixed point set of this involution.

This relation generalizes to an arbitrary surface.[10,11] The Teichmüller space of $\Sigma$ is the fixed point set of the Teichmüller space of $\overline{\Sigma}$ under the action of the involution

$$J \mapsto -I_* \circ J \circ I_* \tag{4.1}$$

To obtain the region of integration in (3.1), one must take the quotient of this fixed point set by the action of the mapping class group of $\Sigma$, which can be viewed as a subgroup of the mapping class group of $\overline{\Sigma}$.

CONCLUSIONS

We have reduced the Polyakov path integral for an open or unoriented surface $\Sigma$ to an expression involving known quantities on its closed oriented double $\overline{\Sigma}$. The amplitude is an integral over a fundamental region of a real slice of the Teichmüller space of $\overline{\Sigma}$; the integrand is essentially a square root of the integrand for $\overline{\Sigma}$, up to explicitly known correction factors. The path integral can thus be expressed in terms of well-understood functions such as theta functions on $\overline{\Sigma}$. As in the case of the closed oriented world sheet, the result is not pretty, but it allows one to do calculations.

For instance, it is now possible to write an explicit expression for the three closed string scattering amplitude (the three-holed sphere or "pair of pants") in terms of genus two theta functions. As a check, we have shown that the resulting expression reduces, in the limit that one boundary curve shrinks to a point, to the correct expression for the closed string propagator (the cylinder) with a vertex operator insertion.

We hope that these results will provide some useful insight into closed string field theory. For instance, one should be able to obtain the field theory vertex by amputating the external propagators from the three closed string scattering amplitude. To do so, a better understanding of the role of external state ghosts, and in particular of ghost zero modes, is needed. Work on this problem is in progress.

ACKNOWLEDGEMENTS

This work was supported in part by NSF Grants PHY-8605978, PHY-8215249, and PHY-8318635, DOE Grant DOE-AC02-76ERO-2220, the Robert A. Welch Foundation, and the Center for Relativity of the University of Texas.

## REFERENCES

1. Yu. Manin, *Phys.Lett.* **172B** (1986) 184
2. M. Dugan, Berkeley preprint UCB-PTH-87/07
3. A. Cohen, G. Moore, P. Nelson, and J. Polchinski, *Nucl.Phys.* **B281** (1987) 127
4. S. Blau, S. Carlip, M. Clements, S. Della Pietra, and V. Della Pietra, Texas preprint UTTG-14-87
5. C. Lovelace, *Phys.Lett.* **32B** (1970) 703
6. C.P. Burgess and T.R. Morris, *Nucl.Phys.* **B291** (1987) 256
7. M. Schiffer and D. Spencer, *Functionals on Finite Riemann Surfaces*, Princeton University Press (1954)
8. S. Della Pietra and V. Della Pietra, to appear
9. O. Alvarez, *Nucl.Phys.* **B216** (1983) 125
10. C.J. Earle, in *Advances in the Theory of Riemann Surfaces*, 1969 Stony Brook Conference, Princeton University Press (1971)
11. M. Sepälä, *Teichmüller Spaces of Klein Surfaces*, Annales Academice Scientiarum Fennica, Mathematica Dissertationes, Helsinki (1978)

# Chapter VI
# Loops

# TWO LOOP PARTITION FUNCTION IN (COMPACTIFIED) HETEROTIC STRING VACUA

Joseph J. Atick*

Institute for Advanced Study, Princeton, NJ 08540

and

Ashoke Sen†‡

Stanford Linear Accelerator Center
Stanford University, Stanford, CA 94305

ABSTRACT

Two loop partition function for the heterotic string theory, compactified on any background preserving space-time supersymmetry at the string tree level, is explicitly calculated. This includes $E_8 \times E_8$ or $SO(32)$ heterotic string theory in ten dimensional flat space-time.

The issue of non-renormalization theorems and vacuum stability in compactified fermionic string theories has been the focus of considerable attention for the last few years. However, genuine progress in these directions has been hampered by the complexity of fermionic string perturbation theory. Until recently most of the progress has been achieved through general symmetry and effective lagrangian considerations [1]. Though illuminating, these arguments unfortunately do not expose the interplay between the relevant world-sheet dynamics and their space time manifestation, nor do they provide a check of the internal consistency of string theory. It is therefore very important at this stage to try to establish the validity of the non-renormalization theorems and understand vacuum stability directly through explicit string perturbative calculations.

The complexity of fermionic string perturbation theory beyond one loop can be essentially traced to the presence of the supermoduli [2]. These are modes of the world-sheet gravitino that cannot be gauged away by any world-sheet symmetry. On a world-sheet of genus $g \geq 2$ there are $2g-2$ of them. These are anologous to the moduli or the modes of the graviton that cannot be gauged away. Just as in the latter case, we have to integrate over them in the path integral, and the added

⋆ Work supported by Department of Energy, contract DE-AC02-76ER02220

† Work supported by the Department of Energy, contract DE-AC03-76SF00515.

‡ Address after March 1, 1988: Theoretical Physics Group, Tata Institute of Fundamental Research, Homi Bhabha Road, Bombay 400005, India.

complexity lies roughly speaking in finding the correct integration measure. In some natural setting these modes of the gravitino and graviton can be thought of as the odd and even coordinates of the supermoduli space parametrizing superconformally inequivalent super Riemann surfaces [3]. At this moment, however, not much is known about this space or the structures that can exist over it. Consequently a measure [4] on it is not yet known in any explicitness that would enable us to carry immediate perturbative calculations of string amplitudes.

An alternate approach to the problem has been to perform the integration over the odd moduli in advance in the path integral at the expense of introducing new operator insertions on the world-sheet. Although it may deprive us of important insights which could only be gained by working on supermoduli space, this procedure has the potential of being explicit: The path integral is now carried out on ordinary Riemann surfaces with spin structures, with which one is certainly more familiar. An associated problem in the path integral is posed by the zero modes of the commuting superconformal ghosts. On a world-sheet of genus $g \geq 2$ there are $2g-2$ zero modes for the superconformal ghost $\beta_{z\theta}$, the counting being the same as that for the gravitino zero modes. These have to be removed in a natural way before one can render the path integral well defined.

Recently Verlinde and Verlinde [5] have written down a path integral expression for the fermionic string measure after the integration over the supermoduli and soaking up of the ghost zero modes have been performed. Unfortunately, this does not yet imply that we possess a practical formalism to all orders in the perturbative expansion. As was pointed out in ref.[6], fermionic string perturbation seems to exhibit an inherent ambiguity steming essentially from ambiguities [7] in defining integration over the variables of a Grassmann algebra on a non-compact space—in this case the Grassmann valued coordinates of the supermoduli space. These ambiguities show up in the formalism of ref. [5] through the fact that the answer seems to depend on the choice of basis of the super-Beltrami differentials in terms of which we expand the gravitino field. The ambiguity is a total derivative in the moduli space which does not integrate to zero in general. At genus two nevertheless there exits a natural resolution of these ambiguities. Through various considerations of world-sheet supersymmetry, modular invariance, and decoupling of unphysical states it was shown in ref.[6] that the basis of super-Beltrami differentials at the boundary $\Delta_1$ ( where the surface degenerates into two tori ) has to go to delta functions concentrated at the two nodes. The analogous statement at arbitrary genus is currently not known.

In this talk we shall use string perturbation to address the question of vacuum stability. In particular we shall see how to explicitly calculate the two loop cosmological constant in arbitrary supersymmetry preserving backgrounds. We shall see that it is zero except for compactifications which possess a Fayet-Iliopoulos $D$-term at one loop [8,9]. In that case we shall find that there is an induced cosmological constant given by the square of the one loop coefficient of the Fayet-Iliopoulos terms. This cosmological constant was first calculated in ref. [10].

To start our discussion we shall write down an expression for the two loop heterotic string partition function following ref.[5],

$$W = \int D[XBC]\Big(\prod_{i=1}^{6} dm_i\Big)\Big(\prod_{a=1}^{2} d\varsigma^a\Big)\Big(\prod_{i=1}^{6}(\eta_i \mid B)\Big)$$
$$\Big(\prod_{a=1}^{2} \delta((\chi_a \mid B))\Big) e^{-S(X,B,C,m_i,\varsigma^a)}, \tag{1}$$

where $\{m_i\}$ are the six moduli for the genus two surface and $\{\varsigma^a\}$ are the two supermoduli [2]; $B$ denotes the reparametrization ghosts $b_{zz}$, $\bar{b}_{\bar{z}\bar{z}}$ and the super-reparametrization ghost $\beta_{z\theta}$. Similarly $C$ stands for the reparametrization ghost fields $c^z$, $\bar{c}^{\bar{z}}$ as well as the super-reparametrization ghost field $\gamma^\theta$. $X$ denotes the set of all the matter fields. The inner products $(\eta_i \mid B)$, $\delta((\chi_a \mid B))$ in (1), which are there to soak up the various ghost zero modes, are defined as follows:

$$\begin{aligned}(\eta_i \mid B) &= \int d^2z\{\eta_{i\bar{z}}{}^{z} b_{zz} + \bar{\eta}_{iz}{}^{\bar{z}} \bar{b}_{\bar{z}\bar{z}} + \eta_{i\bar{z}}{}^{\theta}\beta_{z\theta}\} \\ (\chi_a \mid B) &= \int d^2z \chi_{a\bar{z}}{}^{\theta}\beta_{z\theta}\end{aligned} \tag{2}$$

where $\eta_i$, $\chi_a$ form a basis for the super-Beltrami differentials and are defined through the following equations:

$$\begin{aligned}\eta_{i\bar{z}}{}^{z} &= g_{z\bar{z}}\frac{\delta g^{zz}}{\delta m_i}, \qquad \bar{\eta}_{iz}{}^{\bar{z}} = g_{z\bar{z}}\frac{\delta g^{\bar{z}\bar{z}}}{\delta m_i}, \\ \eta_{i\bar{z}}{}^{\theta} &= \frac{\delta \chi_{\bar{z}}^{\theta}}{\delta m_i}, \qquad \chi_{a\bar{z}}{}^{\theta} = \frac{\delta \chi_{\bar{z}}^{\theta}}{\delta \varsigma^a}.\end{aligned} \tag{3}$$

and we have assumed that the world-sheet metric $g^{\alpha\beta}(m_i)$ is independent of the odd coordinates. At this stage we can carry out the integration over the odd moduli. It is easy to see that the expression we arrive at is given by:

$$W = \int D[XBC]\prod_{i=1}^{6} dm_i e^{-S_0}\prod_{a=1}^{2}\delta((\chi_a \mid B))$$
$$\Big[\prod_{a=1}^{2}\{(\chi_a \mid T) + \frac{\partial}{\partial \varsigma^a}\}\prod_{i=1}^{6}(\eta_i \mid B)\Big]_{\varsigma^a=0}, \tag{4}$$

The above expression can be made more explicit through a particular choice of basis for the super-Beltrami differentials $\chi_{a\bar{z}}^{\theta}$ given by:

$$\chi_{a\bar{z}}^{\theta} = \delta^{(2)}(z - z_a) \quad (a = 1, 2) \tag{5}$$

where $\{z_a\}$ are apriori arbitrary points on the Riemann surface. Also we shall group the six real moduli into three complex moduli $(m_i, m_{\bar{i}})$ and choose the metric such that $\eta_{\bar{i}\bar{z}}{}^{z} = \bar{\eta}_{iz}{}^{\bar{z}} = 0$. In this basis the above expression takes the following form[6]:

$$
\begin{aligned}
W = \int \prod_{i=1}^{3}[dm_i d\bar{m}_i] D[XBC] e^{-S_0} \prod_{i=1}^{3}(\bar{\eta}_{\bar{i}} \mid \bar{b})\xi(z_0) \Big[ & Y(z_1)Y(z_2)\prod_{i=1}^{3}(\eta_i \mid b) \\
& + Y(z_1)\partial\xi(z_2)\sum_{j=1}^{3}(-1)^{j+1}\frac{\partial z_2}{\partial m_j}\prod_{i\neq j}(\eta_i \mid b) \\
& + Y(z_2)\partial\xi(z_1)\sum_{j=1}^{3}(-1)^{j+1}\frac{\partial z_1}{\partial m_j}\prod_{i\neq j}(\eta_i \mid b) \\
& - \partial\xi(z_1)\partial\xi(z_2)\sum_{j=1}^{3}\sum_{k>j}(-1)^{j+k}\prod_{i\neq j,k}(\eta_i \mid b)\left(\frac{\partial z_1}{\partial m_j}\frac{\partial z_2}{\partial m_k} - \frac{\partial z_1}{\partial m_k}\frac{\partial z_2}{\partial m_j}\right)\Big]
\end{aligned}
\tag{6}
$$

where $Y(z_i)$ is the picture changing operator given by[11]:

$$
Y =: e^{\phi}T_F := c\partial\xi + e^{\phi}T_F^{matter} - \frac{1}{4}\{\partial\eta e^{2\phi}b + \partial(\eta e^{2\phi}b)\} = \{Q_B, \xi\} \tag{7}
$$

with $\xi$, $\phi$, $\eta$ related to the superconformal ghosts through the bosonization prescription: $\beta = \partial\xi e^{-\phi}$, $\gamma = \eta e^{\phi}$. $Q_B$ is the BRST charge. The factor of $\xi(z_0)$ is needed to soak up the $\xi$ zero mode. The final answer is independent of $z_0$. In writing down eq.(6) we have dropped terms which vanish by $(b, \bar{b})$ ghost charge conservation.

It is perhaps worth emphasizing at this point that $(z, \bar{z})$ denotes a coordinate system such that $g^{zz} = g^{\bar{z}\bar{z}} = 0$ everywhere on the Riemann surface at the particular point $\{m_i\}$ in the moduli space where we are evaluating the string integrand. Thus the choice of coordinates $(z, \bar{z})$ varies with the moduli. In defining $\frac{\partial z_a}{\partial m_i}$ or the various derivatives appearing in (3) we must keep the coordinate system fixed. In other words, after we choose the specific coordinate system $(z, \bar{z})$ we evaluate $z_a(m_i + \delta m_i)$, $g^{zz}(m_i+\delta m_i)$, $g^{\bar{z}\bar{z}}(m_i+\delta m_i)$, $\chi_{\bar{z}}{}^{\theta}(m_i+\delta m_i)$ in this coordinate system and take $\delta m_i \to 0$ limit of appropriate ratios to calculate $\frac{\partial z_a}{\partial m_i}$ and the various super Beltrami differentials. As has become clear from the analysis of ref.[6,12,13], the choice $\frac{\partial z_a}{\partial m_i} = 0$ is not in general consistent with modular invariance. So we shall not drop these terms from our analysis. Furthermore $z_a(m, \bar{m})$ is in general not a holomorphic function of the moduli. Terms with $\frac{\partial z_a}{\partial \bar{m}_i}$ though drop out from expression (6) by ghost charge conservation of the $\bar{b}$ system. We should also point out that in view of the results of ref. [6] the only constraint we shall impose on $z_a(m_i, \bar{m}_i)$ is that at the boundary $\Delta_1$ of the moduli space the points $z_1$ and $z_2$ should coincide with the nodes $p_1, p_2$ on the tori $T_1, T_2$ respectively.

Before we evaluate the cosmological constant consider the following amplitude:

$$
\Lambda = \int_{M-\{z_1,z_2\}} d^2y\langle\partial X(y)\bar{\partial}X(y)\rangle \tag{8}
$$

with the correlator defined using the path integral in expression (6), and the $y$ integration runs over the Riemann surface with the points $z_1, z_2$ removed. The correlator in (8) is given by: $(Im\Omega)^{-1}_{ij}\omega_i(y)\bar{\omega}_j(\bar{y})\langle\ I\ \rangle$ where $\omega_i$ are the normalized abelian differentials and $\Omega$ is the period matrix, plus terms which come from the contraction of $\partial X(y)$ with $\partial X(z_1)$ in $Y(z_1)$ and $\bar{\partial} X(\bar{y})$ with $\partial X(z_2)$ in $Y(z_2)$ and vice versa. It is easy to see that the $y$ dependence in these latter correlators is given by a total derivative of the form: $\partial_y(\langle\partial X(z_1)X(y)\rangle\langle\partial X(z_2)\bar{\partial}X(y)\rangle + (z_1 \leftrightarrow z_2))$. This total derivative could contribute at the boundary of the $y$ integral, namely at $(z_1, z_2)$, iff the correlator in question develops a pole of the form $(\bar{y} - \bar{z}_a)^{-1}$. By examining the correlator in question we can see that such poles do not arise. Consequently these total derivatives integrate to zero. Using the fact that $\int d^2y\omega_i(y)\bar{\omega}_j(\bar{y}) = (Im\Omega)_{ij}$, we immediately conclude that on a given genus surface, $\Lambda$ as defined in (8) is the partition function $\langle\ I\ \rangle$ up to an overall numerical factor.

At this stage we can express $\partial X\bar{\partial}X$ as the contour integral of the suprsymmetry current around the dilatino vertex:

$$\delta^{\beta}_{\alpha}\langle\partial X(y)\bar{\partial}X(y)\rangle = \oint \frac{dx}{2\pi i}\langle J_{\alpha}(x)V^{\beta}(y)\rangle \tag{9}$$

where $J_{\alpha}(= e^{-\frac{\phi}{2}}\hat{S}^{+}S_{\alpha})$ is the +ve chirality 4-dimensional supersymmetry current, and $V^{\alpha}(y)$ is given by:

$$\begin{aligned} V^{\alpha}(y) =& \bar{\partial}X^{\mu}(\gamma_{\mu})^{\alpha\dot{\beta}}[e^{\frac{1}{2}\phi}\partial X^{\nu}(\gamma_{\nu})_{\dot{\beta}\gamma}\hat{S}^{-}S^{\gamma} \\ &+ \frac{1}{2}e^{\frac{3}{2}\phi}\eta b\hat{S}^{-}S_{\dot{\beta}} + e^{\frac{1}{2}\phi}S_{\dot{\beta}}\lim_{w\to z}(w-z)^{1/2}T_F^{\rm int}(w)\hat{S}^{-}(z)] \end{aligned} \tag{10}$$

in the notation of ref. [10]. We can now deform the supersymmetry contour and attempt to shrink it to a point. If the supersymmetry current $J_{\alpha}(x)$ had only the physical pole at $x = y$ then contour deformation would lead to zero answer for (9). We would then conclude that the cosmological constant is zero in all theories which possess a space-time super current $J_{\alpha}(x)$[14]. However as was first pointed out in ref. [5] the supersymmetry current possesses spurious poles– poles not dictated by the operator product expansion. The origin of those poles was explained in ref.[10]. Their implication to contour deformation is that (9) may be expressed as the sum of residues at the spurious poles. Let $\{r_l\}$ denote the location of these poles on the surface. For $g = 2$ it turns out that $l = 1, \cdots 8$ (in general there are $2^{2g-2}g$ of these on a genus g surface [10]) and they are given in this case by the zeros of the function $f(x) = \prod_{\delta}\vartheta[\delta](-\frac{1}{2}\vec{x} + \frac{1}{2}\vec{y} + \sum_{a=1}^{2}\vec{z}_a - 2\vec{\Delta})$ in the $x$-plane. After contour deformation the expression for the cosmological constant takes the form:

$$\Lambda = -\int [\prod_i dm_i d\bar{m}_i] \sum_l \int d^2y \oint_{r_l} \frac{dx}{2\pi i} D[XBC] e^{-S_0} J_\alpha(x) V^\alpha(y) \prod_{i=1}^{3} (\bar{\eta}_{\bar{i}} \mid \bar{b}) \xi(z_0)$$

$$\Big[ Y(z_1) Y(z_2) \prod_i (\eta_i \mid b) + Y(z_1) \partial\xi(z_2) \sum_{j=1}^{3} (-)^{j+1} \frac{\partial z_2}{\partial m_j} \prod_{i \neq j} (\eta_i \mid b)$$

$$+ Y(z_2) \partial\xi(z_1) \sum_{j=1}^{3} (-)^{j+1} \frac{\partial z_1}{\partial m_j} \prod_{i \neq j} (\eta_i \mid b) \Big] \qquad (11)$$

In expression (11) we have dropped terms that vanish by independent ghost charge conservation in the holomorphic and the anti-holomorphic sectors. Also, no sum over $\alpha$ is implied in this equation. We shall now show that (11) is a total derivative on the moduli space. Let us first observe that replacing $z_1$ by $\tilde{z}_1$, the spurious poles will shift to $\{r_l'\}$. By choosing $\tilde{z}_1$ appropriately we can ensure that $\{r_l\} \bigcap \{r_l'\} = 0$. Consequently in equation (11) for the cosmological constant we can replace $Y(z_1)$ and $\partial\xi(z_1)\frac{\partial z_1}{\partial m_j}$ by $Y(z_1) - Y(\tilde{z}_1)$ and $\{\partial\xi(z_1)\frac{\partial z_1}{\partial m_j} - \partial\xi(\tilde{z}_1)\frac{\partial \tilde{z}_1}{\partial m_j}\}$ respectively, without changing the answer for the $\Lambda$.

Now we write

$$Y(z_1) - Y(\tilde{z}_1) = \oint \frac{dz}{2\pi i} J_{BRST}(z)(\xi(z_1) - \xi(\tilde{z}_1)) \qquad (12)$$

and deform the BRST contour and try to shrink it to a point in the first two terms in eq. (11). The obstructions to this deformation are the poles in the argument of the BRST current at the locations of $V^\alpha(y)$, $\partial\xi(z_2)$, $(\eta_i \mid b)$ and $\xi(z_0)$. The latter residue vanishes identically due to the absence of a $\xi$ mode in the correlator that can be used to soak up the zero mode. Had we not carried out the above subtraction we would have had to worry about this residue. $V^\alpha(y)$ in (10) is BRST invariant up to total derivative in $y$. Consequently the residue of the BRST current at the location $y$ will be a total derivative in $y$. This could contribute at the boundary $(z_1, z_2)$ of the $y$ integration only if the integrand develops a $(\bar{y} - \bar{z}_a)^{-1}$ pole. Remembering that the only $\bar{y}$ dependence in the problem comes through the $\bar{\partial}X(y)$ factor in (10) we can see that no such singularity exists. Finally the residue of $J_{BRST}$ at $\partial\xi(z_2)$ is $\partial Y(z_2)$ and at the location of $(\eta_i \mid b)$ is given by $(\eta_i \mid T)$, where $T$ is the stress tensor on the world sheet. The latter insertion in the correlator may in turn be expressed as $-\frac{\partial S_0}{\partial m_i}$. Combining these results together, we may express $\Lambda$ as,

$$\Lambda = \int [\prod_{i=1}^{3} dm_i d\bar{m}_i] \sum_j \frac{\partial}{\partial m_j} M_j \qquad (13)$$

where the density $M_j$ is given by:

$$M_j = \int d^2y \oint_{r_l} \frac{dx}{2\pi i} \int D[XBC] e^{-S_0} \xi(\tilde{z}_1) \xi(z_1) (-1)^j$$

$$Y(z_2) \prod_{i \neq j} (\eta_i \mid b) \prod_{i=1}^{3} (\bar{\eta}_{\bar{i}} \mid \bar{b}) J_\alpha(x) V^\alpha(y) \qquad (14)$$

In writing down eq. (14) we have dropped terms that vanish by $(b, \bar{b})$ ghost charge conservation, and have set $z_0 = \tilde{z}_1$. Notice that so far we have been working entirely in a model independent setting. We have carried out world-sheet manipulations in a manner that is independent of the background fields except that they must allow the existence of a spacetime supersymmetry current at the string tree level.

Being a total derivative in moduli, expression (14) can be written as a boundary term. The boundary of genus two moduli space has two distinct components denoted by $\Delta_0$ and $\Delta_1$. The first is where the Riemann surface degenerates by pinching a nontrivial homology cycle. This is conformally equivalent to one of the handles becoming infinitely long. It is not difficult to see that no contribution can result from this boundary* : If $s$ stands for the length of the handle in some appropriate coordinates, then $\Delta_0$ corresponds to $s \to \infty$. The string integrand in these coordinates behaves like $\frac{1}{s^5} e^{-m^2 s}$. The polynomial factor comes from the degeneration of the $(det Im\Omega)^{-5}$ factor in the measure and $e^{-m^2 s}$ is the effect of space-time propagation of the string modes in the long tube of length $s$. In the absence of tachyons the limit $s \to \infty$ yields a vanishing contribution.

The other boundary $\Delta_1$ is where the Riemann surface degenerates by pinching a trivial homology cycle and can be pictured as a genus two surface breaking up into two tori $T_1$ and $T_2$ connected by a thin (long) tube. We shall next see that $\Lambda$ receives a contribution from this boundary: In the neighborhood of $\Delta_1$ we can parametrize the moduli space by $(\tau_1 = \Omega_{11}, \tau_2 = \Omega_{22}, t = \Omega_{12})$ where $\tau_1$, $\tau_2$ are the modular parameters of the tori $T_1$ and $T_2$ respectively. $t$ can be thought of as the pluming fixture variable (see for e.g. [17]) and can be written as $t = re^{i\theta}$ where $r$ is the radius of the cylinder connecting the two tori. The boundary in question corresponds to $t \to 0$. Since the boundary is a point of measure zero the integral

$$\Lambda = \int dt d\bar{t} \frac{\partial}{\partial t} M_t(t, \bar{t}) \tag{15}$$

would be non-vanishing on the boundary only if

$$\lim_{t \to 0} M_t(t, \bar{t}) \sim \frac{1}{t}. \tag{16}$$

We shall now briefly sketch the calculation of $M_t$ (for details see ref.[10]). The fastest way to determine the behaviour of $M_t$ near the boundary is to use the factorization theorem [15]†

$$\langle A_1(z_1) A_2(z_2) \rangle_{g=2} \sim \sum_{\Phi} \langle A_1(z_1) \Phi(p_1) \rangle_{T_1} \langle \Phi^\dagger(p_2) A_2(z_2) \rangle_{T_2} t^{h_\Phi} \bar{t}^{\bar{h}_\Phi} \tag{17}$$

where $(h_\Phi, \bar{h}_\Phi)$ are the conformal dimensions of the field $\Phi$ propagating in the narrow tube connecting the two tori; $p_1$, $p_2$ are the nodes on $T_1$ and $T_2$ respectively.

---

* J. J. A would like to thank G. Moore for discussions on this point

† One of course can calculate all the ghost correlators and the free matter part at two loops explicitly in terms of $\vartheta$-functions and prime forms and then the effect of these correlators on the behaviour of $M$ near the boundary can be inferred using well known formulae for degeneration of $\vartheta$-functions [16]. Nevertheless one would still have to use the factorization theorem to exhibit the contribution of the interacting part of the matter correlators. In ref.[10] the analysis is carried along these lines.

In view of the results of ref. [6] we shall take in the correlator in eq. (14) $z_1$ and $z_2$ to lie on $T_1$ and $T_2$ respectively. Ultimately we have to take the $z_1 \to p_1$ and $z_2 \to p_2$ limit. We shall also take $\tilde{z}_1$ to lie on $T_2$ for convenience, although the final result may be shown to be manifestly independent of the location of $\tilde{z}_1$. The $y$ integration runs over the tori $T_1$ and $T_2$. Let us take for definiteness $y$ on $T_1$. It is not hard to prove that the other case gives the same contribution [10]. In this case, the positions $r_l$ of the spurious poles in the $t \to 0$ limit may be found by analyzing the behavior of $\vartheta[\delta](-\frac{1}{2}\vec{x} + \frac{1}{2}\vec{y} + \sum_{a=1}^{2} \vec{z}_a - 2\vec{\Delta})$ in this limit. It can be seen that four of these poles lie on $T_1$, and the other four lie on $T_2$[10].

Let us first take $x$ on $T_2$. Application of the factorization theorem to the correlator in eq. (14) yields:

$$
\begin{aligned}
&t^{h_\Phi}\bar{t}^{\bar{h}_\Phi-2}\overline{\eta(\tau_1)}^2\,\overline{\eta(\tau_2)}^2\langle\xi(z_1)b(w_1)V^\alpha(y)\Phi(p_1)\rangle_{T_1} \\
&\quad\langle\Phi^\dagger(p_2)b(w_2)e^{-\phi(x)/2}\hat{S}^+(x)S_\alpha\xi(\tilde{z}_1)Y(z_2)\rangle_{T_2}
\end{aligned}
\tag{18}
$$

where we have explicitly exhibited the contribution from the factorization of the antiholomorphic ghost determinant. We also used the fact that $(\eta_1 \mid b)_0$, $(\eta_2 \mid b)_0$ in the limit $t \to 0$ reduce to $b(w_1)$, $b(w_2)$ inserted at arbitrary points $w_1$, $w_2$ on $T_1$, $T_2$ respectively. Using various ghost charge conservations and the fact that the operator $\Phi$ has to have dimension (0,1) in order to contribute at the boundary we can determine what $\Phi$ has to be:

$$
\Phi(p_1) = c(p_1)e^{-\phi(p_1)/2}\hat{S}^+(p_1)S_\alpha(p_1)U(\bar{z}) \tag{19}
$$

where $U(\bar{z})$ is any operator of conformal dimension $(0,1)$ and neutral under all ghost charges. The only such operators in the heterotic string are associated with gauge currents (recall that for every dim. $(0,1)$ operator we can construct the vertex operator for a massless gauge boson by adjoining it with $(\partial X + ik \cdot \psi\psi)e^{ik\cdot X}$). It is clear that the only nonvanishing contribution to that matrix element could come from the $(0,1)$ operators associated with the abelian factors of the unbroken gauge group. We now have to compute the one loop matrix elements appearing in eq.(18), with $\Phi$ as given in eq. (19). The superconformal ghost correlator can be calculated readily. In that one finds that there exists a simple pole in $x$ on $T_2$. This pole accounts for four of the spurios poles on the genus two surface before degeneration since it exists on $T_2$ for each spin structure on $T_1$. Finally the one-loop matrix elements of the interacting matter fields can be computed by applying the results of ref. [9], and the answer can be exhibited entirely in terms of properties of the massless spectrum. From this one can easily calculate the residue of the correlator at the spurious poles.

The other four poles in the $x$ plane lie on the torus $T_1$. The contribution to $M_t$ from these poles may be analyzed by taking $x$ on the torus $T_1$, and analyzing the resulting correlator using factorization theorem. The relevant intermediate operator $\Phi(p_1)$ is given by $c(p_1) : \xi(p_1)\eta(p_1) : U(p_1)$. We find a simple pole on $T_1$, whose position is independent of the spin structure on the torus $T_2$. This accounts for the other four poles. The residue at this pole however can be seen to vanish after summing over spin structures on $T_2$.

Combining all the results, we find that the cosmological constant at two loops is given by:

$$\Lambda \sim \sum_a c^{(a)} c^{(a)} \tag{20}$$

where $c^{(a)}$ is the coefficient of the Fayet-Iliopoulos $D$ term associated with the a'th abelian factor in the gauge group, induced at one loop [8,9]. More explicitly it is given by

$$c^{(a)} = \frac{g}{192\pi^2} \sum_i n_i q_i^{(a)} h_i \tag{21}$$

where $n_i$ is the number of massless fermions (bosons) with chirality $h_i$ and $U^{(a)}(1)$ charge $q_i^{(a)}$. Needless to say this means that the cosmological constant at two loops vanishes for all string vacua which have tree-level supersymmetry and which do not develop a one loop Fayet-Iliopoulos $D$-term. These include among other things flat space-time and the standard compactifications of the $E_8 \times E_8$ heterotic string on Calabi-Yau backgrounds.

It is important to see whether one can push string perturbation theory at higher loops to the same level of explicitness that we have witnessed at two loops. One obstacle in that direction seems to be the ambiguities alluded to earlier. Another interesting question is the connection between the boundary terms in the moduli space and the vev's of the auxiliary fields. In our analysis above we found that the two loop boundary term was related to the square of the vev of the auxialiary $D$ term. One is tempted at this stage to conjecture a correlation between the two. It is therefore important to try to calculate within string perturbation directly the vev's of the auxiliary fields, *e.g.* $\langle F \rangle$ and $\langle D \rangle$ at higher loops. At one loop it can be shown through some variant of contour deformation arguments that $\langle F \rangle = 0$. However at higher genus no such statement exists so far. It is therefore safe to say that the $F$-term non renormalization theorems at higher loops have not yet explicitly been established in string perturbation theory.

Acknowledgements: We would like to thank the organizers of this conference for their hospitality. J. J. A would like to thank M. Dine, G. Moore, N. Seiberg and E. Witten for illuminating discussions.

## References

1. M. Dine and N. Seiberg, Phys. Rev. Lett. **57** (1986) 2625.
2. G. Moore, P. Nelson and J. Polchinski, Phys. Lett. **B169** (1986) 47; E. D'Hoker and D. Phong, **B278** (1986) 225; G. Moore and P. Nelson, Nucl. Phys. **B274** (1986) 509; S. Giddings and P. Nelson, Harvard preprint HUTP-87/A062 (1987).
3. D. Friedan, in Unified String Theories, eds. M. Green and D. Gross (World Scientific, 1986); L. Crane and J. Rabin, Commun. Math. Phys., to appear; P. Freund and J. Rabin, EFI-87-22-CHICAGO;
4. M. Baranov and A. Shvarts, Pisma ZETF **42** (1985) 340 (JETP Lett. **42** (1986) 419); M. Baranov, Yu Manin, I. Frolov and A. Shvarts, Yad. Phys. **43** (1986) 1053 (Sov. J. Nucl. Phys. **43** (1986) 670); M. Baranov and A. Shvarts, Niels Bohr Inst. preprint (1987).
5. E. Verlinde and H. Verlinde, Phys. Lett. **B192** (1987) 95.

6. J. J. Atick, J. M. Rabin and A. Sen, preprint IASSNS-HEP-87/45 (SLAC-PUB-4420).
7. B. de Witt, Supermanifolds, Cambridge Univ. Press (1984); A. Rogers, J. Math. Phys. **26** (1985) 385, **27** (1986) 710; J. Rabin, Physica **15D** (1985) 65 (Proceedings of the Workshop on Supersymmetry in Physics, held at Los Alamos, N. Mex., Dec., 1983); M. Rothstein, Trans. AMS **299** (1987) 387.
8. M. Dine, N. Seiberg and E. Witten, Nucl. Phys. **B289** (1987) 589; M. Dine, I. Ichinose and N. Seiberg, Nucl. Phys. **B293** (1987) 253.
9. J. J. Atick, L. Dixon and A. Sen, Nucl. Phys. **B292** (1987) 109.
10. J. J. Atick and A. Sen, SLAC-PUB-4292, to appear in Nucl. Phys. B.
11. D. Friedan, E. Martinec and S. Shenker, Nucl. Phys. **B271** (1986) 93.
12. G. Moore and A. Morozov, preprint IASSNS-HEP-87/47.
13. J. J. Atick, G. Moore and A. Sen, in preparation.
14. E. Martinec, Phys. Lett. **B171** (1986) 189.
15. D. Friedan and S. Shenker, Nucl. Phys. **B281** (1987) 509; Phys. Lett. **175B** (1986) 287.
16. J. D. Fay,Theta Functions on Riemann Surfaces, Springer Notes in Mathematics, Vol.352 (Springer, Berlin, 1973); D. Mumford, Tata Lectures on Theta, Vols.I,II (Birkhauser, Basel, 1983).
17. P. Nelson, Phys. Reports **149** (1987) 304.

# STRING PERTURBATION THEORY AND EFFECTIVE LAGRANGIANS

Igor Klebanov*

Stanford Linear Accelerator Center
Stanford University, Stanford, CA 94305

ABSTRACT

We isolate logarithmic divergences from bosonic string amplitudes on a disc. These divergences are compared with 'tadpole' divergences in the effective field theory with a cosmological term, which also contains an effective potential for the dilaton. Also, corrections to $\beta$-functions are compared with variations of the effective action. In both cases we find an inconsistency between the two. This is a serious problem which could undermine our ability to remove divergences from the bosonic string.

This talk is based on the work done in collaboration with W. Fischler and L. Susskind.[1]

Since the recent resurgence of interest in the theory of strings as the most fundamental objects in physics, a new approach to string dynamics has emerged. This approach, which did not figure during the earlier golden age of strings in the seventies, was pioneered in the papers by Callan, Martinec, Perry and Friedan,[2] Sen[3] and Lovelace.[4] These authors start by writing down a renormalizable two-dimensional field theory describing propagation of strings in classical backgrounds which can be viewed as condensates of massless string modes. If conformal invariance is considered to be a fundamental principle of string theory, then the vanishing of the $\beta$-functions of the 2-d field theory should insure consistent string propagation. Therefore we find that strings can propagate not only in flat 26-dimensional space, but also in curved space with some classical antisymmetric tensor and dilaton backgrounds. The $\beta$-functions can be computed in the loop expansion of the 2-d field theory. The resulting expressions are functions of background fields in spacetime with increasing number of derivatives, depending on how many field theory loops have been taken into account.

The amazing feature of the equations resulting from setting all the $\beta$-functions to zero is that they turn out to be variations of a generally covariant space-time action functional depending on the massless background fields. This equivalence was discovered and tested to low orders in the derivative expansion by a number of authors whose work relied primarily on the sigma-model background field method.[2,5,6] Although a general proof of equivalence between conformal invariance conditions of the world-sheet theory and variational equations of a spacetime effective action is yet

* Work supported by the Department of Energy, contract DE–AC03–76SF00515.

to be constructed, recent work by a number of authors[7] has made important steps towards such a proof.

A crucial property of the effective action found via the sigma-model route is that it generates string scattering amplitudes for the massless modes. In other words, it is the effective action in the standard field theory sense. An immediate question that comes to mind is how to incorporate into our effective action the effects of string loop dynamics. Naively this appears to pose serious problems for the $\beta$-function method because of the following simple argument. The $\beta$-functions are sensitive only to the short distance effects on the string world sheet. Therefore they are independent of the world sheet topology: the standard field theory $\beta$-functions computed on a sphere (string tree level), a torus (one string loop), etc. are all going to be the same. If we believe that conformal invariance generates string equations of motion, this would imply that the effective action is not renormalized by string loop effects, which is in direct contradiction with non-vanishing of scattering amplitudes beyond tree level.

A natural resolution of this apparent paradox was proposed by Fischler and Susskind.[8] The basic observation is that beyond tree level the Polyakov functional integral prescription for S-matrix calculations contains more integrations than at the tree level. In order to calculate an $n$-particle amplitude one is instructed to integrate over the modular parameters of the surface with $n$ punctures. For example, the fact that is crucial for this talk is that a disc with $n$ punctures has more modular parameters than a sphere with n puncures. A more familiar statement is that on a sphere SL(2,C) invariance allows us to fix locations of 3 vertex operators, while on a disc SL(2,R) invariance allows for fixing only one closed string vertex, and the angular coordinate of the other. The extra integrations that need to be carried out on a disc give rise to extra logarithmic divergences beyond those encountered at string tree level. These divergences give rise to the loop corrections to $\beta$-functions. Therefore, for applications to strings, one is no longer interested in an ordinary two-dimensional field theory, but rather in a peculiar combination of 2-d field theories defined on world sheets with different topology and supplied with a prescription for integration over the moduli of these world sheets.

After having identified the source of the loop corrections to the string equations of motion it is important to check that the resulting equations follow the pattern discovered at the string tree level. Namely, they should be equivalent to variations of the loop corrected effective action whose form is dictated by general covariance and simple counting of the powers of the string coupling constant. In this talk I will present a calculation of the simplest consistency check, the one that concerns the cosmological term which is simultaneously the effective potential for the dilaton. The result of our calculation is rather perplexing: the loop corrections to $\beta$-functions do not turn out to be consistent with the expected form of the effective action! This may be a serious problem that could primarily affect our understanding of the behaviour of cosmological constant in string theory. I should mention that a conclusion opposite to ours have been reached in a recent preprint by Callan et. al.[9] Since the methods used there are quite different from ours, we do not fully understand the nature of this discrepancy.

Let me now proceed to a more formal explanation of the problem. The propagation of the closed bosonic string in gravitational and dilaton backgrounds is governed by the following 2-d action:[10,2]

$$S = \frac{1}{4\pi\alpha'}\int d^2\sigma\sqrt{\gamma}(\gamma^{\mu\nu}g_{ij}(X)\partial_\mu X^i\partial_\nu X^j - \frac{\alpha'}{2}\phi(X)R^{(2)}) - \frac{1}{4\pi}\oint ds\phi(X)\kappa \quad (1)$$

where $\gamma_{\mu\nu}(\sigma_1,\sigma_2)$ is the world sheet metric, $g_{ij}(X)$ is the 26-dimensional metric, and $\phi(X)$ is the dilaton field which couples to the world sheet and boundary curvatures. The fields $g_{ij}(X)$ and $\phi(X)$ can be thought of as an infinite collection of couplings in the 2-d field theory. These couplings become renormalized and satisfy renormalization group equations

$$\frac{\partial g_{ij}}{\partial \log\lambda} = \beta_{ij}(g_{ij},\phi) \quad (2)$$

$$\frac{\partial \phi}{\partial \log\lambda} = \beta_\phi(g_{ij},\phi) \quad (3)$$

Bypassing certain technicalities associated with the choice of renormalization scheme, the conformal invariance conditions are

$$\beta_{ij} = \beta_\phi = 0 \quad (4)$$

Remarkably, these equations can be derived from a generally covariant action[2]

$$I = \int d^{26}X\sqrt{g}e^\phi(R + (\partial\phi)^2 + O(\alpha')) \quad (5)$$

The $O(\alpha')$ corrections in (5) are higher derivative terms where each additional pair of derivatives introduces a factor of $\alpha'$.[2,5]

The processes leading to the effective action of (5) are string tree graphs calculated on a spherical world sheet. The factor $\exp(\phi)$ is understood as follows: the path integral for the string has a factor

$$\exp\left(\frac{\phi_0}{8\pi}\int d^2\sigma\sqrt{\gamma}R^{(2)}\right) = \exp(\phi_0(1-g)) \quad (6)$$

where $\phi_0$ is the zero momentum part of $\phi$ and $g$ is the genus of the world sheet. For a sphere ($g = 0$) the effective action must be weighted by $\exp(\phi_0)$, and locality requires that this be replaced by $\exp(\phi)$.

Let us now consider corrections to the action (5) due to processes with a small hole in the world sheet. Such processes occur in the theory of coupled open and closed strings. Since the disc has genus 1/2, the action term with the least possible number of derivatives is

$$\delta I = \int d^{26}X\sqrt{g}\exp(\phi/2)J \quad (7)$$

where $J$ is a constant.

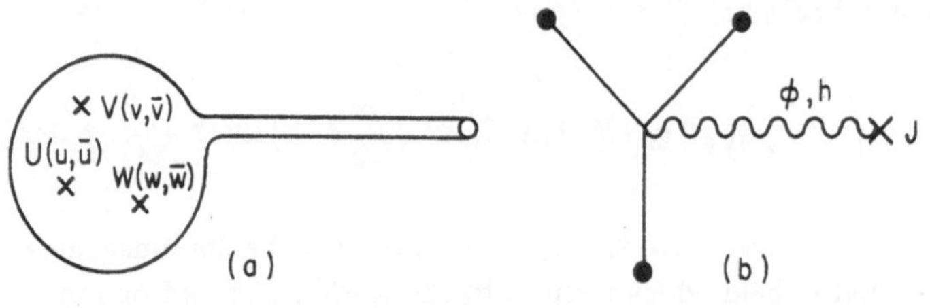


**Figure 1.** The tadpole configurations a) in string theory, b) in field theory.

To understand the leading corrections to closed string $\beta$-functions we must consider closed-string scattering amplitudes on a disc. We will represent the disc by a sphere with a hole and for simplicity look at three-particle scattering since it illustrates the basic point. As usual, the sphere is stereographically projected onto the complex plane. We may integrate over positions of two of the vertex operators. Alternatively we can hold the vertices fixed and integrate over the radius $a$ and location $z$ of the hole. Then the tadpole divergences occur in the small hole limit which is conformally equivalent to a sphere with a long tube attached (fig. 1a)). After this change of integration variables, the amplitude is

$$\exp(\phi_0/2)\int daa^{-3}\int d^2z \mid v-w\mid^2\mid v-u\mid^2\mid u-w\mid^2 < V(v,\bar{v})U(u,\bar{u})W(w,\bar{w}) > \quad (8)$$

where the expectation value is computed on a sphere with a hole of radius $a$ centered at $z$, and $V$,$U$,$W$ are vertex operators for emission of arbitrary closed string physical states. The Green's function needed to compute (8) is given by

$$< X(u,\bar{u})X(w,\bar{w}) >= -\log\mid u-w\mid^2 - \log\mid 1-\frac{a^2}{(z-u)(\bar{z}-\bar{w})}\mid^2 \quad (9)$$

Expanding (9) for small $a$ gives

$$< X(u,\bar{u})X(w,\bar{w}) >= -\log\mid u-w\mid^2 - \frac{a^2}{(z-u)(\bar{z}-\bar{w})} - \frac{a^2}{(z-w)(\bar{z}-\bar{u})} \quad (10)$$

The $O(a^2)$ contribution to the propagator is identical to the effect of the operator $:\partial X_i\bar{\partial}X^i:$ inserted at the point $z$ (normal ordering means that we drop the self-contraction). Thus, to obtain the coefficient of the logarithmic divergence in the amplitude (8), we expand the integrand to order $a^2$ and integrate over $a$:[11]

$$e^{\phi_0/2}\int_\lambda\frac{da}{a}\int d^2z \mid v-w\mid^2\mid v-u\mid^2\mid u-w\mid^2 \left\langle V(v,\bar{v})U(u,\bar{u})W(w,\bar{w}):\partial X_i\bar{\partial}X^i:(z,\bar{z})\right\rangle \quad (11)$$

where the expectation is now evaluated on a sphere. This logarithmic divergence in the disc amplitudes provides the leading correction to the tree level $\beta$-functions for the closed string modes. To find these corrections, we need to decompose the operator insertion that replaces a small hole on a sphere in terms of the operators that enter the two-dimensional action (1). Using dimensional regularization, de Alwis has found that[12]

$$: \partial X_i \bar{\partial} X^i := \partial X_i \bar{\partial} X^i + \frac{d}{8}\sqrt{\gamma}\, R^{(2)} \tag{12}$$

where we have set $\alpha' = 2$. Recalling the logarithmically divergent counterterms that are necessary on a sphere, we find that the $\beta$-functions corrected by small holes are:

$$\beta_{ij} = R_{ij} - \nabla_i \nabla_j \phi + J g_{ij} \exp(-\phi/2) + \ldots$$

$$\beta_\phi = -1/2 \nabla^2 \phi - 1/2(\partial\phi)^2 - \frac{d}{2} J \exp(-\phi/2) + \ldots \tag{13}$$

$J$ is the normalization factor of the insertion that we did not bother to fix. A subtle point in the formulae (13) is that the $\beta$-function terms $- \nabla_i \nabla_j \phi$ and $-1/2(\partial\phi)^2$ are absent unless we implement a divergent shift of the sigma model target space variables: $X^i \to X^i + \log \lambda \partial^i \phi$.[5] This particular choice of renormalization scheme is necessary even at tree level to insure that the $\beta$-functions are equivalent to variations of the effective action. However, it is easy to check that the leading corrections arising from small holes cannot be shifted by different choices of renormalization scheme on a sphere (they do not contain any spacetime derivatives).

We observe now that, even though the first two terms in (13) are equivalent to variations of a spacetime action, the small hole corrections violate this equivalence. They cannot be obtained by adding to the action the 'cosmological term', which is fixed up to normalization by general covariance and counting of string coupling constants. This is the basic result of our work. To make our discussion of this important point as clear as possible, let us enumerate all the zero-momentum vertex operators relevant for our problem. To find the correct soft graviton and soft dilaton operators it is convenient to carry out a rescaling on the couplings of the two-dimensional theory, $g_{ij} \to g_{ij} \exp(-2\phi/d - 2)$ which, as will be shown later, is necessary to identify the physical dilaton.[10] Then we can read off the vertex operators as the coefficients of terms linear in dilaton $\phi$ and graviton $h_{ij}$ in the 2-d action (1) (we will state all operators up to overall normalization constants). Soft graviton emission is given by an insertion of $\partial X^i \bar{\partial} X^j$ while the soft dilaton emission is produced by

$$\partial X_i \bar{\partial} X^i + \frac{d-2}{8}\sqrt{\gamma}\, R^{(2)} =: \partial X_i \bar{\partial} X^i : -\frac{1}{4}\sqrt{\gamma}\, R^{(2)} \tag{14}$$

We found that the operator insertion that must replace a small hole to satisfy the desired equivalence with the effective action is

$$: \partial X_i \bar{\partial} X^i : + \frac{1}{4}\sqrt{\gamma}\, R^{(2)} \tag{15}$$

Please note that, although an accidental conspiracy of factors makes the operator (15) appear as a soft dilaton operator with a flipped sign, their origin is very different. Actually, the operator (15) must be a linear combination of a physical operator (dilaton) and an unphysical operator (the trace of graviton). This is so because the cosmological constant term creates a tadpole for the trace of graviton. If one thinks of tadpoles as injecting the operators onto the sphere and substitutes correct tadpole and propagator factors, one recovers the precise mixture of the dilaton and graviton operators that must replace the hole for consistency with the effective action. This procedure is an alternative to finding precisely which combinations of the variations of the effective action should be equal to the $\beta$-functions and it gives identical results. Now it should be clear to you why the old statement of Ademollo et al.[13] that the logarithmic divergences should be proportional to soft dilaton emission amplitudes cannot be compatible with the idea of effective action.

As elucidated in the previous discussion, the precise insertion we are finding by a direct string-theoretic argument is $:\partial X_i \bar{\partial} X^i:$. It disagrees with what is necessary for consistency of the effective action as well as with the Ademollo et al. theorem.[13] Thus, in order to find agreement with the effective action, we need to find an additional renormalization of the curvature coupling, which is topological when integrated over the sphere. Since identification of such terms is notoriously subtle, we have carried out a test of our results.

We calculated the logarithmically divergent part of a three-graviton amplitude on a disc and compared it with tadpole diagrams of the effective field theory, specified by the action $I + \delta I$. The string calculation amounts to carrying out the integration over the position $z$ and radius $a$ of the hole in (8). For each radius $a$, the $z$ integration covers the whole plane excluding only those regions which would cause one of the vertex operators to be inside the hole. This exercise turns out to have a simple answer: the logarithmically divergent part of the amplitude is proportional to $\sqrt{\alpha'}\partial/\partial\sqrt{\alpha'}A_{tree}(\rho_i, k_i)$ where $\sqrt{\alpha'}\partial/\partial\sqrt{\alpha'}$ effectively counts the power of momenta in a given term of the corresponding tree amplitude for three gravitons.

This answer is to be compared with the effective field theory divergences of the form $1/k^2|_{k^2=0}$ which arise from the tadpole graphs of fig. 1b). We identify this divergence with the logarithmic divergence in the world sheet cut-off:

$$\frac{1}{k^2} = \int_{\lambda}^{1} da a^{k^2-1}$$

A very important feature of the field theory calculation is the presence of a tadpole for an unphysical state, the trace of graviton. Contrary to what is sometimes said in the literature, propagation of this state into vacuum provides an additional source of divergence.

In order to calculate the dilaton contribution to the graphs of fig. 1b), we recall the soft dilaton theorem. The emission amplitude for a zero-momentum dilaton is given by[13]

$$A_\phi = -\frac{1}{d-2}\left(\sqrt{\alpha'}\frac{\partial}{\partial\sqrt{\alpha'}} - 2\right) A \tag{16}$$

where $A$ is the amplitude without the soft dilaton. Multiplying this by the propagator and tadpole factors from the conveniently rescaled action ($g_{ij} \to g_{ij}\exp(-2\phi/$

$d-2)$)

$$I = \int d^{26}X\sqrt{g}\left(R - \frac{(\partial\phi)^2}{d-2} + \ldots\right) + \int d^{26}X\sqrt{g}\exp\left(\phi\left(-\frac{2}{d-2} - \frac{1}{2}\right)\right) J \quad (17)$$

we obtain the dilaton contribution to the divergence. Note that in (17), as opposed to (5), the dilaton has a standard kinetic term (up to a factor). The mixing between dilaton and graviton has been eliminated. This identifies the dilaton field in (17) as proportional to the physical dilaton.[2,10]

Since the gravitational couplings are determined by general covariance, we can derive a similar theorem for the graviton contribution to the divergence (we found it convenient to work in the standard harmonic gauge). Adding the two, we find that the net divergence in the amplitude is proportional to $(\sqrt{\alpha'}\partial/\partial\sqrt{\alpha'} + 2)A_{tree}$. The string and field theory answers disagree! As expected, the missing $\beta$-function insertion $\sim \sqrt{\gamma}R^{(2)}$ translates into a missing logarithmic divergence proportional to the tree amplitude. The reason is that an integrated curvature insertion on a sphere produces an answer proportional to the tree amplitude. Therefore, our results, although inconsistent with what we expected to find, possess some internal consistency.

I have no time to dwell on quite a few attempts we made to find the missing term. None of them have led to clear-cut results. I would like to add a note of caution, however. The amplitudes that we considered are divergent. At this time there is no universally applicable prescription for regularizing divergences in string theory. Although the 'small fixture' regularization that we introduced is physically plausible and is easy to carry out, it is not inconceivable that some other scheme will identify the extra term needed for consistency. Such a scheme may be available only in the framework of closed string field theory. Alternatively one could look at string amplitudes directly in non-flat backgrounds. By adjusting the background to eliminate all divergences we would then find the loop-corrected equations of motion. At this point both of these approaches appear to be difficult.

I would like to conclude my talk by mentioning a possibly interesting extension of our results. The problem we have found occurs on a disc when all the closed string vertex operators approach each other. Let us consider an equivalent calculation on a surface of arbitrary topology. The situation when all the vertices approach each other can be conformally mapped into a sphere with a small fixture of any genus. Using multipole expansion for the Green's function on an arbitrary surface we can show that the term quadratic in the size of the fixture is proportional to an insertion of the same operator $:\partial X \cdot \bar{\partial} X:$ replacing the fixture on a sphere. As we argued previously, the effective action requires that the operator replacing the fixture depend on its genus. This argument would generalize the inconsistency we are finding to a surface of arbitrary genus. Unfortunately it is hard to make this argument precise due to existence of overlapping divergences in all higher-order calculations. For example, if we calculate closed string amplitudes on an annulus, there are going to be $(\log\lambda)^2$ divergences due to the fact that each hole gives rise to a tadpole. Thus, a new apparatus of stringy renormalization is needed for subtracting higher order divergences and exposing the $\beta$-functions. We know of at least one case of a higher genus calculation, however, where this is not necessary. This is the recently investigated case of $D$-term supersymmetry breaking in string theory.[14] The first place where the dilaton tadpole shows up is a double torus. Due to the arguments stated above, we expect a careful identification of the insertion on a sphere to reveal a mismatch with the effective action.

The only topology where the insertion turns out to be consistent with the effective action considerations is a torus. This may have to do with the fact that only in this case the integrated curvature vanishes (the tadpole for a state that couples to $R^{(2)}$ is zero). We should remark here that, in a theory of closed bosonic strings only, the torus provides the leading contribution to tadpoles. We have carried out an explicit calculation on a torus and convinced ourselves that the insertion that must replace a small handle is $: \partial X \cdot \bar{\partial} X :$, the same as for a small hole. For the case of genus 1 this insertion turns out to be consistent with the effective action considerations which proceed in complete analogy with our calculations for genus 1/2. However, as explained above, we expect trouble once we move on to genus 2.

Let me point out that the small hole limit in the sense of the above paragraph is quite different from the small hole limit of the annulus diagram of the open string theory. If one inserts open string vertex operators on the outer boundary of an annulus, there is a logarithmic divergence that occurs in the limit when the radius of the inner boundary vanishes. In this case, however, this variable is a modular parameter of an empty annulus.[15] Therefore the ratio of the determinants needed to calculate the zero-point function depends on the radius of the hole. This brings in an extra term into the operator that must replace the hole. It is not easy, however, to identify this extra term as a curvature insertion. A reasonable thing to do is to carry out a comparison of the logarithmically divergent part of the amplitude for three gauge bosons with the effective field theory. This work is in progress now (the issue is complicated by the presence of open string tachyons in the bosonic model).[16] However, a similar comparison of the four gauge boson amplitude in open superstring theory (for a gauge group not equal to SO(32)) appears to yield agreement with the effective action including the tadpole terms. Therefore, we conjecture that the problem we found does not afflict open string amplitudes but is only present for pure closed string amplitudes. It is suggestive to compare the situation with A. Strominger's associativity anomaly which occurs only in the tadpole limit in pure closed string amplitudes.[17] Perhaps, clarifying the relation between our calculation and the associativity anomaly could shed some further light on the problem.

## ACKNOWLEDGEMENTS

I thank the organizers of the Boulder Workshop on Superstrings for giving me a chance to present this work in a stimulating environment. I am indebted to L. Susskind and W. Fischler for many hours of conversations during which most of the ideas relevant to this talk were developed. I also thank T. Banks, C. Callan, C. Lovelace, V. Periwal, M. Peskin, J. Polchinski, B. Ratra, N. Seiberg and A. Sen for helpful conversations. I am grateful to A. Sen and C. Wendt for reading the manuscript.

## REFERENCES


1. W. Fischler, I. Klebanov and L. Susskind, SLAC preprint SLAC-PUB-4298, April 1987
2. C. Callan, E. Martinec, M. Perry and D. Friedan, *Nucl. Phys.* **B262** (1985) 593
3. A. Sen, *Phys. Rev.* **D32** (1985) 2102;, *Phys. Rev. Lett.* **55** (1985) 1846
4. C. Lovelace, *Nucl. Phys.* **B273** (1986) 413
5. C. Callan, I. Klebanov, M. Perry, *Nucl. Phys.* **B272** (1986) 111
6. R. Metsaev and A. Tseytlin, Lebedev Inst. preprint Print-87-0184
7. A.M. Polyakov, Talk at the 1986 Congress of Mathematics, Berkeley, SLAC-TRANS-222
   A. B. Zamolodchikov, *JETP Lett.* **43** (1986) 731
   T. Banks and E. Martinec, preprint EFI-87-12
   B. Sathiapalan, preprint UCLA/87/TEP/17
   R. Brustein, D. Nemeschansky and S. Yankielowicz, preprint USC-87/004
8. W. Fischler and L. Susskind, *Phys. Lett.* **171B** (1986) 383, *Phys. Lett.* **173B** (1986) 262
9. C. Callan, C. Lovelace, C. Nappi and S. Yost, Princeton preprint PUPT-1045
10. E. Fradkin and A. Tseytlin, *Nucl. Phys.* **B261** (1985) 1;, *Phys. Lett.* **158B** (1985) 316
11. S.R. Das and S.J.Rey, *Phys. Lett.* **186B** (1987) 328
12. S. de Alwis, *Phys. Lett.* **168B** (1986) 59
13. M. Ademollo, R. d'Auria, F. Gliozzi, E. Napolitano, S. Sciuto, and P. di Vecchia, *Nucl. Phys.* **B94** (1975) 221
14. M. Dine, N. Seiberg and E. Witten, *Nucl. Phys.* **B289** (1987) 589
    M. Dine, I. Ichinose and N. Seiberg, preprint IASSNS-HEP-87/17
    J. Atick, L. Dixon and A. Sen, *Nucl. Phys.* **B292** (1987) 109
    J. Atick and A. Sen, SLAC preprint SLAC-PUB-4292, April 1987
15. C. Callan, C. Lovelace, C. Nappi and S. Yost, *Nucl. Phys.* **B290** (1987) 1
16. I. Klebanov, Work in progress
17. A. Strominger, Lecture notes at ICTP School on Superstrings, Trieste, April 1987

# Chapter VII
# Anomalies and Chiral Bosonization

# CHIRAL BOSONIZATION

Rafael I. Nepomechie

Department of Physics

University of Miami, Coral Gables, Florida 33124

## INTRODUCTION AND SUMMARY

One of the remarkable features of physics in 1+1 dimensions is the quantum equivalence of a Dirac Fermion and a scalar [1]. This phenomenon ("Abelian Bosonization") has been generalized to the non-Abelian case by Witten [2]. In both the Abelian and non-Abelian cases, the models are "non-chiral" — i.e., they describe both left-movers and right-movers.

In the talk delivered at the Workshop, I began with a review of non-chiral non-Abelian Bosonization, following Ref. [3]. Since this material is already published, I do not repeat it here. Rather, I report only on the second half of the talk, which was on chiral Bosonization: the Fermi - Bose equivalence of chiral models, which have (say) only left-movers. This is new work which I have done with L. Mezincescu; it is based on results presented in Ref. [4].

First, we discuss the Abelian case. A classical covariant model describing a chiral scalar has been found by Siegel [5]. Although the action contains an additional field, there is a gauge invariance which guarantees that the only propagating field is the chiral scalar. When the model is quantized, this gauge invariance in general becomes anomalous [6]. One way of cancelling the anomaly has been proposed in Ref. [6]: namely, making a suitable modification of the Siegel action. To determine whether the modified Siegel model is equivalent to a model of a single Weyl Fermion, we follow Ref. [3], and construct the corresponding current and stress-energy generating functionals. We find that the two models indeed have identical current correlation functions. However, the modified Siegel model cannot be coupled to gravity in a straightforward manner. In particular, we find the correct couplings only at the linearized level. Hence, it is not yet clear whether the model's stress-tensor correlation functions are the same as those of a Weyl Fermion.

There is a second, more direct, way of cancelling the Siegel anomaly: namely, increasing the number of chiral scalars to 26. In contrast with the first approach, there is now no difficulty in coupling to gravity.

Next, we describe a non-Abelian generalization of the Siegel model. Upon quantization, the Siegel invariance of this model also becomes anomalous. To cancel the anomaly, one must add to the model additional chiral fields. As in the Abelian case, the anomaly cancellation condition is precisely the string no-ghost condition.

ABELIAN CASE

**1. Fermi model**

Fermi-Bose equivalence can be discussed conveniently in terms of the generating functionals of the corresponding models. In order to construct the generating functional of the Fermi model, we consider the action for a single complex Weyl Fermion field $\psi(x) = \frac{1}{2}(1-\gamma_5)\psi(x)$, which is coupled to a background Abelian gauge field $A_\mu(x)$ and gravity (zweibein) $e_\mu{}^a(x)$,

$$S_F = \int d^2x\, e\, i\bar{\psi} e^\mu{}_a \gamma^a (\nabla_\mu - iA_\mu)\, \tfrac{1}{2}(1-\gamma_5)\psi\,, \tag{1a}$$

$$= \int d^2x\, e\, i\bar{\psi}\gamma^- D_-\psi\,. \tag{1b}$$

In writing the second line, we have used the notations *

$$x^\pm = \frac{1}{\sqrt{2}}\left(x^0 \pm x^1\right), \tag{2a}$$

$$D_- = e^\mu{}_- D_\mu = e^\mu{}_-(\nabla_\mu - iA_\mu)\,, \tag{2b}$$

as well as the identity

$$\gamma^a\gamma_5 = \epsilon^{ab}\gamma_b\,. \tag{2c}$$

By construction, this classical action is invariant under local gauge, local Lorentz, and general-coordinate tranoformations.

Functional integration defines the effective action $W_F$, the generating functional of the connected current and stress-energy correlation functions:

$$\exp\{iW_F\} = \int [d\psi] \exp\left\{iS_F[\psi, A_\mu, e_\mu{}^a]\right\}, \tag{3}$$

We shall always normalize the measure (in this case, $[d\psi]$) so that the functional integral is unity in the absence of the external potentials.

---

* We use a Minkowski metric, with $\eta_{00} = -1$, $\eta_{11} = 1$; also, $\epsilon^{01} = 1$. In light-cone coordinates, $\eta^{+-} = -1 = \epsilon^{+-}$. The $2\times 2$ dimensional Dirac matrices satisfy $\{\gamma^a, \gamma^b\} = -2\eta^{ab}$, $\gamma_5 = \gamma^0\gamma^1$.

It is well-known that $W_F$ is not invariant under local gauge transformations. Furthermore, as discussed in [7], it is possible to maintain local Lorentz symmetry, so that $W_F$ is in fact a functional of the metric tensor $g_{\mu\nu}$, rather than the zweibein $e_\mu{}^a$. However, $W_F$ is then not invariant under general-coordinate transformations [7,8]. Indeed, finite local counterterms can be added to the effective action, so that the anomalous variations under infinitesimal gauge and general-coordinate transformations are given by

$$\delta_\theta W_F = -\frac{1}{2\pi}\int d^2x\, \theta\partial_+ A^+ , \qquad \delta A_\mu = \partial_\mu \theta , \tag{4a}$$

$$\delta_\epsilon W_F = -\frac{1}{48\pi}\int d^2x\, h^{++}\partial_+^3 \epsilon^+ , \qquad \delta h_{\mu\nu} = \partial_\mu \epsilon_\nu + \partial_\nu \epsilon_\mu . \tag{4b}$$

Here we have set $g_{\mu\nu} = \eta_{\mu\nu} + h_{\mu\nu}$, and we have worked to lowest order in the background fields $A_\mu$ and $h_{\mu\nu}$. The higher-order terms follow from the Wess-Zumino consistency conditions [9,7].

From Eqs. (4), we see that the linearized effective action depends only on $A^+$ and $h^{++}$. (The same is true for the classical action, as can easily be seen by linearizing (1b) and using the free $\psi$ equations of motion.) Hence, Eqs. (4) uniquely determine the linearized effective action. Moreover, the full, non-linear effective action $W_F[A_\mu, g_{\mu\nu}]$ is completely determined by the Ward identities (4) together with the Wess-Zumino consistency conditions.

## 2. Bose model

We have seen that the effective action for a Weyl Fermion in $1+1$ dimensions is completely determined by the Ward identities (4). Hence, if we find a Bose model which obeys the same Ward identities, then we shall have established that the effective actions of the Fermi and Bose models are equal.

As a candidate for the Bose model, we first consider the classical action [5]

$$S = \int d^2x\; -\frac{1}{2}\partial_\mu\phi\partial^\mu\phi + \frac{1}{8}\left(\eta^\mu{}_\nu - \epsilon^\mu{}_\nu\right)\left(\eta^\rho{}_\sigma - \epsilon^\rho{}_\sigma\right)\lambda^{\nu\sigma}\partial_\mu\phi\partial_\rho\phi \tag{5a}$$

$$= \int d^2x\; \partial_+\phi\partial_-\phi + \frac{1}{2}\lambda^{--}\left(\partial_-\phi\right)^2 , \tag{5b}$$

where $\phi(x)$ is a real scalar field, and $\lambda^{\mu\nu}(x)$ is a real symmetric tensor field. Varying $\lambda^{--}$ and $\phi$ yields the classical equations of motion

$$\frac{1}{2}\left(\partial_-\phi\right)^2 = 0 , \tag{6a}$$

$$2\left(\partial_+\partial_-\phi\right) + \partial_-\left(\lambda^{--}\partial_-\phi\right) = 0 , \tag{6b}$$

respectively. The first equation implies $\partial_-\phi = 0$ ; and so the second equation is automatically satisfied. Since $\lambda^{--}$ drops out of the field equations, it must be a gauge degree of freedom. Indeed, the action (5) has the ("Siegel") invariance [5]

$$\delta\phi = \frac{1}{2}(\eta^{\mu\nu} - \epsilon^{\mu\nu})\xi_\nu\partial_\mu\phi = \xi^-\partial_-\phi , \tag{7a}$$

$$\delta\lambda^{--} = -2\partial_+\xi^- - \lambda^{--}\overleftrightarrow{\partial}_-\xi^- , \tag{7b}$$

where $\xi^\mu(x)$ is an infinitesimal vector. Thus, classically this model describes one chiral scalar, with corresponding current $\partial_+\phi$.

In order to quantize the model, it is necessary to fix the Siegel invariance. Choosing the gauge $\lambda^{--} = 0$ leads to the ghost Lagrangian [6]

$$\mathcal{L}_{\text{ghost}} = b^{++}\left(-2\partial_+c^- - \lambda^{--}\overleftrightarrow{\partial}_-c^-\right) . \tag{8}$$

Consider the generating functional $W[\lambda^{--}]$, defined by functional integration over $\phi$ and the ghosts:

$$\exp\{iW[\lambda^{--}]\} = \int [d\phi][d(\text{ghosts})]\exp\{i\left(S[\phi,\lambda^{--}] + S_{\text{ghost}}\right)\} . \tag{9}$$

To lowest order in $\lambda^{--}$, the generating functional is given by

$$\begin{aligned} W[\lambda^{--}] &= \frac{1}{2}\int d^2x \int d^2y \left.\frac{\delta^2 W}{\delta\lambda^{--}(x)\delta\lambda^{--}(y)}\right|_{\lambda=0}\lambda^{--}(x)\lambda^{--}(y) \\ &= \frac{1}{8}\int d^2x\int d^2y\, i\langle T^* U_{--}(x)U_{--}(y)\rangle\, \lambda^{--}(x)\lambda^{--}(y), \end{aligned} \tag{10}$$

where

$$U_{--}(x) = 2\frac{\delta S}{\delta\lambda^{--}(x)} = (\partial_-\phi)^2 - 2[2b^{++}\partial_-c^- + (\partial_-b^{++})\,c^-] . \tag{11}$$

By straightforward computation, one finds

$$i\langle T^* U_{--}(x)U_{--}(y)\rangle = \frac{1}{24\pi}(1-26)\,\frac{\partial_-^3}{\partial_+}\delta(x-y), \tag{12}$$

where the contribution $-26$ comes from the ghosts. Hence, under a Siegel transformation,

$$\delta_\xi W[\lambda^{--}] = \frac{25}{48\pi}\int d^2x\, \lambda^{--}\partial_-^3\xi^- \neq 0 . \tag{13}$$

That is, the Siegel symmetry is anomalous [6]. If the model is to describe only a chiral scalar, this anomaly must be cancelled.

We know of two ways of cancelling the Siegel anomaly. The first, proposed in Ref. [6], is to modify the Siegel action (5) by adding a new term which is linear in $\phi$:

$$S = \int d^2x\, \{\partial_+\phi\partial_-\phi + \frac{1}{2}\lambda^{--}(\partial_-\phi)^2 - \alpha\lambda^{--}\partial_-^2\phi\} . \tag{14}$$

The parameter $\alpha$ is yet to be determined. Under the transformations

$$\delta\phi = \xi^-\partial_-\phi + \alpha\partial_-\xi^- , \tag{15a}$$

$$\delta\lambda^{--} = -2\partial_+\xi^- - \lambda^{--}\overleftrightarrow{\partial}_-\xi^- , \tag{15b}$$

the action is not invariant:

$$\delta_\xi S = -\alpha^2 \int d^2x\, \lambda^{--}\partial_-^3\xi^- . \tag{16}$$

That is, the model has a tree-level Siegel "anomaly." Hence, this anomaly can be made to cancel against the one-loop anomaly (13), by choosing

$$\alpha^2 = 25/48\pi . \tag{17}$$

In this way, the generating functional $W[\lambda^{--}]$ becomes Siegel invariant. Now the functional integral over $\lambda^{--}$ is trivial, since nothing depends on $\lambda^{--}$ [6].

The modified Siegel model (14), (17) is clearly a candidate for a (Bosonized) Weyl Fermion. To determine whether this is indeed the case, we try to couple the model to backgrounds $A_\mu$ and $g_{\mu\nu}$ in such a way that the corresponding effective action obeys the Ward identities of the Fermion theory.

Since $\phi$ is real, it must shift under a gauge transformation:

$$\delta\phi = \theta , \qquad \delta A_\mu = \partial_\mu\theta . \tag{18}$$

The coupling of the model (14) to a background gauge field is described by the action

$$S = \frac{1}{4\pi}\int d^2x\, \{\partial_+\phi\partial_-\phi + \frac{1}{2}\lambda^{--}(D_-\phi)^2 - \alpha\lambda^{--}D_-^2\phi - 2\partial_+\phi A_-\} , \tag{19}$$

where $D_\mu\phi = \partial_\mu\phi - A_\mu$. Indeed, under a gauge transformation (18), this action has the "anomalous" response

$$\delta_\theta S = -\frac{1}{2\pi}\int d^2x\, \theta\partial_+ A^+ . \tag{20}$$

In (19), we have rescaled the action by thc factor $1/4\pi$, so that the anomaly (20) coincides with that of the Fermi model (4a). Moreover, under the covariantized Siegel transformations

$$\delta\phi = \xi^- D_-\phi + \alpha\partial_-\xi^- , \tag{21a}$$

$$\delta\lambda^{--} = -\,2\partial_+\xi^- - \lambda^{--}\overleftrightarrow{\partial}_-\xi^- , \tag{21b}$$

we see that

$$\delta_\xi S = -\frac{\alpha^2}{4\pi}\int d^2x\, \lambda^{--}\partial_-^3\xi^- . \tag{22}$$

The quantum theory is described by the Bosonic effective action $W_B$ defined by

$$\exp\{iW_B[\lambda^{--},A^+]\} = \int [d\phi][d(\text{ghosts})]\exp\{i\,(S[\phi,\lambda^{--},A^+] + S_{\text{ghost}})\} . \tag{23}$$

By suitably adjusting the value of $\alpha$, it follows that

$$\delta_\xi W_B = 0\,, \tag{24a}$$

$$\delta_\theta W_B = -\frac{1}{2\pi}\int d^2x\ \theta\partial_+ A^+\,. \tag{24b}$$

(There is no additional contribution to the gauge anomaly (24b) from the functional integral measure.) Since the Fermion and Boson effective actions have the same gauge variations, the two models have identical current correlation functions.

Coupling to background gravity is more difficult. With the help of zweibeins, the original (unmodified) Siegel action (5) can be made general-coordinate invariant [10,6]:

$$S = \int d^2x\ \sqrt{-g}\{-\frac{1}{2}g^{\mu\nu}\partial_\mu\phi\partial_\nu\phi + \frac{1}{2}\lambda^{--}e^\mu{}_-e^\nu{}_-\partial_\mu\phi\partial_\nu\phi\}\,. \tag{25}$$

This action is also invariant under the Siegel transformations

$$\delta\phi = \xi^- D_-\phi\,, \tag{26a}$$

$$\delta\lambda^{--} = -2D_+\xi^- - \lambda^{--}\overleftrightarrow{D}_-\xi^-\,, \tag{26b}$$

where $\xi^- = e_\mu{}^-\xi^\mu$, $D_\pm = e^\mu{}_\pm D_\mu$, and $D_\mu$ is the general-coordinate covariant derivative.

Upon quantization, the Siegel symmetry becomes anomalous. In the approach of Ref. [6], the action must reduce to (14) in the flat-space limit; hence, one might guess that Siegel invariance can be restored by adding the term $-\alpha\lambda^{--}D_-^2\phi$ to the Lagrangian (25). However, as we shall see, this is incorrect.

Let us consider the simpler problem of determining the *linearized* couplings to gravity. That is, we set $g_{\mu\nu} = \eta_{\mu\nu} + h_{\mu\nu}$, and work to leading order in the background fields $\lambda^{--}$ and $h^{\mu\nu}$. As a trial, we take the action

$$S = \int d^2x\ \{\partial_+\phi\partial_-\phi + \frac{1}{2}h^{++}(\partial_+\phi)^2 + \frac{1}{2}h^{--}(\partial_-\phi)^2 + \frac{1}{2}\lambda^{--}(\partial_-\phi)^2 \\ -\alpha\lambda^{--}\partial_-^2\phi - \beta h^{--}\partial_-^2\phi\}\,. \tag{27}$$

This is the linearization of (25), plus the term (with coefficient $\alpha$) introduced in [6], plus a new term (with coefficient $\beta$) which is also linear in $\phi$. This new term will be needed to cancel a curved-space Siegel anomaly. Correspondingly, the Siegel transformations are given by the linearization of (26):

$$\delta\phi = \xi^-\partial_-\phi\,, \tag{28a}$$

$$\delta\lambda^{--} = -2\partial_+\xi^- - \lambda^{--}\overleftrightarrow{\partial}_-\xi^- - h^{--}\overleftrightarrow{\partial}_-\xi^-\,. \tag{28b}$$

As before, we quantize in the gauge $\lambda^{--} = 0$. From the expression (28b) for $\delta\lambda^{--}$, we see that $h^{--}$ couples to the ghosts in the same way as $\lambda^{--}$. Consider now the effective action $W[\lambda^{--}, h^{\mu\nu}]$, found by functionally integrating over $\phi$ and the ghosts:

$$\exp\{iW[\lambda^{--}, h^{\mu\nu}]\} = \int [d\phi][d(\text{ghosts})]\exp\{i\left(S[\phi, \lambda^{--}, h^{\mu\nu}] + S_{\text{ghost}}\right)\}\,. \tag{29}$$

To second order in the background fields,

$$W[\lambda^{--},h^{\mu\nu}] = \frac{1}{8}\int d^2x \int d^2y \; \{i\langle T^* U_{--}(x) U_{--}(y)\rangle \; \lambda^{--}(x)\lambda^{--}(y)$$
$$+2i\langle T^* U_{--}(x) T_{\mu\nu}(y)\rangle \; \lambda^{--}(x) h^{\mu\nu}(y) + i\langle T^* T_{\mu\nu}(x) T_{\rho\sigma}(y)\rangle \; h^{\mu\nu}(x) h^{\rho\sigma}(y)\}\,, \tag{30}$$

where

$$\begin{aligned} U_{--} =& 2\frac{\delta S}{\delta\lambda^{--}} = (\partial_-\phi)^2 - 2\alpha\partial_-^2\phi + \text{ghosts}\,, \\ T_{--} =& 2\frac{\delta S}{\delta h^{--}} = (\partial_-\phi)^2 - 2\beta\partial_-^2\phi + \text{ghosts}\,, \\ T_{++} =& 2\frac{\delta S}{\delta h^{++}} = (\partial_+\phi)^2\,, \qquad T_{+-} = 0\,. \end{aligned} \tag{31}$$

In order that $W$ be Siegel invariant, the terms in (30) involving $\lambda^{--}\lambda^{--}$ and $\lambda^{--}h^{\mu\nu}$ must be local. We have already seen that the $\lambda^{--}\lambda^{--}$ term is non-local, unless $\alpha^2 = 25/48\pi$. Similarly, it is easy to see that

$$i\langle T^* U_{--}(x) T_{--}(y)\rangle = \left(\frac{1}{24\pi} + 2\alpha\beta - \frac{26}{24\pi}\right) \frac{\partial_-^3}{\partial_+}\delta(x-y)\,. \tag{32}$$

Hence, we require $\beta = \alpha \neq 0$; this explains the need for the new term in (27). Moreover, with this choice for $\beta$, it follows that

$$i\langle T^* T_{--}(x) T_{--}(y)\rangle = \left(\frac{1}{24\pi} + 2\beta^2 - \frac{26}{24\pi}\right) \frac{\partial_-^3}{\partial_+}\delta(x-y) = 0\,. \tag{33}$$

Finally,

$$i\langle T^* T_{++}(x) T_{++}(y)\rangle = \frac{1}{24\pi}\,\frac{\partial_+^3}{\partial_-}\delta(x-y)\,. \tag{34}$$

Putting all this together, we see that $W[\lambda^{--},h^{\mu\nu}]$ satisfies

$$\delta_\xi W = 0\,, \tag{35a}$$

$$\delta_\epsilon W = -\frac{1}{48\pi}\int d^2x \; h^{++}\partial_+^3\epsilon^+\,, \qquad \delta h_{\mu\nu} = \partial_\mu\epsilon_\nu + \partial_\nu\epsilon_\mu\,. \tag{35b}$$

That is, $W$ is Siegel invariant, and has the same anomalous variation under general-coordinate transformations as the effective action of a Weyl Fermion, at the linearized level.

To couple to both gauge and linearized gravity backgrounds is now easy:

i. Starting with Eq. (27), replace $\partial_-\phi$ by $D_-\phi = (\partial_-\phi - A_-)$;

ii. add the term $\partial_+\phi A^+$; and

iii. rescale the action by $1/4\pi$.

Clearly, the corresponding effective action $W_B[\lambda^{--}, A^+, h^{++}]$ is Siegel invariant, and obeys the same gauge and gravitational Ward identities as the Weyl Fermion effective action (4). Hence, at the *linearized* level, the two effective actions are equal

$$W_B[\lambda^{--} = 0, A^+, h^{++}] = W_F[A^+, h^{++}]\,. \tag{36}$$

This is still not a demonstration of quantum equivalence of the two models, which would require equality of the effective actions at the full non-linear level. Because of the term $-\beta h^{--}\partial_-^2\phi$ in the Bosonic action (27), it is not clear how the couplings to gravity can be generalized to the full non-linear level.

Thus far, we have discussed the modification of the Siegel action proposed in Ref. [6] to cancel the Siegel anomaly. There is a second way to cancel this anomaly, which avoids the difficulties with coupling to gravity. Instead of a single chiral scalar, we consider a set of $d$ such scalars $\phi^\alpha$, $\alpha = 1, \ldots, d$, with action

$$S = \frac{1}{4\pi}\int d^2x\ \partial_+\phi^\alpha\partial_-\phi^\alpha + \frac{1}{2}\lambda^{--}\partial_-\phi^\alpha\partial_-\phi^\alpha\,. \tag{37}$$

From (12), we see that the Siegel anomaly is proportional to $d-26$; hence, for $d = 26$, the anomaly is absent. Making use of our previous results, it is now straightforward to couple this model to gauge and gravitational backgrounds:

$$\begin{aligned} S = \frac{1}{4\pi}\int d^2x\ \sqrt{-g}\{-\frac{1}{2}g^{\mu\nu}\partial_\mu\phi^\alpha\partial_\nu\phi^\alpha + \frac{1}{2}\lambda^{--}e^\mu{}_-e^\nu{}_-(D_\mu\phi)^\alpha(D_\nu\phi)^\alpha \\ +(g^{\mu\nu} + \frac{\epsilon^{\mu\nu}}{\sqrt{-g}})\partial_\mu\phi^\alpha A^\alpha_\nu\}\,. \end{aligned} \tag{38}$$

where $(D_\mu\phi)^\alpha = \partial_\mu\phi^\alpha - A^\alpha_\mu$. Hence, the corresponding effective action obeys the Ward identities and consistency conditions for a set of 26 Weyl Fermions. We conclude that the two models have identical effective actions, and are therefore equivalent.

NON-ABELIAN CASE

## 1. Bose model

There exists a non-Abelian generalization of the Siegel model (5), which describes a free chiral Lie-algebra-valued current. The action is given by [4]

$$S = -\frac{1}{8\pi}\int d^2x\ tr\{\partial_+gg^{-1}\partial_-gg^{-1} + \frac{1}{2}\lambda^{--}\left(g^{-1}\partial_-g\right)^2\} - \frac{1}{24\pi}\int_{\mathcal{N}} tr\left(g^{-1}dg\right)^3, \tag{39}$$

where $g(x)$ is a matrix field in some real, orthogonal representation of a compact, semi-simple group $G$. Moreover, $\mathcal{N}$ is a three-dimensional manifold, whose boundary is spacetime. The action therefore consists of the Wess-Zumino-Witten (WZW) action [2], with an additional coupling to the field $\lambda^{--}(x)$. To see that this model indeed describes only one (chiral) current, consider the $\lambda^{--}$ equation of motion,

$$\frac{1}{2}\,tr\left(g^{-1}\partial_-g\right)^2 = 0\,. \tag{40}$$

Since the group $G$ is compact, this implies

$$g^{-1}\partial_-g = 0\,. \tag{41}$$

Hence, only a single field $\lambda^{--}$ is needed to set an entire Lie-algebra-valued current to vanish. The remaining current which the model describes is $\partial_+ gg^{-1}$.

As in the Abelian case, the action (39) does not provide dynamics for the field $\lambda^{--}$, as it is a gauge degree of freedom. The gauge transformations under which the action is invariant are

$$\delta g = \xi^- \partial_- g\,, \tag{42a}$$

$$\delta\lambda^{--} = -\,2\partial_+\xi^- - \lambda^{--}\overleftrightarrow{\partial}_-\xi^-\,. \tag{42b}$$

These are a direct generalization of the Siegel transformation laws (7).

It should be noted that the Wess-Zumino term in the action (39) is separately invariant under the transformations (42). As we shall see, the coefficient of the Wess-Zumino term is determined by the requirement that the model should couple correctly to a background vector gauge field. It should also be noted that, on setting $g = exp(i\sigma_2\phi) \in SO(2)$, the action (39) reduces to the Siegel action (5), rescaled by the factor $1/4\pi$.

When this model (39) is quantized, the Siegel symmetry (42) becomes anomalous, as in the Abelian case. However, in order to cancel the anomaly, we cannot generalize the approach of Ref. [6], which consists of adding another term to the action. Indeed, the term one would add is

$$-\alpha\lambda^{--}\ tr\ \partial_-\left(g^{-1}\partial_- g\right),$$

which is identically zero, since $g^{-1}\partial_- g$ is Lie-algebra-valued. Hence, in order to cancel the Siegel anomaly, we must follow the second approach : namely, we must add to the model additional chiral fields.

As an example, consider the model describing a chiral Lie-algebra-valued current $\partial_+ gg^{-1}$ and a set of $d$ chiral Abelian currents $\partial_+\phi^\alpha$, $\alpha = 1,\ldots, d$, with corresponding actions (39) and (37), respectively. Again we fix the Siegel symmetry with the gauge choice $\lambda^{--} = 0$ , and we consider the generating functional $W|\lambda^{--}|$ defined by

$$\exp\{iW[\lambda^{--}]\} = \int [d\phi^\alpha][dg][d(\text{ghosts})]\exp\{i\left(S[\phi^\alpha,\ \lambda^{--}] + S[g,\ \lambda^{--}] + S_{\text{ghost}}\right)\}\,. \tag{43}$$

The generating functional can again be expanded as in Eq. (10). For the case that the group $G$ is simple, the two-point function of $U_{--} \equiv 2\,\delta S/\delta\lambda^{--}$ is given by

$$i\langle T^* U_{--}(x)U_{--}(y)\rangle = \frac{1}{24\pi}\left(d + c_G - 26\right)\frac{\partial_-^3}{\partial_+}\delta(x-y)\,, \tag{44a}$$

where

$$c_G = d_G/\left(1 + C_A/\kappa\right)\,. \tag{44b}$$

Here, $d_G = \dim G$, and $\kappa$ and $C_A$ are defined in terms of the generators $T_a$ of $G$ as follows:

$$[T_a, T_b] = i f_{abc} T_c \quad , \quad T_a^\dagger = T_a$$
$$tr(T_a T_b) = \kappa \delta_{ab} \quad , \quad f_{acd} f_{bcd} = C_A \delta_{ab} \, . \tag{45}$$

(See Refs. 11 and 3.) For a level-one representation of a simply-laced group, the quantity $c_G$ is equal to the rank of the group. If $G$ is a direct product of simple groups, $c_G$ is given by a sum of terms (44b), one such term for each simple factor.

From (44), we learn that the Siegel anomaly is absent, provided that the condition

$$d + c_G - 26 = 0 \tag{46}$$

is satisfied. This is precisely the no-ghost condition for a string on a group manifold [11]. Moreover, we observe that for a chiral Lie-algebra-valued current, there is a restriction

$$c_G \leq 26 \, . \tag{47}$$

In contrast, for the non-chiral case [2], there is no such restriction.

Now let us consider the coupling of the model (39) to a background (anti-Hermitian) non-Abelian gauge field $A_\mu$. The correct action is given by

$$S = -\frac{1}{8\pi} \int d^2x \; tr\{\partial_+ g g^{-1} \partial_- g g^{-1} + \frac{1}{2} \lambda^{--} \left(g^{-1} D_- g\right)^2 - 2 \partial_+ g g^{-1} A_-\}$$
$$- \frac{1}{24\pi} \int_{\mathcal{N}} tr \left(g^{-1} dg\right)^3 \, , \tag{48}$$

where $D_\mu g = \partial_\mu g - A_\mu g$. Indeed, under a gauge transformation

$$\delta g = a g \, , \tag{49a}$$
$$\delta A_\mu = \partial_\mu a + [a, A_\mu] \, , \tag{49b}$$

the action (48) has the anomalous response

$$\delta_a S = -\frac{1}{4\pi} \int d^2x \; tr \; a \partial_+ A_- \, . \tag{50}$$

As we shall see below, this coincides with the gauge anomaly of a set of Majorana-Weyl Fermions $\psi_-$, transforming as $\delta\psi_- = a\psi_-$. The (properly normalized) Wess-Zumino term in (48) is crucial for obtaining this result. Moreover, the action is invariant under the covariantized Siegel transformations, with $\delta g = \xi^- D_- g$, and $\delta\lambda^{--}$ as in (42b).

It is easy to couple also to background gravity:

$$S = -\frac{1}{8\pi} \int d^2x \, \sqrt{-g} \; tr \; \{-\frac{1}{2} g^{\mu\nu} \partial_\mu g g^{-1} \partial_\nu g g^{-1} + \frac{1}{2} \lambda^{--} e^\mu{}_- e^\nu{}_- g^{-1} D_\mu g g^{-1} D_\nu g$$
$$+ (g^{\mu\nu} + \frac{\epsilon^{\mu\nu}}{\sqrt{-g}}) \partial_\mu g g^{-1} A_\nu\} - \frac{1}{24\pi} \int_{\mathcal{N}} tr \left(g^{-1} dg\right)^3 \, . \tag{51}$$

Functional integration over the dynamical variables $g$, the "spectator" chiral fields $\phi^\alpha, \alpha = 1, \ldots, d$, and the ghosts, leads to the generating functional $W_B[\lambda^{--}, A_\mu, g_{\mu\nu}]$, which at the linearized level satisfies

$$\delta_\xi W_B = 0\,, \tag{52a}$$

$$\delta_a W_B = \frac{1}{4\pi}\int d^2x\ tr\ a\partial_+ A^+\,, \tag{52b}$$

$$\delta_\epsilon W_B = -\frac{1}{48\pi}(d + c_G)\int d^2x\ h^{++}\partial_+^3\epsilon^+\,. \tag{52c}$$

The higher-order terms follow from the Wess-Zumino consistency conditions. As in the Abelian case, these Ward identities completely determine $W_B$.

**2. Fermi model**

What is the corresponding Fermionized model, which reproduces the Ward identities given by Eqs. (52) ? Naively, one would suppose that it consists of a set of Majorana-Weyl spinors $\psi$ (the group index carried by the spinor is suppressed), as well as the $d$ chiral "spectators" $\phi^\alpha$. (We are interested here only in non-Abelian Bosonization; thus, we do not Fermionize the spectators, although of course we could if we so wished.) However, it is easy to see that this would be incorrect. The chiral spectators $\phi^\alpha$ couple to the Siegel field $\lambda^{--}$, as in (37). Hence, in order to avoid a Siegel anomaly, the spinors $\psi$ must also couple to $\lambda^{--}$. But, it is not possible to achieve such a coupling: $\lambda^{--}$ couples to $T_{--}$; and for Weyl spinors $\psi$ obeying $\psi = \frac{1}{2}(1-\gamma_5)\psi$, $\partial_-\psi = 0$, the stress-tensor component $T_{--}$ is identically zero!

The resolution of this puzzle is to take the spinors $\psi$ to be *Majorana*, instead of Majorana-Weyl. Now $T_{--}$ does not vanish, and a coupling to $\lambda^{--}$ can be achieved. Indeed, consider the model

$$S = \int d^2x\ \frac{i}{2}\{\bar\psi\gamma^\mu\partial_\mu\psi - \frac{1}{2}\lambda^{--}\bar\psi\gamma_-\partial_-\psi\} \tag{53a}$$

$$= i\frac{\sqrt{2}}{2}\int d^2x\ \{\psi_+^T\partial_+\psi_+ + \psi_-^T\partial_-\psi_- + \frac{1}{2}\lambda^{--}\psi_+^T\partial_-\psi_+\}\,, \tag{53b}$$

where $\psi_\pm = \frac{1}{2}(1 \pm \gamma_5)\psi$.

Observe that only $\psi_+$ couples to the Siegel field $\lambda^{--}$. The classical equations can be shown to imply $\psi_+ =$ constant; only $\psi_-$ propagates, with $\partial_-\psi_- = 0$. Correspondingly, the action is invariant under the transformations

$$\delta\psi_- = 0\,, \tag{54a}$$

$$\delta\psi_+ = \xi^-\partial_-\psi_+ + \frac{1}{2}(\partial_-\xi^-)\psi_+\,, \tag{54b}$$

$$\delta\lambda^{--} = -2\partial_+\xi^- - \lambda^{--}\overleftrightarrow{\partial}_-\xi^-\,. \tag{54c}$$

Remarkably, similar models have already been considered [5, 12], although for entirely different reasons (namely, supersymmetry.)

For convenience, we choose the Fermi and Bose models to contain the same representation of the group $G$. Hence, if the matrix field $g(x)$ has dimension $N \times N$, then the (real, orthogonal) group representation carried by $\psi(x)$ has dimension $N$. For the model consisting of the $N$ Majorana spinors with action given by (53), plus the $d$ chiral spectators $\phi^\alpha$ and the ghosts, we find that

$$i\langle T^* U_{--}(x) U_{--}(y)\rangle = \frac{1}{24\pi}\left(d + \frac{N}{2} - 26\right) \frac{\partial_-^3}{\partial_+}\delta(x-y) . \tag{55}$$

Comparing with the corresponding results (44), (46) for the Bosonic model, we see that the Siegel anomaly cancels for

$$c_G = \frac{N}{2} . \tag{56}$$

This condition also appears in the case of non-chiral Bosonization. (See, e.g., Ref. [3].)

It is completely straightforward to couple the model (53) to background gauge and gravitational fields. Corresponding to the gauge transformation (49), the Fermions transform according to

$$\delta\psi_- = a\psi_- , \qquad \delta\psi_+ = 0 . \tag{57}$$

Hence, the gauge and general-coordinate invariant action is

$$S = \int d^2x \; e \; \frac{i}{2}\{\bar\psi e^\mu{}_a \gamma^a [\partial_\mu - \tfrac{1}{2}(1-\gamma_5)A_\mu]\psi - \frac{1}{2}\lambda^{--}\bar\psi\gamma_- e^\mu{}_- \partial_\mu \psi\} . \tag{58}$$

Functional integration over the Fermion fields, the spectators, and the ghosts now leads to the generating functional $W_F[\lambda^{--}, A_\mu, g_{\mu\nu}]$, which at the linearized level satisfies the Ward identities

$$\delta_\xi W_F = 0 , \tag{59a}$$

$$\delta_a W_F = \frac{1}{4\pi}\int d^2x \; tr \; a\partial_+ A^+ , \tag{59b}$$

$$\delta_\epsilon W_F = -\frac{1}{48\pi}\left(d + \frac{N}{2}\right)\int d^2x \; h^{++}\partial_+^3 \epsilon^+ . \tag{59c}$$

Comparing with the Bose Ward identities (52), we conclude that the Bose and Fermi generating functionals are equal,

$$W_B[\lambda^{--}, A_\mu, g_{\mu\nu}] = W_F[\lambda^{--}, A_\mu, g_{\mu\nu}] , \tag{60}$$

provided that the condition (56) is satisfied.

DISCUSSION

We have seen that chiral Bosonization is more restrictive than non-chiral Bosonization. In particular, only certain multiplets of chiral fields (e.g., 26 chiral scalars) can be readily Fermionized. Remarkably, the condition that the Siegel anomaly should vanish is automatically satisfied in string theory, since it coincides with the no-ghost condition. Indeed, it would appear that chiral Bosonization is natural only within a string context.

One can contemplate a variety of applications of chiral Bosonization to string theory. For instance, consider the Fermionic formulation of the $E_8 \times E_8$ heterotic string [10], in which the $O(16) \times O(16)$ subgroup is realized linearly on a set of 32 free Majorana-Weyl Fermions. Generalizing the arguments presented here, it should be possible to demonstrate the equivalence of this Fermionic system to a Bose $E_8 \times E_8$ sigma model, as in Eq. (39). This Bosonic formulation has the advantage that the full $E_8 \times E_8$ symmetry is realized linearly. Another application of chiral Bosonization is the demonstration [13] of the equivalence of the Ramond-Neveu-Schwarz and the Green-Schwarz superstring formalisms.

We have been concerned only with local properties of corresponding Fermi and Bose models, and have therefore taken the two-dimensional spacetime (worldsheet) to be a plane. Recently, there has been much interest in Abelian Bosonization on worldsheets which are higher genus Riemann surfaces, both for the non-chiral [14] and chiral [15] cases. These investigations have relied on operator methods. It would be interesting to extend the present Lagrangian analysis to such worldsheets of more complicated topology.

ACKNOWLEDGMENTS

The work described here, based on Ref. [4], was done in collaboration with L. Mezincescu. I am grateful to the organizers of the Workshop, P.G.O. Freund and K.T. Mahanthappa, for the opportunity to present these results. This work was supported in part by the National Science Foundation under Grant No. PHY-87 03390.

REFERENCES

1. T. Skyrme, Proc. R. Soc. **A262** (1961) 237, J. Math. Phys. **12** (1971) 1735; S. Coleman, Phys. Rev. **D11** (1975) 2088; S. Mandelstam, Phys. Rev. **D11** (1975) 3026.
2. E. Witten, Comm. Math. Phys. **92** (1984) 455.
3. L.S. Brown and R.I. Nepomechie, Phys. Rev. **D35** (1987) 3239; L.S. Brown, G.J. Goldberg, C.P. Rim, and R.I. Nepomechie, Phys. Rev. **D36** (1987) 551.
4. L. Mezincescu and R.I. Nepomechie, "Critical Dimensions for Chiral Bosons," UMTG-140
5. W. Siegel, Nucl. Phys. **B238** (1984) 307.
6. C. Imbimbo and A. Schwimmer, Phys. Lett. **193B** (1987) 455; J. Labastida and M. Pernici, "On the BRST Quantization of Chiral Bosons," IASSNS-HEP-87/29.
7. W. Bardeen and B. Zumino, Nucl. Phys. **B244** (1984) 421.
8. L. Alvarez-Gaumé and E. Witten, Nucl. Phys. **B234** (1983) 269.

9. J. Wess and B. Zumino, Phys. Lett. **37B** (1971) 95.
10. D. Gross, J. Harvey, E. Martinec, and R. Rohm, Nucl. Phys. **B256** (1985) 253; Nucl. Phys. **B267** (1986) 75.
11. V. Knizhnik and A. Zamolodchikov, Nucl. Phys. **B 247** (1984) 83; P. Goddard and D. Olive, Nucl. Phys. **B 257** [FS14] (1985) 226; I. Todorov, Phys. Lett. **153 B** (1985) 77; D. Nemeschansky and Y. Yankielowicz, Phys. Rev. Lett. **54** (1985) 620; S. Jain, R. Shankar, and S. Wadia, Phys. Rev. **D 32** (1985) 2713; E. Bergshoeff, S. Randjbar - Daemi, A. Salam, H. Sarmadi, and E. Sezgin, Nucl. Phys. **B269** (1986) 77; A. Redlich and H. Schnitzer, Phys. Lett. **167B** (1986) 315; A. Redlich, H. Schnitzer, and K. Tsokos, Nucl. Phys. **B291** (1987) 429; D. Gepner and E. Witten, Nucl. Phys. **B278** (1986) 493; A. Eastaugh, L. Mezincescu, E. Sezgin and P. van Nieuwenhuizen, Phys. Rev. Lett. **57** (1986) 29; A. Ceresole, A. Lerda, P. Pizzochero, and P. van Nieuwenhuizen, Phys. Lett. **B** ; R.I. Nepomechie, Phys. Rev. **D33** (1986) 3670; M. Duff, B. Nilsson, C. Pope, and N. Warner, Phys. Lett. **171B** (1986) 170.
12. M.T. Grisaru, L. Mezincescu, and P.K. Townsend, Phys. Lett. **179B** (1986) 247; R. Brooks, S.J. Gates, and F. Muhammad, "Unidexterous Superspace: The flax of (super)strings."
13. E. Witten, in: Fourth Workshop on Grand Unification, ed. A. Weldon, P. Langacker and P. Steinhardt (Birkhäuser, Basel, 1983); P. Goddard, D.I. Olive, and A. Schwimmer, Phys. Lett. **157B** (1985) 393; R.I. Nepomechie, Phys. Lett. **178B** (1986) 207; *ibid* **180B** (1986) 423.
14. L. Alvarez-Gaumé, J.B. Bost, G. Moore, P. Nelson, and C. Vafa, Phys. Lett. **178B** (1986) 41; "Bosonization on Higher Genus Riemann Surfaces," HUTP-86/A062.
15. T. Eguchi and H. Ooguri, Phys. Lett. **187B** (1987) 127; E. Verlinde and H. Verlinde, Nucl. Phys. **B288** (1987) 357.

# SUPERSTRING ANOMALIES

Paul H. Frampton
Institute of Field Physics
Department of Physics and Astronomy
University of North Carolina
Chapel Hill, NC 27514

## ABSTRACT

The two-loop anomalies of the SO(32) open superstring are partially analyzed.* In particular, it is shown that there is no direct generalization of the Adler-Bardeen theorem since one counterterm which was absent at one loop level becomes necessary at two-loops. It is also shown how cancellation of the leading gauge anomaly at two loops involves diagrams with closed-string intermediate states.

In four spacetime dimensions, gauge theories coupled to chiral fermions have the property that the chiral anomaly calculated at one-loop order is not changed by higher-loop corrections(2). This remarkable fact can be understood in several ways: Feynman diagrams, path integrals, or differential geometry. In higher numbers of spacetime dimensions, there are at least two reasons to suspect that the Adler-Bardeen (AB) theorem may not generalize: first, there are more invariants involved than in four dimensions; second, the Feynman diagram for the one-loop anomaly is such that the simplest radiative corrections are not all equivalent to propagator and vertex corrections as they are for the triangle graph in four dimensions.

Nevertheless, in closed superstring theories there exist arguments analogous to the AB theorem. In E(8)xE(8) heterotic string theory there is modular invariance to all orders and anomalies which are absent do not appear

*This work was done in collaboration with T. W. Kephart and T.-C. Yuan. (Ref.1)

at higher genus since they would correspond to a violation of modular invariance[3]. In such a closed string theory there is only one diagram for each number of loops (genus). For the O(32) open superstring[4], on the other hand, life is more complicated: there are several diagrams at each order and no generalized Adler-Bardeen theorem is to be expected since there is no full modular invariance.

In the open superstring theory at one loop, using STr and $\Lambda$ for symmetized trace and generators in the adjoint of O(N), and Str and $\lambda$ in the defining representation of O(N) one finds for the hexagon diagram the well-known relationship

$$\mathrm{STr}\,\Lambda^6 = (N-32)\mathrm{Str}\lambda^6 + 15S(\mathrm{tr}\lambda^4\mathrm{tr}\lambda^2) \qquad (1)$$

and, in particular, the leading gauge anomaly vanishes if N=32.

At the two-loop order there are two classes of diagrams:

Class A: diagrams with only OPEN internal propagators.

Class B: diagrams with CLOSED internal propagators.

The Class A diagrams can all be represented with respect to a skeleton diagram (Fig. 1)

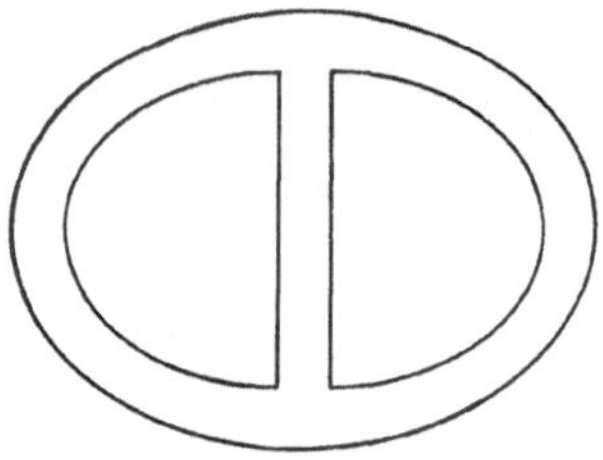

Figure I.

To this skeleton, one obtains the general hexagon diagram by attaching six external legs to the three propagators. The six may be partitioned in seven ways 600, 510, 411, 420, 330, 321, 222. One then sums over all possible twists (a +ve sign for no twist, a -ve sign for a twist). For each partition (abc), the sum over twists gives

$$V_A^{abc} = \sum_i H_{A_i}^{abc} Z_{A_i}^{abc} \qquad (2)$$

where $H_{A_i}^{abc}$ is a group theoretic factor which is easy to calculate and which gives useful information.

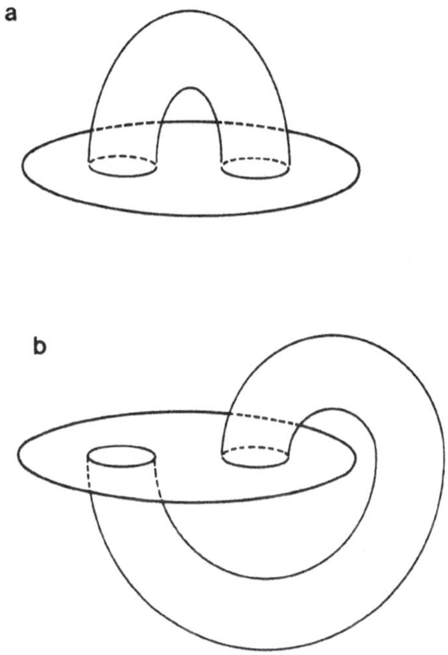


Figure 2.

For Class B diagrams there may only one internal closed string propagator (because $\kappa \sim g^2$). There are two such diagrams: Fig. 2a shows the orientable pot lid and Fig. 2b shows the non-orientable Klein pot-lid. Each of these two contributions factors as $V_B = H_B Z_B$ where $H_B$ is the group factor.

At one-loop level, the Green-Schwarz mechanism depends on the antisymmetric tensor field $B_{\mu\nu}$ in the non-leading anomaly, $(\mathrm{tr}F^2)(\mathrm{tr}F^4)$. At two-loops, even the cancellation of the leading gauge anomaly ($\mathrm{tr}F^6$) needs the presence of the supergravity multiplet in diagrams of Class B.

Now we can look at the results of the calculations of the group theory factors. The Class B diagrams both give $N\mathrm{tr}\lambda^n$ for O(N) and n=4 or 6 for d=6 or 10 respectively. The Class A diagrams give the results shown in Table I for the d=6 box diagram, using standard calculational methods (5,6).

Table I

| $H^{abc}$ | $\mathrm{tr}\lambda^4$ | $(\mathrm{tr}\lambda^2)^2$ |
|---|---|---|
| $H^{400}_{(A)}$ | 2(N-2)(N-8) | 6(N-2) |
| $H^{310}_{(A)}$ | (N-4)(N-8) | 3(N-4) |
| $H^{220}_{(A)}$ | $N^2$-10N+32 | 3N-14 |
| $H^{211}_{(A)}$ | -8(N-4) | 4(N-4) |
| $H_{(A)}$ | $4N^2$-50N+128 | 8(2N-7) |
| $H_{(B)}$ | 2N | 0 |
| $H^{(box)}=H_{(A)}+H_{(B)}$ | 4(N-4)(N-8) | 8(2N-7) |

We see that the $\mathrm{tr}\lambda^4$ term does not vanish for any N of O(N) for either Class A or Class B separately but does for the sum H when N=8, the value of N for which the one-loop anomaly cancelled. (The other solution N=4 must be irrelevant since for this choice of N there is a one-loop anomaly).

The $\mathrm{tr}\lambda^4$ terms for $H^{400}_{(A)}$ and $H^{310}_{(A)}$ both vanish for N=8; this is because they are a propagator insertion, and vertex insertion, respectively to the one-loop graph.

The Class A contributions to the two-loop hexagon anomaly give the polynomial factors listed in Table II. Class B terms and the total $H^{(hex)}$ are also given.

Table II

| $H^{abc}_{(A)}$ | $tr\lambda^6$ | $tr\lambda^2 tr\lambda^4$ | $(tr\lambda^2)^3$ |
|---|---|---|---|
| $H^{600}_{(A)}$ | $2(N-2)(N-32)$ | $30(N-2)$ | 0 |
| $H^{510}_{(A)}$ | $(N-4)(N-32)$ | $15(N-4)$ | 0 |
| $H^{420}_{(A)}$ | $N^2-22N + 128$ | $9N-74$ | 6 |
| $H^{330}_{(A)}$ | $N^2-18N + 128$ | $6(N-13)$ | 9 |
| $H^{411}_{(A)}$ | $-20N + 128$ | $2(5N-38)$ | 6 |
| $H^{321}_{(A)}$ | $-14N + 128$ | $7N-82$ | 9 |
| $H^{222}_{(A)}$ | $-12N + 128$ | $6(N-14)$ | 10 |
| $H_{(A)}$ | $5N^2-190N + 896$ | $83N-514$ | 40 |
| $H_{(B)}$ | $2N$ | 0 | 0 |
| $H^{(hex)} = H_{(A)} + H_{(B)}$ | $(5N-28)(N-32)$ | $83N - 514$ | 40 |

The coefficient of the $tr\lambda^6 \sim trF^6$ term factors. Thus we conclude that the leading gauge anomaly vanishes for an SO(32) gauge group. It is also evident that $B_{\mu\nu}$ field is needed if this leading anomaly coefficient at two-loops is to vanish for SO(32).

The group factor $40(tr\lambda^2)^3$ can be generated by an anomaly $\sim (trF^2)^3$. Such a term can not appear in one-loop group traces of adjoints but can always appear in two and higher loops. This is the only new tensor invariant available beyond one-loop in D=10. To cancel this non-leading term a further counterterm of the form $B(trF^2)^2$ needs to be introduced. A sketch of the complete two-loop counterterms is given at the end.

A final remark concerning $H^{(hex)}$ is in order. In analogy with previous work in string theory, we assume all seven class (A) and the two class (B) diagrams must be added with equal weight to preserve unitarity[7,8]. It is gratifying that when this is done in D dimensions we get precisely the same $(N - 2^{D/2})$ factor to cancel the leading gauge anomaly as was obtained at one loop.

Assuming that the leading gravitational anomaly cancels at one loop and fixing N=32 for simplicity we find (up to an overall normalization) the 12-form associated with the total two-loop anomaly must take the form

$$\begin{aligned} I_{12} = \; & a_1 trF^4 trF^2 + b_1(trF^2)^3 \\ & + a_2 trR^4 trR^2 + b_2(trR^2)^3 \\ & + a_3 trR^2 trF^4 + b_3 trR^2(trF^2)^2 \\ & + a_4 trF^2 trR^4 + b_4 trF^2(trR^2)^2 \end{aligned} \tag{3}$$

with $a_1 = 2142$ and $b_1 = 40$. The consistent anomaly is

$$G \sim \int I'_{10} \tag{4}$$

where $I_{12} = dI_{11}$ and $\delta I_{11} = dI'_{10}$ have been used. One must then find a counterterm $S_c$ to add to the action such that $\delta\delta_c = -G$, otherwise the theory is anomalous. Choosing a minimal form

$$\begin{aligned} S_c \sim \int [ & a_1 B\, trF^4 + b_1\, B(trF^2)^2 + a_2 B\, trR^4 + b_2\, B(trR^2)^2 \\ & + \Delta\, B\, trF^2 trR^2 - \Delta \overset{o}{\omega}_{L3} \overset{o}{\omega}_{Y3}\, trR^2 - \Delta \overset{o}{\omega}_{Y3} \omega_{L3} trF^2 ] \end{aligned} \tag{5}$$

gives the required cancellation provided that, in Eq. (5),

$$a_1 = a_3 \tag{6a}$$

$$a_2 = a_4 \tag{6b}$$

$$\Delta = (b_3 - b_1) = (b_4 - b_2) \tag{6c}$$

Since we have computed only $a_1$ and $b_1$ we have made only a consistency check. The fact that $b_1 \neq 0$ at two-loops (recall $b_1 = 0$ at one loop) the counterterm $B(trF^2)^2$ appears first only at this order. A similar new term is likely to appear in the O(32) heterotic string, presumably from a different part of the integration region for the double torus. For E(8) x E(8) there is a proportionality between $trF^4$ and $(trF^2)^2$ so this distinction between counterterms does not exist.

In summary, we have shown there exists no leading gauge anomaly at two loops for the SO(32) open superstring and that a new counterterm $B(trF^2)^2$ must be added to cancel the anomaly coming from the twelve-form $(trF^2)^3$ for which we have calculated the non-vanishing coefficient at two loops. A full analysis of the field theory part of the two-loop anomalies will undoubtedly be quite difficult but would certainly help clarify the details of the SO(32) open superstring.

References


1. P. H. Frampton, T. W. Kephart and T.-C. Yuan, UNC-Chapel Hill Report No. IFP-299-UNC (August 1987), to appear in Phys. Rev. Lett. (October 1987).

2. S. L. Adler and W. A. Bardeen, Phys. Rev. 182, 1517 (1969).

3. E. Witten in "Anomalies, Geometry, and Topology" eds. W. A .Bardeen and A. R. White, World Scientific (1985) p. 61.

4. M. B. Green and J. H. Schwarz, Phys. Lett. 149B, 117(1984).

5. P. H. Frampton and T. W. Kephart, Phys. Rev. Lett. 50, 1343, 1347 (1983).

6. P. H. Frampton and T. W. Kephart, Phys. Rev. D28, 1010 (1983); Phys. Lett. 131B, 80 (1983).

7. P. H. Frampton, P. Goddard and D. Wray, Nuov. Cim. 3A, 755 (1971).

8. D. J. Gross, A. Neveu, J. Scherk and J. H. Schwarz, Phys. Rev. D2, 697 (1970).

# TOPOLOGY, SUPERSPACE, AND HETEROTIC GRAVITATIONAL ANOMALIES

Burt A. Ovrut*†

Department of Physics
University of Pennsylvania
Philadelphia, Pa. 19104

ABSTRACT

The (1,0) superdiffeomorphic and Lorentz anomalies are constructed using cohomology in heterotic superspace. The superfield gravitational Wess-Zumino term is presented, and the relationship between superdiffeomorphic and Lorentz anomalies discussed.

In this paper I would like to discuss a topological approach to heterotic superdiffeomorphic and Lorentz anomalies developed in collaboration with J. Louis and R. Garreis [1]. This work is carried out using the superfield formulation of (1,0) supergravity [2] and, hence, world sheet supersymmetry is manifest. Although topology, and in particular cohomology, have been successfully applied [3] to the study of anomalies involving component fields, previous attempts to apply cohomology to superspace [4] were beset with difficulties. These difficulties arose, primarily, from the lack of a Stokes theorem for superspace. However, it was shown in [1] that for (1,0) superspace there is a modified notion of Stokes theorem which enables topological techniques to be employed. Using these techniques, a satisfactory superspace theory of gauge and gravitational anomalies can be developed. It is interesting to note that this modified notion of Stokes theorem is intimately related to the choice of torsion constraints [2] in (1,0) superspace.

We begin by briefly reviewing the formalism of (1,0) supermanifolds. A point in superspace is written as

$$z^M = (x^+, x^-, \theta^{+1}) \tag{1}$$

The geometry is determined by the supervielbeins $E_M{}^A, E_A{}^M$ and the Lorentz connection $\phi_{MA}{}^B$. The associated gauge groups are

1) superdiffeomorphisms - parameters $\xi^M$

*Work supported in part by the Department of Energy Contract Number DOE-AC02-76-ERO-3071 and NATO Grant Number 86/0684.

†Invited talk at the NATO Workshop on Superstrings, Boulder, Colorado, July 27-August 1, 1987.

2) Lorentz transformations - parameter L and generator

$$\lambda_A{}^B = \left(\begin{array}{cc|c} 1 & & \\ & -1 & \\ \hline & & 1/2 \end{array}\right) \tag{2}$$

Under superdiffeomorphisms

$$\begin{aligned} \delta_D E_M{}^A &= -\xi^L \partial_L E_M{}^A - (\partial_M \xi^L) E_L{}^A \\ \delta_D E_A{}^M &= -\xi^L \partial_L E_A{}^M + E_A{}^L (\partial_L \xi^M) \\ \delta_D \phi_M &= -\xi^L \partial_L \phi_M - (\partial_M \xi^L) \phi_L \end{aligned} \tag{3}$$

Under Lorentz transformations

$$\begin{aligned} \delta_L E_M{}^A &= L E_M{}^B \lambda_B{}^A \\ \delta_L E_A{}^M &= -L \lambda_A{}^B E_B{}^M \\ \delta_L \phi_M &= -\partial_M L \end{aligned} \tag{4}$$

It is clear from (3) and (4) that, thus far, the supergravity multiplet is highly reducible. The torsion superfields, $T_{BC}{}^A$, and the Lorentz curvature superfields $R_{CDA}{}^B (= R_{CD}\lambda_A{}^B)$ can be constructed from the vielbein and connection. The supergravity multiplet can be rendered irreducible by imposing the following constraints

$$\begin{aligned} T_{+1+1}{}^a &= -2i\delta^{+a} \\ T_{bc}{}^a &= T_{+1+1}{}^{+1} = 0 \\ T_{-+1}{}^a &= 0 \end{aligned} \tag{5}$$

The Bianchi identities then imply

$$\begin{aligned} T_{a+1}{}^{+1} &= T_{++1}{}^a = 0 \\ R_{++1} &= R_{+1+1} = 0 \\ R_{-+1} &= -2iT_{+-}{}^{+1} \\ R_{+-} &= 2\mathcal{D}_{+1} T_{+-}{}^{+1} \end{aligned} \tag{6}$$

Some of the consequences of (5) and (6) are

1) $\phi_M$ are determined from $E_M{}^A$, $E_A{}^M$;

2) $E_+{}^M$ are determined from $E_-{}^M$, $E_{+1}{}^M$.

Define $H_A{}^B$ by

$$\delta E_M{}^A = E_M{}^B H_B{}^A \tag{7}$$

Then constraints (5) are left invariant for independent $H_{+1}{}^A$, $H_-{}^a$ as long as

$$H_-^{\ +1} = -\frac{i}{2}(\mathcal{D}_{+1}H_-^{\ +} - \mathcal{D}_-H_{+1}^{\ \ +})$$

$$H_+^{\ +} = -i\,(\mathcal{D}_{+1}H_{+1}^{\ \ +} - 2iH_{+1}^{\ \ +1}) \tag{8}$$

$$H_+^{\ -} = -i\mathcal{D}_{+1}H_{+1}^{\ \ -}$$

Super-Weyl transformations are defined as those transformations that leave the torsion constraints invariant and which satisfy

$$\delta_W E_M^{\ a} = \ell E_M^{\ a} \tag{9}$$

Under super-Weyl transformations

$$\begin{aligned}
\delta_W E_M^{\ a} &= \ell E_M^{\ a} \\
\delta_W E_M^{\ +1} &= \frac{\ell}{2} E_M^{\ +1} - iE_M^{\ +}(\partial_{+1}\ell) \\
\delta_W E_a^{\ M} &= -\ell E_a^{\ M} + i(\partial_{+1}\ell)\delta_{a+}E_{+1}^{\ \ M} \\
\delta_W E_{+1}^{\ \ M} &= -\frac{\ell}{2} E_{+1}^{\ \ M} \\
\delta_W \phi_M &= N_a E_M^{\ a}(\partial_a\ell) + E_M^{\ +1}(\partial_{+1}\ell)
\end{aligned} \tag{10}$$

The density superfield is

$$E_{+1}^{-1} = (\det\,(E_a^{\ m})^{-1})(E_{+1}^{\ \ +1} - E_{+1}^{\ \ m}(E_a^{\ m})^{-1}E_a^{\ +1}) \tag{11}$$

Under the above transformations

$$\begin{aligned}
\delta_D E_{+1}^{-1} &= -(-)^m\partial_M(\xi^M E_{+1}^{-1}) \\
\delta_L E_{+1}^{-1} &= -\frac{L}{2} E_{+1}^{-1} \\
\delta_W E_{+1}^{-1} &= \frac{3\ell}{2} E_{+1}^{-1}
\end{aligned} \tag{12}$$

Denote $dx^+\,dx^-\,d\theta^{+1}$ by dz. Then the superspace integration measure is $dz\,E_{+1}^{-1}$. The (1,0) supergravity Einstein action is given by

$$\begin{aligned}
S_{SG} &= \int dz\,E_{+1}^{-1}\,2T_{|\ \ }^{\ +1} \\
&= \int dz\,E_{+1}^{-1}\,i(\mathcal{D}_-\phi_{+1} - \mathcal{D}_{+1}\phi_-) \\
&= 0
\end{aligned} \tag{13}$$

Having reviewed the formalism of (1,0) superspace we can now construct the sigma model for the heterotic superstring [5]. The matter superfields are

1) $X^\mu$ - $\mu = 0,\ldots,D-1$

2) $\psi_{-1}^{I}$ - $I = 1,\ldots,N$

Note that -1 is equivalent to $-+1$. Under the above transformations

$$\delta_D X^\mu = -\xi^L \partial_L X^\mu \quad , \quad \delta_D \psi^I_{-1} = -\xi^L \partial_L \psi^I_{-1}$$
$$\delta_L X^\mu = 0 \quad , \quad \delta_L \psi^I_{-1} = \frac{L}{2} \psi^I_{-1} \tag{14}$$
$$\delta_W X^\mu = 0 \quad , \quad \delta_W \psi^I_{-1} = -\frac{\ell}{2} \psi^I_{-1}$$

The most general action that is superdiffeomorphic, Lorentz, and super-Weyl invariant is

$$S = S_X + S_\psi \tag{15}$$

where

$$S_X = \int dz\, E_{+1}^{-1} [-i(\mathcal{D}_{+1} X^\mu)(\mathcal{D}_- X_\mu)]$$
$$S_\psi = \int dz\, E_{+1}^{-1} [-\psi^I_{-1}(\mathcal{D}_{+1} \psi^I_{-1})] \tag{16}$$

We have taken $G_{\mu\nu} = \eta_{\mu\nu}$, $B_{\mu\nu} = 0$, $g_{IJ} = \delta_{IJ}$, and $A_{+1K}{}^I = 0$ in (16).

We now discuss the superdiffeomorphic anomaly associated with the heterotic string. Prior to gauge fixing the effective action, $W[E_A{}^M]$, is given by

$$e^{iW} = N \int [dX^\mu][d\psi^I_{-1}]\, e^{i(S_X + S_\psi)} \tag{17}$$

The superdiffeomorphic variation of W can be written as

$$\delta_D W = \int dz\, E_{+1}^{-1} \Lambda_P{}^M G_{D-1M}{}^P \tag{18}$$

where

$$\Lambda_P{}^M = \partial_P \xi^M \tag{19}$$

Note that $\delta_D W$ vanishes if and only if $G_{D-1M}{}^P$ vanishes. Therefore, $G_{D-1M}{}^P$ is the superdiffeomorphic anomaly. What is the functional structure of $G_{D-1M}{}^P$? Remarkably, this can be determined topologically. This is done in three steps.

1) BRST transformations:

Let $\xi$ be a constant, anticommuting parameter and define superdiffeomorphic ghost superfields by $\xi^M = \xi C^M$. Note that $C^\pm$ and $C^{+1}$ are anticommuting and commuting superfields respectively. The BRST transformations of $E_A{}^M$, $X^\mu$, and $\psi^I_{-1}$ are obtained from superdiffeomorphisms by replacing $\xi^M$ with $\xi C^M$. Define BRST generator, $\Sigma_D$, by

$$\Sigma_{BRST}\phi_i = \xi \Sigma_D \phi_i \tag{20}$$

Note that $\Sigma_D$ is an anticommuting operator. Hence,

$$\Sigma_D^2 = 0 \tag{21}$$

It follows from the $\delta_D$ variations above that

$$\Sigma_D E_A{}^M = -C^L \partial_L E_A{}^M + (-)^a E_A{}^L (\partial_L C^M)$$
$$\Sigma_D X^\mu = -C^L \partial_L X^\mu \tag{22}$$
$$\Sigma_D \psi^I_{-1} = -C^L \partial_L \psi^I_{-1}$$

Define

$$C_P{}^M = (-)^p \partial_P C^M \tag{23}$$

Then

$$\Sigma_D C_P{}^M = - C^L \partial_L C_P{}^M - (-)^{p+\ell} C_P{}^L C_L{}^M \tag{24}$$

Also

$$\Sigma_D E_{+1}^{-1} = - \partial_L (C^L E_{+1}^{-1}) \tag{25}$$

2) Wess-Zumino consistency condition:

Replacing $\xi^M$ by $\xi C^M$ in $\delta_D W$ implies

$$\Sigma_D W = - \int dz\ E_{+1}^{-1} C_P{}^M G_{D-1M}{}^P \tag{26}$$

Operating with $\Sigma_D$ yields

$$\int dz\ \Sigma_D (E_{+1}^{-1} C_P{}^M G_{D-1M}{}^P) = 0 \tag{27}$$

Can $\Sigma_D$ be brought through $E_{+1}^{-1}$? Decompose $\Sigma_D = \Sigma_\ell + \Sigma_g$ where

$$\begin{aligned}
&\Sigma_\ell E_A{}^M = - C^L \partial_L E_A{}^M + (-)^a E_A{}^L (\partial_L C^M) \quad , \quad \Sigma_g E_A{}^M = 0 \\
&\Sigma_\ell X^\mu = - C^L \partial_L X^\mu \ , \ \Sigma_g X^\mu = 0 \\
&\Sigma_\ell \psi_{-1}^I = - C^L \partial_L \psi_{-1}^I , \ \ \Sigma_g \psi_{-1}^I = 0 \\
&\Sigma_\ell C_P{}^M = - C^L \partial_L C_P{}^M , \ \ \Sigma_g C_P{}^M = - (-)^{p+\ell} C_P{}^L C_L{}^M
\end{aligned} \tag{28}$$

Note that

$$\Sigma_\ell E_{+1}^{-1} = - \partial_L (C^L E_{+1}^{-1}), \ \Sigma_g E_{+1}^{-1} = 0 \tag{29}$$

It can be shown that $\Sigma_g^2 = 0$. Assume that $G_{D-1M}{}^P$ behaves like a Lie algebra valued scalar superfield. That is

$$\Sigma_\ell G_{D-1M}{}^P = - C^L \partial_L G_{D-1M}{}^P \tag{30}$$

Then

$$\begin{aligned}
\int dz\ \Sigma_\ell (E_{+1}^{-1}\ C_P{}^M G_{D-1M}{}^P) &= - \int dz\ \partial_P (C^P E_{+1}^{-1}\ C_M{}^L G_{D-1L}{}^M) \\
&= 0
\end{aligned} \tag{31}$$

Therefore

$$\int dz\ E_{+1}^{-1}\ \Sigma_g (C_P{}^M G_{D-1M}{}^P) = 0 \tag{32}$$

This is the Wess-Zumino consistency condition for the superdiffeomorphic anomaly. Note that (32) is linear in the ghost fields and $G_{D-1M}{}^P$ has Lorentz charge $-+1$.

3) Topological solution of the consistency condition:

Let $\Gamma_{NM}{}^R$ be the superfield Christoffel connection. Define $\Gamma_M{}^R = dy^N\ \Gamma_{NM}{}^R$. Differential $dy^N$ is chosen so that $\Gamma_M{}^R$ has the same statistics as $C_M{}^R$. Under BRST transformations take

$$\Sigma_\ell \, \Gamma_M{}^R = dy^N(-)^n(C^L\partial_L\Gamma_{NM}{}^R + C_N{}^L\Gamma_{LM}{}^R)$$
$$\Sigma_g \, \Gamma_M{}^R = dC_M{}^R + (-)^{1+m+\ell}(C_M{}^L\Gamma_L{}^R + \Gamma_M{}^L C_L{}^R) \tag{33}$$

The curvature two-form is defined by

$$R_P{}^Q = d\Gamma_P{}^Q + (-)^{1+p+r}\,\Gamma_P{}^R\Gamma_R{}^Q \tag{34}$$

Consider the three-form

$$\omega_3^0(\Gamma) = (-)^q\Gamma_Q{}^P R_P{}^Q + \frac{1}{3}(-)^{r+p+q}\,\Gamma_Q{}^P\Gamma_P{}^R\Gamma_R{}^Q \tag{35}$$

It can be shown that

$$(-)^q R_Q{}^P R_P{}^Q = d\omega_3^0(\Gamma) \tag{36}$$

Now define $\tilde{\Gamma}_M{}^R = \Gamma_M{}^R - C_M{}^R$. Furthermore, let $\Delta_D = d + \Sigma_g$. Note that $\Delta_D^2 = 0$. The associated curvature two-form is

$$\tilde{R}_P{}^Q = \Delta_D\,\tilde{\Gamma}_P{}^Q + (-)^{1+p+r}\,\tilde{\Gamma}_P{}^R\tilde{\Gamma}_R{}^Q \tag{37}$$

Since the algebra of $\Delta_D$, $\tilde{\Gamma}_P{}^Q$ is the same as d, $\Gamma_P{}^Q$ implies

$$(-)^q\tilde{R}_Q{}^P\tilde{R}_P{}^Q = \Delta_D\,\omega_3(\tilde{\Gamma}) \tag{38}$$

where

$$\omega_3(\tilde{\Gamma}) = (-)^q\tilde{\Gamma}_Q{}^P\tilde{R}_P{}^Q + \frac{1}{3}(-)^{r+p+q}\,\tilde{\Gamma}_Q{}^P\tilde{\Gamma}_P{}^R\tilde{\Gamma}_R{}^Q \tag{39}$$

However, one can show that $\tilde{R}_P{}^Q = R_P{}^Q$. Therefore

$$d\omega_3^0(\Gamma) = \Delta_D\omega_3(\tilde{\Gamma}) \tag{40}$$

Now $\omega_3(\tilde{\Gamma})$ can be expanded as $\omega_3(\tilde{\Gamma}) = \omega_3^0 + \omega_2^1 + \omega_1^2 + \omega_0^3$ where, for example

$$\omega_2^1 = -(-)^q C_Q{}^P d\Gamma_P{}^Q \tag{41}$$

Comparing the left and right of (40) and equating terms of identical form and ghost number yields the descent equations. The equation of interest is

$$\Sigma_g\,\omega_2^1 = -\,d\omega_1^2 \tag{42}$$

One cannot integrate superforms over superspace. Therefore, consider the component form of (42)

$$\Sigma_g\,\omega_{2AB}^1 = \mathcal{D}_A\,\omega_{1B}^2 - (-)^{ab}\mathcal{D}_B\omega_{1A}^2 + T_{AB}{}^D\,\omega_{1D}^2 \tag{43}$$

Integrating over superspace yields

$$\int dz\; E_{+1}^{-1}\,\Sigma_g\,\omega_{2AB}^1 = \int dz\; E_{+1}^{-1}\,T_{AB}{}^D\,\omega_{2D}^1 \tag{44}$$

Note that the torsions $T_{AB}{}^D$ are an obstruction to Stokes theorem. However, recall that $G_{D-1M}{}^P$ has Lorentz charge $-+1$ and that $T_{-+1}{}^D = 0$ for any D. Therefore

$$\int dz\; E_{+1}^{-1}\,\Sigma_g\,\omega_{2-+1}^1 = 0 \tag{45}$$

and, hence, $\omega^1_{2-+1}$ is a non-trivial solution of the Wess-Zumino consistency condition. The $-+1$ component of $\omega^1_2$ is

$$\omega^1_{2-+1} = (-)^{1+p} C_Q{}^P E_{-1}{}^{NM} \partial_M \Gamma_{NP}{}^Q \qquad (46)$$

where $E_{-1}{}^{NM} = (-)^n E_-{}^N E_{+1}{}^M - E_{+1}{}^N E_-{}^M$. Comparing (45) with (32) implies

$$G_{D-1M}{}^P \propto (-)^m E_{-1}{}^{NL} \partial_L \Gamma_{NM}{}^P \qquad (47)$$

The constant of proportionality can be found by a one-loop supergraph calculation. The result is

$$G_{D-1M}{}^P = -\frac{1}{168\pi} (D-N)(-)^m E_{-1}{}^{NL} \partial_L \Gamma_{NM}{}^P \qquad (48)$$

The Lorentz anomaly also has a topological solution. We find that

$$G_{L-1B}{}^A \propto (-)^b E_{-1}{}^{NM} \partial_M \phi_{NB}{}^A \qquad (49)$$

The superdiffeomorphic and Lorentz anomalies are closely related. To see this define superfields $\not{H}_M{}^N$ by

$$(e^{\not{H}})_M{}^N = E_A{}^M \delta_A{}^N \qquad (50)$$

and let

$$\Gamma_t = (e^{-t\not{H}})\Gamma(e^{t\not{H}}) - (e^{-t\not{H}})d(e^{t\not{H}}) \qquad (51)$$

where $0 \leq t \leq 1$. Then the Wess-Zumino term for the superdiffeomorphic anomaly is

$$S[E,\Gamma] = \int_0^1 dt \int dz\, E_{+1}^{-1} \not{H}_Q{}^P G_{D-1}[\Gamma_t]_P{}^Q \qquad (52)$$

where

$$G_{D-1}[\Gamma_t]_P{}^Q = -\frac{1}{168\pi} (D-N)(-)^m E_{-1}{}^{NL} \partial_L \Gamma_{tNM}{}^P \qquad (53)$$

Decompose $\delta_D = \mathcal{L}_\xi + T_\Lambda$ where $\Lambda_M{}^N = \partial_M \xi^N$. It follows that

$$\mathcal{L}_\xi \Gamma_{tNM}{}^R = -\xi^L \partial_L \Gamma_{tNM}{}^R - (\partial_N \xi^L)\Gamma_{tLM}{}^R$$
$$T_\Lambda \Gamma_t = [\Gamma_t, \Lambda_t] - d\Lambda_t \qquad (54)$$

where

$$\Lambda_t = (e^{-t\not{H}})\Lambda(e^{t\not{H}}) + (e^{-t\not{H}})T_\Lambda(e^{t\not{H}}) \qquad (55)$$

Also note that $\Gamma_o = \Gamma$, $\Lambda_o = \Lambda$, and $\Lambda_1 = 0$. Consider $\delta_D S[E,\Gamma]$. After a long calculation we find

$$\delta_D S[E,\Gamma] = -\int dz\, E_{+1}^{-1} \Lambda_Q{}^P G_{D-1}[\Gamma]_P{}^Q \qquad (56)$$

It follows that

$$\delta_D(W + S[E,\Gamma]) = 0 \qquad (57)$$

Hence, by adding Wess-Zumino term (52) to the action the superdiffeomorphic anomaly can be made to vanish. However, there is now a Lorentz anomaly, as we now show. The relation between the Christoffel and Lorentz connections is

$$\Gamma_{NL}{}^{M} = -(-)^{n(a+\ell)}(E_L{}^A \partial_N E_A{}^M - E_L{}^A \phi_{NA}{}^B E_B{}^M) \tag{58}$$

Defining

$$\phi_\tau = (e^{-\tau \not{H}})\phi(e^{\tau \not{H}}) - (e^{-\tau \not{H}})d(e^{\tau \not{H}}) \tag{59}$$

where $\tau = t - 1$. Then

$$\Gamma_{tP}{}^{Q} = \delta_P{}^A \delta_B{}^Q \phi_{\tau A}{}^B \tag{60}$$

The superdiffeomorphic Wess-Zumino term can now be written as

$$S[E,\Gamma] = -\int_{-1}^{0} d\tau \int dz\; E_{+1}^{-1} \not{H}_B{}^A G_{L-1}[\phi_\tau]_A{}^B \tag{61}$$

One can now calculate $\delta_L S[E,\Gamma]$. We find that

$$\delta_L(W + S[E,\Gamma]) = \int dz\; E_{+1}^{-1}\, L_A{}^B G_{L-1}[\phi]_B{}^A \tag{62}$$

where the Lorentz anomaly is given by

$$G_{L-1}[\phi]_B{}^A = \frac{1}{168\pi}(D-N)(-)^a E_{-1}{}^{NM} \partial_M \phi_{NB}{}^A \tag{63}$$

Since there is no longer a superdiffeomorphic anomaly we can now fix this gauge, thus introducing ghost superfields into the action. The contribution of these ghosts to the Lorentz anomaly can be calculated with supergraphs. We find that the entire Lorentz anomaly is

$$G_{L-1B}{}^{A} = \frac{1}{168\pi}(D-N+22)(-)^a E_{-1}{}^{NM} \partial_M \phi_{NB}{}^A \tag{64}$$

Finally, the super-Weyl anomaly can also be calculated topologically. Setting $D - N + 22 = 0$, we find that the condition that the super-Weyl anomaly vanish is $D - 10 = 0$. Hence, we find the well known result that the heterotic string has no superdiffeomorphic, Lorentz, or super-Weyl anomalies as long as $D = 10$ and $N = 32$.

## References


[1] R. Garreis, J. Louis, and B. Ovrut, Phys. Lett. B (1987), to appear; Univ. of Pennsylvania preprint UPR-0337T, June, 1987.

[2] M. Evans and B. Ovrut, Phys. Lett. 174B, (1986) 63; Phys. Lett. 175B, (1986) 145; Phys. Lett. 184B (1987) 153; Phys. Lett. 186B (1987) 134; Phys. Rev. D35 (1987) 3045.
R. Brooks, F. Muhammad, and S. Gates, Nuc. Phys. B268 (1986) 599.
G. Moore and P. Nelson, Nuc. Phys. B274 (1986) 509.

[3] See B. Zumino, Chiral Anomalies and Differential Geometry, Relativity, Groups and Topology II, Les Houches (1983) and L. Baulieu, Phys. Reports C129 (1985) 1, and references therein.

[4] G. Girardi, R. Grimm, and R. Stora, Phys. Lett. 156B (1985) 203.
L. Bonora, P. Pasti, and M. Tonin, Phys. Lett. 156B (1985) 341; Nuc. Phys. B252 (1985) 458.

[5] D. Gross, J. Harvey, E. Martinec, and R. Rohm, Phys. Rev. Lett. 54 (1985) 502; Nuc. Phys. B256 (1985) 253; Nuc. Phys. B267 (1986) 75.

# ANOMALY-FREE THEORIES REMAIN ANOMALY-FREE AFTER COMPACTIFICATION WITH NO ISOMETRIES

J.E. Bjorkman and Y. Tosa

Department of Physics
University of Colorado
Boulder, Co. 80309

## INTRODUCTION

It has long been known that the classical gauge symmetries of a theory may be broken by quantum effects. These quantum anomalies prevent conservation of the gauge currents, which ultimately leads to a loss of unitarity. Therefore, it is extremely important that a theory have no anomalies.

The local anomaly for any given field is related to the number of zero modes of the wave operator for that field in $D+2$ dimensions, where D is the dimension of space-time [1-3]. The number of zero modes is determined topologically by the Atiyah-Singer index theorem [4]. When the original space is compactified, its topology is changed. This in turn changes the number of zero modes; hence, compactification may induce new anomalies that were not present in the original theory.

In general, one must check that a particular compactification scheme does not generate any new anomalies. Witten [5]; Green, Schwarz and West [6]; and Das and Kwon [7] have shown that the specific case of anomaly-free $D=10$, $N=1$ supergravity remains anomaly-free after compactification to $D=4$, 6 and 8. This leads one to ask if this is a general property of supergravity theories.

We have shown that, as long as the compact manifold has no isometries, then there is a simple relationship between the anomaly prior to compactification and the anomaly after compactification. Using this relationship we then demonstrated that most anomaly-free theories remain anomaly-free after compactification, subject to a few restrictions on the compactification scheme [8].

## COMPACTIFICATION

We now turn to a discussion of how the original theory is compactified. Since there can be no local anomaly for odd dimensions, we only need to consider the even dimensional cases. Therefore, we will start with a space M of dimension $D = 2n$. This is reduced to a space $M'$ of dimension $D' = 2n'$ with an internal compact space K of dimension $D_0 = 2n_0$. We use a prime to denote variables in the space $M'$ and a subscript 0 to denote variables in the compact space K. We will see later that in order to relate the anomaly after compactification to the original anomaly, the compact space can have no isometries; therefore, compactification leads to

$$M \to M' \times K. \tag{1}$$

We assume there is a gauge group G defined on the space M, and that after compactification, the associated field strength F will in general acquire a vev, $\langle F \rangle$, on K; thereby breaking the original gauge group G. Given $\langle F \rangle$, we can find a subgroup $G_0$ of G which contains $\langle F \rangle$. Then we can find a maximal subgroup $G'$ of G which commutes with $G_0$ such that

$$G \supset G' \times G_0. \tag{2}$$

We assume G, $G'$ and $G_0$ are all simple groups and the associated field strengths are F, $F'$, and $F_0$, respectively. We also assume that the background field $F_0$ can only be a function of the space K and that $F'$ is a function only of the space $M'$.

ANOMALIES AND COMPACTIFICATION

To get an idea how the anomaly after compactification might be related to the original anomaly, consider the following observations:

i. After compactification, the anomaly of a given field, i, in $M'$ is related to a differential form (I-form) in $D'+2$ dimensions.

$$\delta \Gamma'_i \sim I^i_{2n'+2},$$

where $\mathrm{ind}\,\not{D}^i = \int I^i_{2n'+2}$ is the number of zero modes of the wave operator.

ii. The numbers of fields after compactification are given by indices on K. We may understand this from the following: The wave operator for any field decomposes into a sum of wave operators on $M'$ and K [9].

$$\not{D}^i = \not{D}^i_{M'} + \not{D}^i_K$$

The eigenvalues of $\not{D}_K$ act as masses for the fields in $M'$. Therefore, the number of massless fields in $M'$ is given by the number of zero eigenvalues of $\not{D}_K$. This is determined by the Atiyah-Singer index theorem [1 4].

$$n_i = \mathrm{ind}\,\not{D}^i_K = \int_K I^i_{2n_0}$$

iii. The total anomaly after compactification is a sum of the individual anomalies for each field in $M'$.

$$I^{total}_{2n'+2} = \sum_i n_i I^i_{2n'+2} = \sum_i \left[\int_K I^i_{2n_0}\right] I^i_{2n'+2}$$

iv. Under certain circumstances the I-form for M factorizes.

$$I_{2n+2} \sim \sum_i I^i_{2n_0} I^i_{2n'+2}$$

These observations lead us to suppose that the differential form after compactification is related to the original I-form by

$$I^{total}_{2n'+2} \sim \int_K I_{2n+2}.$$

To see how this works, we consider the individual cases of spin-1/2, spin-3/2, Yang-Mills, and self-dual antisymmetric tensors.

A. Spin-1/2

The simplest case is a spin-1/2 gauge singlet. Its I-form before compactification is

$$I^{1/2}_{2n+2}(1) = c\,\hat{A}(M)\big|_{vol+2}, \tag{3}$$

where $\hat{A}$ is the Dirac genus, and c is a constant which depends on whether or not we have Majorana fermions. We need to know under what conditions the I-form will factorize into a product of I-forms on M′ and K. The Dirac genus belongs to a multiplicative sequence [10]; therefore, for a product manifold the Dirac genus factorizes. In other words,

$$\hat{A}(M) = \hat{A}(M')\hat{A}(K), \tag{4}$$

whenever the curvature tensor is block diagonal, i.e.

$$R = \begin{pmatrix} R' & 0 \\ 0 & R_0 \end{pmatrix}. \tag{5}$$

This will occur as long as the compact space has no isometries. In this case

$$I^{1/2}_{2n+2}(1) = c\,\hat{A}(M')\big|_{vol'+2}\hat{A}(K)\big|_{vol_0} = \left(\frac{c}{c'}\right) I^{1/2}_{2n_0}(1)\, I^{1/2}_{2n'+2}(1). \tag{6}$$

To calculate the anomaly after compactification, we must know how the fields decompose after compactification. The decomposition of a left handed spinor representation, L, of SO(D) branching into (SO(D′), SO($D_0$)) is given by:

$$L \rightarrow (L, L) + (R, R). \tag{7}$$

Therefore, a spin-1/2 field decomposes into

$$\Psi_L \rightarrow (\Psi_L, \Psi_L) + (\Psi_R, \Psi_R). \tag{8}$$

As we observed earlier, the number of fields in M′ is given by the index of the wave operator in K.

$$n_{1/2} = \mathrm{ind}\, \not{D}_K^{1/2} = c_0 \int_K I_{2n_0}^{1/2}(1). \tag{9}$$

The anomaly after compactification is then just the sum of the anomalies for each field in M′.

$$I_{2n'+2}^{\mathrm{total}} = n_{1/2} I_{2n'+2}^{1/2}(1) = \left[c_0 \int_K I_{2n_0}^{1/2}(1)\right] I_{2n'+2}^{1/2}(1). \tag{10}$$

If we compare eq. (10) to eq. (6), we see that the relationship between the final and original anomaly is given by

$$I_{2n'+2}^{\mathrm{total}} = \left(\frac{c_0 c'}{c}\right) \int_K I_{2n+2}^{1/2}(1). \tag{11}$$

B. Spin - 3/2

A slightly more complicated case that factorizes into a sum of products of I-forms is a spin-3/2 field. The I-form for a spin-3/2 field is given by

$$I_{2n+2}^{3/2} = c\,\hat{A}(M)\,[\mathrm{Tr}(e^{iR/2\pi} - 1) + (2n-1)]\big|_{\mathrm{vol}+2}. \tag{12}$$

Recall that the curvature tensor is block diagonal. This implies the simple relation

$$\mathrm{Tr}\, e^{iR/2\pi} = \mathrm{Tr}\, e^{iR'/2\pi} + \mathrm{Tr}\, e^{iR_0/2\pi}. \tag{13}$$

Inserting eqs. (4) and (13) into eq. (12) gives the factorized anomaly

$$I_{2n+2}^{3/2} = \frac{c}{c'}\left\{I_{2n_0}^{1/2}(1)\, I_{2n'+2}^{3/2} + \left[I_{2n_0}^{1/2}(1) + I_{2n_0}^{3/2}\right] I_{2n'+2}^{1/2}(1)\right\}. \tag{14}$$

Now we need to know the decomposition of the gravitino under compactification. For the purpose of counting the index, a gravitino field is given by [9]

$$\Psi_L^{3/2} = V \otimes \Psi_L + \Psi_R. \tag{15}$$

This decomposes as

$$\Psi_L^{3/2} \to (\Psi_L^{3/2}, \Psi_L) + (\Psi_L, \Psi_L^{3/2}) - (\Psi_R, \Psi_L) + (L \leftrightarrow R). \tag{16}$$

The number of spin-1/2 and spin-3/2 fields in $M'$ is then given by

$$n_{3/2} = \operatorname{ind} \not{D}_K^{1/2} = c_0 \int_K I_{2n_0}^{1/2}(1)$$
$$n_{1/2} = \operatorname{ind} \not{D}_K^{3/2} + \operatorname{ind} \not{D}_K^{1/2} = c_0 \int_K \left[ I_{2n_0}^{1/2}(1) + I_{2n_0}^{3/2} \right], \tag{17}$$

so the total anomaly after compactification is

$$I_{2n'+2}^{\text{total}} = n_{3/2} I_{2n'+2}^{3/2} + n_{1/2} I_{2n'+2}^{1/2}(1). \tag{18}$$

Inserting eq. (17) into eq. (14) and comparing with eq. (18) gives the following relation between the original I-form and the I-form after compactification:

$$I_{2n'+2}^{\text{total}} = \left( \frac{c_0 c'}{c} \right) \int_K I_{2n+2}^{3/2}. \tag{19}$$

C. Yang-Mills

The next case is spin-1/2 fields with gauge interactions. Prior to compactification the I-form is:

$$I_{2n+2}^{1/2}(F) = c \hat{A}(M) \operatorname{ch}(F) \Big|_{\text{vol}+2}, \tag{20}$$

where ch(F) is the Chern-form. We now need to know the decomposition of the Chern-form. Any element F of the Lie algebra associated with G decomposes into a sum over irreducible representations of the generators of $G'$ and $G_0$ as follows:

$$F \to \bigoplus_i (F'_i \otimes 1_0 \oplus 1' \otimes F_{0i}). \tag{21}$$

From this we see that

$$\operatorname{Tr} F^n = \sum_i \sum_{k=0}^{n} \binom{n}{k} \operatorname{tr} F_i'^k \operatorname{tr} F_{0i}^{n-k}. \tag{22}$$

Then it is easy to show that the Chern-form factorizes into the following sum over irreducible representations:

$$\operatorname{ch}(F) = e^{iF/2\pi} = \sum_i \operatorname{ch}(F'_i) \operatorname{ch}(F_{0i}). \tag{23}$$

Note, the factorization of the Chern form does not require simple groups, so this formula is still valid even in the presence of U(1) symmetries. Inserting eqs. (4) and (23) into eq. (20) gives the factorized I-form

$$I^{1/2}_{2n+2}(F) = \frac{c}{c'} \sum_i I^{1/2}_{2n_0}(F_{0i}) I^{1/2}_{2n'+2}(F'_i). \qquad (24)$$

The Yang-Mills spin-1/2 fields decompose as

$$\Psi_L(F) \to \sum_i [(\Psi_L(F'_i), \Psi_L(F_{0i})) + (\Psi_R(F'_i), \Psi_R(F_{0i}))], \qquad (25)$$

so the number of fields in M′ that transform under the representation $F'_i$ is

$$n_i = \mathrm{ind}\, \not{D}^{1/2}_K(F_{0i}) = c_0 \int_K I^{1/2}_{2n_0}(F_{0i}). \qquad (26)$$

Therefore, the total I-form after compactification is given by the sum

$$I^{total}_{2n'+2} = \sum_i n_i I^{1/2}_{2n'+2}(F'_i). \qquad (27)$$

Inserting eq. (26) into (24) and comparing with eq. (27) gives

$$I^{total}_{2n'+2} = \left(\frac{c_0 c'}{c}\right) \int_K I^{1/2}_{2n+2}(F). \qquad (28)$$

D. Self-Dual Antisymmetric Tensors

Finally we turn to self-dual antisymmetric tensors (SDA's). The I-form is

$$I^A_{2n+2} = d\, L(M)|_{vol+2}. \qquad (29)$$

Where L is the Hirzebruch polynomial and d is a constant which depends on the representation of the SDA (complex or real).

The Hirzebruch polynomial is also a member of a multiplicative sequence [10], so it factorizes as

$$L(M) = L(M')L(K). \qquad (30)$$

Inserting eq. (30) into eq. (29) gives the factorized I-form,

$$I^A_{2n+2} = \left(\frac{d}{d'}\right) I^A_{2n_0} I^A_{2n'+2}. \qquad (31)$$

Under SO(D) branching into (SO(D′),SO($D_0$)), a self-dual antisymmetric tensor decomposes as

$$\mathrm{SDA_L} \rightarrow (\mathrm{SDA_L}, \mathrm{SDA_L}) + (\mathrm{SDA_R}, \mathrm{SDA_R}) + (\text{real reps.}). \tag{32}$$

Real reps. have no gravitational anomaly, so the only anomalous fields in M′ are the SDA's. The number of SDA's in M′ is given by

$$n_A = \mathrm{ind}\, \not{D}_K^A = d_0 \int_K I^A_{2n_0}. \tag{33}$$

The total I-form is then

$$I^{\text{total}}_{2n'+2} = n_A I^A_{2n'+2}. \tag{34}$$

Inserting eq. (33) into (31) and comparing with eq. (34) gives

$$I^{\text{total}}_{2n'+2} = \left(\frac{d_0 d'}{d}\right) \int_K I^A_{2n+2}. \tag{35}$$

Notice that the proportionality constant is different for SDA's than the previous cases.

## MATTER CONTENT

For each field we have considered, we have seen that the I-form after compactification is proportional to the integral over K of the original I-form. It is now easy to generalize our results to an arbitrary number of fields. We wish to consider as general a theory as possible with spins less than two. We will assume the existence of a gravity sector which may contain any number of gauge singlet fields of spins 0, 1/2, . . . , 2 as well as antisymmetric tensors. We also assume the presence of Yang-Mills matter with any number of fields of spins 0, 1/2, 1.

Now we consider which fields can contribute to the original anomaly. We have ignored the possibility of more complicated fields such as tensor-spinors, so we may assume the only fields which can generate an anomaly are: spin-1/2, spin-3/2, and self-dual antisymmetric tensors. To be specific for the gravity multiplet, let N be the number (left minus right) of gravitinos, $\Psi_L^{3/2}$, $\ell$ the number of spin-1/2 fields, $\Psi_L$, (or other physical gauge singlets such as shadow matter), and m the number of SDA's, $A_L$. Finally, for the Yang-Mills sector, the only fields that generate an anomaly are spin-1/2 fields which transform under some representation F of the gauge group G, $\Psi_L(F)$.

To calculate the anomaly after compactification, we must determine which fields in the original theory can generate anomalies in the dimensionally reduced theory. We have shown that the only fields which generate anomalies in the dimensionally reduced fields are the same fields which generate anomalies in the original theory [8]. The only subtle point is that non-self-dual antisymmetric tensors can decompose into self-dual tensors at lower dimension. So, although they have no anomaly in the original theory, one might worry that non-self dual tensors could produce an anomaly in the compactified theory. Fortunately, they always produce pairs of left and right handed SDA's, so their total anomaly vanishes [8].

## TOTAL ANOMALY

We are now ready to calculate the total anomalies before and after compactification. For the anomaly prior to compactification the total I-form is

$$I_{2n+2}^{total} = N\, I_{2n+2}^{3/2} + \ell\, I_{2n+2}^{1/2}(1) + m\, I_{2n+2}^{A} + I_{2n+2}^{1/2}(F). \quad (36)$$

Similarly, the total I-form in $M'$ is given by

$$I_{2n'+2}^{total} = n_{3/2} I_{2n'+2}^{3/2} + n_{1/2} I_{2n'+2}^{1/2}(1) + \sum_i n_{1/2}^{i} I_{2n'+2}^{1/2}(F'_i) + n_A I_{2n'+2}^{A}, \quad (37)$$

where we have modified the definitions of the n's to include the generalization to an arbitrary number of original fields. If we now integrate $I_{2n+2}$ over K and use eqs. (11), (19), (28) and (35), we get

$$\int_K I_{2n+2}^{total} = \frac{c}{c'c_0}\left[n_{3/2} I_{2n'+2}^{3/2} + n_{1/2} I_{2n'+2}^{1/2}(1) + \sum_i n_{1/2}^{i} I_{2n'+2}^{1/2}(F'_i)\right] + \frac{d}{d'd_0}\left[n_A I_{2n'+2}^{A}\right]. \quad (38)$$

Except for the SDA contribution, this would be proportional to the total $M'$ I-form. Unfortunately, $\frac{c}{c'c_0} \neq \frac{d}{d'd_0}$, so we must require that the SDA contribution vanishes; i.e., either $n_A = 0$ or $I_{2n'+2}^{A} = 0$. This will occur whenever either $m = 0$, $D' = 4k'$, or $D_0 = 4k_0 + 2$. One of these conditions will always be met, unless we start with SDA's in $D = 4k + 2$ and then compactify to $D' = 4k' + 2$. Subject to this requirement, we obtain our main result:

$$I_{2n'+2}^{total} \propto \int_K I_{2n+2}^{total}. \quad (39)$$

Henceforth, we will assume we have no SDA's if we start in $D = 4k + 2$ and compactify to $D' = 4k' + 2$.

## ANOMALY-FREE THEORIES

Since we have a relationship between the anomaly prior to compactification and the anomaly after compactification, we may now address the question of how compactification will change the anomalies in a particular theory. The relevant theories to examine are those constrained to be anomaly-free prior to compactification.

There are two cases we must consider when discussing anomaly-free theories. The first case is when the original theory is chosen such that $I_{2n+2} = 0$. This case is trivial, since $I_{2n'+2} \propto \int_K I_{2n+2} = 0$. Therefore, any theory, whose anomaly vanishes trivially, will remain anomaly-free after compactification to any lower dimension, as long as no isometries are generated for the compact space.

The second case is when the anomaly cancels by the Green-Schwarz mechanism [11]. Note that the anomaly after compactification is determined by the number of zero modes, which is a topological invariant; hence, the Green-Schwarz counter terms can not contribute to the anomaly after compactification. Therefore, we only need to examine the original

I-form, which in the Green-Schwarz case factorizes in the form

$$I_{2n+2} = (\mathrm{Tr}\,R^2 + k\,\mathrm{Tr}\,F^2)X_{2n-2}. \tag{40}$$

We will now demonstrate, subject to certain restrictions, that the I-form after compactification will also factorize. First note that the invariant polynomial $X_{2n-2}$ is a $(2n-2)$-form, which under compactification goes in general to

$$X_{2n-2} = \sum_{k,i} X_k^{\prime i} X_{0\,2n-2-k}^{i}. \tag{41}$$

After integrating $I_{2n+2}$ over K and projecting out the terms proportional to the volume elements, we find that

$$\begin{aligned} I_{2n'+2} \propto\ & (\mathrm{Tr}\,R'^2 + k\,\mathrm{Tr}\,F'^2)\sum_i X_{2n'-2}^{\prime i}\int_K X_{0\,2n_0}^{i} \\ & + \sum_i X_{2n'+2}^{\prime i}\int_K (\mathrm{Tr}\,R_0^2 + k\,\mathrm{Tr}\,F_0^2)X_{0\,2n_0-4}^{i}. \end{aligned} \tag{42}$$

Note that we have assumed that $G'$ and $G_0$ do not contain any U(1) factors. We now observe that the I-form for $M'$ will also factorize, provided that the second term of eq. (42) vanishes.

The first possibility is if a three-form field strength H exists such that $dH_0 = (\mathrm{Tr}\,R_0^2 + k\,\mathrm{Tr}\,F_0^2)$. Since $H^2$ contributes to the vacuum energy it is globally well defined [5]. We also know that $X_{0\,2n_0-4}^{i}$ is a polynomial of $\mathrm{Tr}\,R_0^m$ and $\mathrm{Tr}\,F_0^m$; therefore, it is also globally well defined and closed [2]. Consequently, by partial integration, the second term in eq. (42) vanishes. An example where an H field strength exists is $D = 10$, $N = 1$ supergravity. We have checked the explicit cases presented in refs. [5-7], and our results agree.

The second possibility is to choose the compact space such that $(\mathrm{Tr}\,R_0^2 + k\,\mathrm{Tr}\,F_0^2) = 0$. An example of this is the holonomy compactification scheme devised by Candelas et al. [12]. If either of these two possibilities occurs, then the second term of eq. (42) vanishes giving

$$I_{2n'+2} \propto (\mathrm{Tr}\,R'^2 + k\,\mathrm{Tr}\,F'^2)\sum_i X_{2n'-2}^{\prime i}\int_K X_{0\,2n_0}^{i}. \tag{43}$$

Thus we see, if the I-form in M factorizes and either of these two conditions is met, then the I-form in $M'$ will also factorize.

Whenever the I-form after compactification factorizes, we may cancel the anomaly in $M'$ by the Green-Schwarz mechanism. However, for $D' = 2$, $X_{2n'-2}$ is a zero-form; therefore, the only terms present in the I-form are the leading terms in the anomaly. One can not cancel the leading terms by adding local counter terms to the action [1-3,13], so the Green-Schwarz mechanism breaks down at $D' = 2$. We conclude that if the original anomaly is cancelled by the Green-Schwarz mechanism, then the theory will remain anomaly-free after compactification, provided that: 1) Neither isometries nor U(1) gauge symmetries are generated; 2) There is a field strength H such that $dH_0 = (\mathrm{Tr}\,R_0^2 + k\,\mathrm{Tr}\,F_0^2)$, or the compact space is chosen such that $(\mathrm{Tr}R_0^2 + k\,\mathrm{Tr}F_0^2) = 0$; 3) The space $M'$ has dimension $D' > 2$.

## CONCLUSIONS

We have seen that if we compactify any supergravity theory with no isometries, then the I-form after compactification is proportional to the integral over the compact space of the I-form prior to compactification, subject to the following restriction on self-dual antisymmetric tensors: If the theory contains SDA's, then it must not be compactified from $D = 4k + 2$ to $D' = 4k' + 2$.

With this result, we were able to show that anomaly-free theories remain anomaly-free after compactification, subject to some constraints on the compactification. In particular, there were two cases:

i. $I_{2n+2} = 0$. Here the anomaly vanishes trivially prior to compactification. Subject to the constraints on isometries and SDA's, it remains anomaly-free after compactification to any lower dimension.

ii. $I_{2n+2} = (\mathrm{Tr}R^2 + k\,\mathrm{Tr}F^2)X_{2n-2}$. Here the I-form factorizes, such that the anomaly cancels by the Green-Schwarz mechanism. Again subject to the constraint on self-dual antisymmetric tensors, the I-form after compactification will factorize and cancel by the Green-Schwarz mechanism as long as: 1) Neither isometries nor U(1) gauge symmetries are generated; 2) There is a field strength H such that $dH_0 = (\mathrm{Tr}R_0^2 + k\,\mathrm{Tr}F_0^2)$, or the compact space is chosen such that $(\mathrm{Tr}R_0^2 + k\,\mathrm{Tr}F_0^2) = 0$; 3) The space $M'$ has dimension $D' > 2$.

## ACKNOWLEDGEMENTS

We wish to thank the organizers of the conference for their hospitality. This work was supported in part by DOE contract # DE-AC02-86ER40253.

## REFERENCES

[1] L. Alvarez-Gaumé and E. Witten, Nucl. Phys. B234 (1983) 269;
[2] B. Zumino, W. Yong-Shi and A. Zee, Nucl. Phys. B239 (1984) 477;
[3] L. Alvarez-Gaumé and P. Ginsparg, Annl. Phys. 161 (1985) 423.
[4] M.F. Atiyah and I.M. Singer, Ann. of Math 87 (1968) 485, 546; 93 (1971) 1,119,139.
[5] E. Witten, Phys. Lett. 149B (1984) 351.
[6] M.B. Green, J.H. Schwarz and P.C. West, Nucl. Phys. B254 (1985) 327.
[7] A. Das and Y.H. Kwon, Phys. Rev. D35 (1987) 1508.
[8] J.E. Bjorkman and Y. Tosa, Colo-Hep-144, (to be published in Physics Letters B).
[9] E. Witten, "Fermion Quantum Numbers in Kaluza-Klein Theory", in Shelter Island II, edited by R. Jackiw, (M.I.T. Press, Cambridge, 1985) p.227.
[10] F. Hirzebruch, Topological Methods in Algebraic Geometry, (Springer Verlag, New York, 1978) pp.8-16.
[11] M.B. Green and J.H. Schwarz, Phys. Lett. 194B (1984) 117; Nucl. Phys. B255 (1985) 93.
[12] P. Candelas, G.T. Horowitz, A. Strominger and E. Witten, Nucl. Phys. B258 (1985) 46.
[13] L. Baulieu, Phys. Lett. 167B (1986) 56.

MANY NEW ANOMALY-FREE THEORIES IN HIGHER DIMENSIONS

Yasunari Tosa
Department of Physics
University of Colorado
Campus Box 390
Boulder, CO 80309

Susumu Okubo
Department of Physics & Astronomy
University of Rochester
Rochester, NY 14627

## I. INTRODUCTION

I report on the works done in collaboration with Susumu Okubo.[1,2]

One of many surprises in superstring theories is the anomaly-free property. People initially believed that the (local) anomaly-free constraints were so strong that no other theories would exist, except those superstring theories with the gauge group $E_8 \otimes E_8$ or SO(32) formulated in D=10 space-time.[3] Thus, the belief of the ultimate theory being found made a quite stir in the physics community in the past. However, we show in this talk that there are infinitely many anomaly-free configurations in various space-time dimensions. Unfortunately, the anomaly-free condition constrains only the fermionic sector of a theory and thus we are still far from constructiong a real physical theory. However, we believe that there may be new structures beyond superstrings which use some of our new anomaly-free solutions.

Because the anomaly-free property is required for the consistency of the gauge principle, any physical theories must satisfy this condition. Therefore, any theories which are to be discovered in future such as supermembrane theories etc must be a subset of anomaly-free theories. Fortunately, we can write down the condition for the anomaly-free property for *field theory limit*, without referring to the underlying physical structures. Thus, we have set out to find *all* possible anomaly-free theories under certain assumptions. We do not assume the underlying structure to be string-type. Our search is a systematic one, which employs Lie algebra trace-identities as much as possible. Thus, our search needs no fluke, but patience. After completion of our works, we were notified the fact that using string-like structures (modular invariance), one can also generate many anomaly-free solutions.[4] Within string theories, people also have found alternatives in various dimensions.[5]

Our work was motivated by Thierry-Mieg,[6] who has extended the search for anomaly-free theories beyond D=10 and found solutions in D=18 and 26. Schellekens[7] has found a solution in D=14. We will find these solutions among *infinitely* many solutions. Actually, finding many solutions was a great surprize to us, since at the beginning of this investigation we thought to prove the uniquness of a few solutions.

## II. FERMIONIC CONTENT AND ANOMALIES

### A. Fermionic content

Since we do not know the underlying theory which yields an anomaly-free field theory, we must fix the fermionic content *by hand*. Once we fix the fermionic content, we can immediately write down the anomaly contribution of these fermions.[8] In this work, we choose the fermionic content which closely immitates the N=1 D=10 supergravity theory. That is, we assume that the massless fermionic sector contains one gravitino and chiral spin 1/2 Y.M. matter fields with *not only* an adjoint rep *but also* multiple copies of non-trivial irreducible representations (irreps) of a *simple* group. The presence of additional irreps besides an adjoint rep is the departure from ordinary superstring theories. If a non-trivial irrep is complex, multiple copies of its conjugate are also allowed. For SO(2n), two different kinds of spinors ($\lambda_n$ and $\lambda_{n-1}$) are allowed. We also find solutions which do not satisfy all the assumptions on the representation. We also discuss cases which do not contain gravitinos. We allow as many Y.M. gauge singlets as required to cancel the pure gravitational anomaly. The allowance of many singlets is another difference between this work and related papers.

Thus, our assumptions are set up as generally as possible for new theories. The assumption of a simple group is the most restrictive one (or the weakest, depending upon a point of view). Various new superstring theories use semi-simple groups.[5] Note that our method is local and thus we cannot, for example, distinguish SO(32) from Spin(32)/$Z_2$.

### B. Anomalies

We look at theories in even-dimensional space-time where chirality is naturally defined. As shown by Alvarez-Gaume and Witten; and by Alvarez-Gaume and Ginsparg,[8] anomalies for a Rarita-Schwinger field and spin 1/2 chiral fermions at $D=2n$ are related to the $(2n+2)$-forms in the following polynomials:

$\hat{A}(R)\mathrm{Tr}\left(\exp\left(\frac{iR}{2\pi}\right)-1\right)+(4k-3)\hat{A}(R)$ for a Rarita-Schwinger field,

$-\ell\hat{A}(R)$ for $\ell$ chiral spin 1/2 (auxiliary and Y.M. singlet) fields,

$\sum\epsilon\hat{A}(R)\mathrm{ch}(F)$ for Yang-Mills spin 1/2 fermions with chirality $\epsilon$,

where $\hat{A}$ is the Dirac genus and ch(F) denotes the Chern character. The Chern character is given by $\mathrm{ch}(F)=\mathrm{Tr}[\exp(iF/(2\pi))$. The summation symbol for Y.M. matter fields implies the sum over various chiralities $\epsilon$ also. Y.M. singlets can be counted as auxiliary fields.

Note that $\hat{A}(R)$ has a generating function:[1,9]

$$\sum_{j=0}^{\infty}\hat{A}_j(R)z^j=\exp\left[\sum_{k=1}^{\infty}\frac{B_k}{4k(2k)!}\frac{\mathrm{Tr}R^{2k}}{(2\pi)^{2k}}z^k\right] \tag{2.1}$$

where $\hat{A}_j(R)$ is a 4j-form and $B_k$ is a Bernoulli number, e.g. $B_1=1/6$, $B_2=$

1/30, $B_3 = 1/42$, $B_4 = 1/30$, $B_5 = 5/66$, $B_6 = 691/2730$, $B_7 = 7/6$, $B_8 = 3617/510$. This exponential form is a generic one for *any* multiplicative sequences.[1,9]

Note that the fundamental difference exists between theories in $D = 4k - 2$ and those in $D = 4k$ : The gravitational part $\hat{A}(R)$ contains only 4m-forms, while the Y.M. part contains $2\ell$-forms. Thus, anomalies for theories in $D = 4k - 2$ have contributions from both pure gravitational particles and Y.M. particles, while anomalies for theories in $D = 4k$ have contributions from only Y.M. particles. Therefore, we must discuss theories in D=4k−2 and D=4k separately, since constraints are completely different as we will see.

## III. CHIRAL FERMIONS IN FOUR DIMENSION

Because we are discussing higher dimensional theories, we must discuss constraints of *chiral* four dimensional theories after compactification. After compactification, the Yang-Mills field usually acquires a vacuum expectation value and the original gauge group is broken. We denote the original group as G and the broken group as $G' \otimes G^0$ where $G^0$ is the part got the vacuum expectation value (vev) and $G'$ is the four dimensional gauge group. Correspondingly, the original rep $\Lambda$ is broken into $\sum_j (\Lambda'_j, \Lambda^0_j)$.

Now, in order to get the chiral four dimensional theory, we must satisfy two conditions for spin 1/2 fermions with a rep $\Lambda'$ for the gauge group $G'$ in four dimension:

i) $n_{1/2}(\Lambda') = n^L_{1/2}(\Lambda') - n^R_{1/2}(\Lambda') \neq 0$.

ii) $n_{1/2}(\Lambda'^*) \neq n_{1/2}(\Lambda')$ if fermions with the complex conjugate rep $\Lambda'^*$ exists.

The second condition comes about, because in four dimension a left-handed fermions with $\Lambda'^*$ can be regarded as a right-handed fermions with $\Lambda'$. The second condition immediately tells us that *the rep $\Lambda'$ for the four dimensional gauge group $G'$ must be complex* and thus $G'$ *must contain* U(1), SU(n) $(n \geq 3)$, SO(4n+2) $(n \geq 2)$, *or* $E_6$, assuming $G'$ to be compact. Witten[10] realized that the first conditions applies to only zero modes and these zero modes are correlated to the zero modes in the compact space. Using the index theorem by Atiyah and Singer, the zero modes in the compact space can be given by the integral of the form given in the previous section over the compact space. Thus, we have the simple consequences:

i) For theories in D=4k-2, both $G'$ and $G^0$ must contain U(1), SU(N) $(N \geq 3)$, or SO(4N+2) $(N \geq 2)$. However, there is no constraints on the original gauge group G. The Yang-Mills F must acquire a vev for the compact space.

ii) For theories in D=4k, both G and $G'$ must contain complex groups listed in i). The Yang-Mills F does not have to acquire a vev.

iii) No four dimensional chiral fermions exist for D=odd theories.

It may be interesting to note that in D=4k-2 the *maximal* subgroup solutions which give four dimensional chiral fermions are only the following:

$$\begin{aligned} SO(4p) &\longrightarrow SO(4q-2)\otimes SO(4(p-q)+2)\\ F_4 &\longrightarrow SU(3)\otimes SU(3)\\ E_7 &\longrightarrow SU(3)\otimes SU(6)\\ E_8 &\longrightarrow SU(3)\otimes E_6\\ E_8 &\longrightarrow SU(5)\otimes SU(5)\end{aligned}$$

Note that $SO(6)\otimes SO(10)$ is not a maximal subgrup of $E_8$, since this subgroup is contained in another maximal subgroup SO(16).

## IV. ANOMALY-FREE SOLUTIONS IN D=4k−2

### A. Anomaly-free constraints in D=4k-2

First, we show that in $D = 4k-2$ the immediate consequence of our assumption on the fermionic content is that *no anomaly-free theories at* $D = 22$ *and beyond* $D > 26$ *exist* in order to have the leading pure gravitational anomaly cancelled. Note that this conclusion is obtained without discussing the cancellation of the rest of the anomalies, i.e. pure Y.M. and mixed anomalies. For theories in $D = 4k$, we do not obtain the limit on the space-time dimension.

For theories with one gravitino, the total anomaly in $D = 4k-2$ is related to the 4k-forms

$$I_k = (\sum \epsilon n_G + 4k - 3 - \ell)\hat{A}_k + \sum_{m=0}^{k-1} \hat{A}_m \left[G_{k-m} + \frac{(-)^{k-m}\sum \epsilon F_{k-m}}{(2(k-m))!}\right] \tag{4.1}$$

where $n_G$ is the dimension of an irrep with chirality $\epsilon$ and

$$G_k = \frac{(-)^k 4k}{B_k} C_k R_k, \qquad F_k = \frac{TrF^{2k}}{(2\pi)^{2k}}, \qquad R_k = \frac{TrR^{2k}}{(2\pi)^{2k}}, \qquad C_k = \frac{B_k}{4k(2k)!},$$

$$\sum_{j=0} \hat{A}_j z^j = \exp(\sum_{m=1} C_m R_m z^m) \quad (\hat{A}_0 = 1).$$

The leading $TrR^{2k}$-term should vanish. Thus, we must have

$$n + (-)^k \frac{4k}{B_k} = 0 \qquad \text{where } n = \sum \epsilon n_G - \ell + 4k - 3. \tag{4.2}$$

We obtain the upperlimit on n,

$$|n| < \sqrt{\frac{k}{\pi}}\left(\frac{\pi e}{k}\right)^{2k}, \tag{4.3}$$

**Table 1.** Solutions for the pure gravitational anomaly cancellation. Thierry-Mieg solutions[7] are the last column.

| D | k | n | $\sum \epsilon n_G(\ell=1)$ |
|---|---|---|---|
| 2 | 1 | 24 | 24 |
| 6 | 2 | −240 | −244 |
| 10 | 3 | 504 | 496 |
| 14 | 4 | −480 | −492 |
| 18 | 5 | 264 | 248 |
| 26 | 7 | 24 | 0 |

using an approximate expression for a Bernoulli number:

$$B_k > 4\sqrt{\pi k}\left(\frac{k}{\pi e}\right)^{2k} \quad \text{(the difference is less than 1\% for } k \geq 5)$$

which can be derived from Euler's identity, relating a Bernoulli number with a Riemann's zeta function.[1] For $k \geq 9$, the right hand side of Eq.(4.3) is less than .66. For $k=8$, $4k/B_k$ is 16320/3617, and for $k=6$, $4k/B_k$ is 65520/691. Hence, the integer solutions for Eq.(4.2) are only those in Table 1.

Following Green-Schwarz,[3] we can cancel the remaining anomalies by adding local counter terms, provided that the rest of the 4k-form terms takes on the form

$$I_k = (R_1 + \alpha)X_{k-1} \quad (k \geq 2) \tag{4.4}$$

where $X_{k-1}$ is a gauge invariant $2(k-1)$-form made out of two forms F and R. The function $\alpha$ is a gauge-invariant Yang-Mills 2-form, which is not necessarily in proportion to $\sum \epsilon F_1$. (In the case of non-supergravity theories, it is not as we will show later.) This way of cancelling the anomalies imposes trace constraints on F. Note the trivial fact that Y.M. singlets do not contribute to these Y.M. traces, but only contribute to the pure gravitational anomaly. Therefore, by suitably adjusting the number of singlets, *the problem of finding anomaly-free solutions reduces to the problem of finding a rep satisfying trace identities in the Lie algebra.* Note that in the case of $k=1$ (D=2), the method above does not work. The only way out is to use U(1) group(s), in order for $\sum \epsilon F_1$ to factorize, since TrF = 0 for a (semi-)simple group.

For the case of theories with one gravitino coupled with Y.M. matter, the coefficient $\alpha$ has a closed expression, which can be found by looking at the coefficients of $R_1 R_{k-1}$ and $R_{k-1}$ terms and is in proportion to $\sum \epsilon F_1$. The result is

$$\alpha = -\frac{6}{(-)^{k-1}\left[\frac{k}{B_k} + \frac{k-1}{B_{k-1}}\right] - \frac{1}{B_1}} \sum \epsilon F_1 = \alpha'(\sum \epsilon F_1) \quad (k \geq 3). \tag{4.5}$$

Thus, $\alpha' = -1/30$ (D=10), $1/42$ (D=14), $-1/30$ (D=18), respectively, which happen to be $-B_2$, $B_3$, $-B_4$. The cases of D=6 and D=26 are exceptions: in D=6, $\alpha$ cannot be fixed, and in D=26 $\alpha$ is overdetermined if $\epsilon F_1 \neq 0$ and $\alpha = 0$ if $\epsilon F_1 = 0$.[1] Hereafter, we omit the summation symbol in front of $\epsilon F_j$.

Table 2. Anomaly-free constraints on the Y.M. fields for theories with one gravitino. Note that $F_k = \mathrm{Tr}F^{2k}/(2\pi)^{2k}$.

| D | k | Y.M. constraints |
|---|---|---|
| 6 | 2 | $\epsilon F_2 = \frac{1}{4}\alpha(6\alpha - \epsilon F_1)$ |
| 10 | 3 | $\epsilon F_3 = \frac{1}{48}(\epsilon F_2)(\epsilon F_1) - \frac{1}{14400}(\epsilon F_1)^3$ |
| 14 | 4 | $\epsilon F_2 = -\frac{1}{196}(\epsilon F_1)^2$ |
| | | $\epsilon F_4 = -\frac{1}{36}(\epsilon F_3)(\epsilon F_1) + \frac{5}{7112448}(\epsilon F_1)^4$ |
| 18 | 5 | $\epsilon F_2 = \frac{1}{100}(\epsilon F_1)^2$ |
| | | $\epsilon F_3 = \frac{1}{7200}(\epsilon F_1)^3$ |
| | | $\epsilon F_5 = \frac{1}{16}(\epsilon F_4)(\epsilon F_1) - \frac{7}{69120000}(\epsilon F_1)^5$ |
| 26 | 7 | $0 = \epsilon F_2 = \epsilon F_3 = \epsilon F_4 = \epsilon F_5 = \epsilon F_7$ and $\alpha = 0$ |

Now, the Y.M. constraints with one gravitino are given in Table.2. The case D=10 is the famous one discovered by Green and Schwarz.[3] Note that the coefficients for the trace constraints are expressed in terms of Bernoulli numbers. Beyond D=10, the pure gravitational anomaly cancellation according to the Green-Schwarz mechanism becomes non-trivial. For D=14, we have two non-trivial equations for n

$$0 = n + \frac{16}{B_4} \text{ for } \mathrm{Tr}R^8, \qquad 0 = \frac{n}{2} + \frac{8}{B_2} \text{ for } (\mathrm{Tr}R^4)^2 \tag{4.6}$$

For D=18, we have

$$0 = n - \frac{20}{B_5} \text{ for } \mathrm{Tr}R^{10}, \qquad 0 = n + \frac{8}{B_2} - \frac{12}{B_3} \text{ for } \mathrm{Tr}R^4\mathrm{Tr}R^6. \tag{4.7}$$

For D=26, we have four independent equations for n and all of them are satisfied by n=24:

$$0 = n - \frac{28}{B_7} \text{ for } \mathrm{Tr}R^{14}, \qquad 0 = n - \frac{20}{B_5} + \frac{8}{B_2} \text{ for } \mathrm{Tr}R^4\mathrm{Tr}R^{10},$$

$$0 = n + \frac{16}{B_4} - \frac{12}{B_3} \text{ for } \mathrm{Tr}R^6\mathrm{Tr}R^8, \qquad 0 = \frac{n}{2} - \frac{6}{B_3} + \frac{8}{B_2} \text{ for } (\mathrm{Tr}R^4)^2\mathrm{Tr}R^6. \tag{4.8}$$

As one can see from Table 1, it seems that finding solutions to these constraints is very difficult. The amazing thing is that we have found solutions to all the dimensions (D=6,10,14,18) and except in the case of D=18 we have found infinitely many solutions, as we will see in the next subsection. For D=26, it seems that no Yang-Mills field is required.

In the case of theories with no gravitinos, the trace constrains are much simpler.[1] The 4k-form in $D = 4k - 2$ is given by

$$I_k = (\sum \epsilon n_G - \ell)\hat{A}_k + \sum_{m=0}^{k-1} \hat{A}_m \frac{(-)^{k-m}}{(2(k-m))!}\epsilon F_{k-m}. \tag{4.9}$$

We require this to factorize as

$$(R_1 + \alpha)(r_{k-1} + r_{k-2} + \ldots + r_1 + r_0) \tag{4.10}$$

where $r_m$ is the term containing m curvature two-forms. Thus, we have

$$\sum \epsilon n_G - \ell = 0 \tag{4.11}$$

$$\epsilon F_1 = \epsilon F_2 = \ldots = \epsilon F_{k-2} = 0 \tag{4.12}$$

$$\epsilon F_k = -\frac{k(2k-1)}{24}\alpha(\epsilon F_{k-1}). \tag{4.13}$$

Note that $\alpha$ is not fixed in this case.

B. Anomaly-free solutions in D=4k-2

The allowance of many shadow matter fields leads us to many new anomaly-free theories. First of all, by adjusting the number of Y.M. singlets, all solutions at D=18 are automatically solutions at D=10. Similarly, all solutions at D=14 and 18 are solutions at D=6, but not in vice versa. What we mean is that a Y.M. matter rep of a group in one dimension may be used in a different dimension and that theory will also be anomaly-free.

In order to make our results concise, we hereafter use the following notation: $\Lambda$ denotes the rep and the symbol [a] denotes the largest integer which does not exceed a. The integers, m and m′, can be negative, unless otherwise specified. Negative integers imply that the chirality for those fields is negative. The symbol, $\lambda_j$ $(1 \leq j \leq n)$, refers to the n fundamental weight system for a simple Lie algebra of rank n. The underlined numbers denote the dimensions of irreps. We call solutions *regular* if the spin connection can be embedded into the gauge group. This condition is needed if one wants to use the holonomy group for dimensional reduction. Here, we have used the Schellekens criterion:[7]

The list of our results for a non-trivial part of Y.M. irreps is the following:[1]

D = 6 (all solutions are regular)

i) $\Lambda$ = adjoint $\oplus$ (any number, any chirality of lowest dim. reps) of $G_2$, $F_4$, $E_6$, $E_7$.

ii) $\Lambda$ = adjoint $\oplus$ m vectors $\ominus$ m′ spinors of SO(N),

where $m + 2^{[\frac{N-9}{2}]} m' = 8 - N$. For $N = 2n$, m′ is the sum of the numbers of two kinds of spinors, $\lambda_n$ and $\lambda_{n-1}$, except N=8 where their numbers must be equal.

iii) $\Lambda$ = adjoint $\ominus$ m $\lambda_j$ of Sp(2n),
where n and m are integer solutions of the following:

$$m = \frac{(n+4)(n-1)(j-1)!\,(2n+1-j)!}{(n+1-j)\,[2n^2+3n+4-3j(2n+2-j)]\,(2n-2)!}.$$

The case where j = 1 is always a solution for an arbitrary n with m = 2(n+4). The largest rank solution with $j \neq 1$ up to Sp(200) is given by (j = 2, m = 2) of Sp(24).

iv) $\Lambda$ = adjoint $\ominus$ m $\lambda_j$ $\ominus$ m′ $\lambda_j^*$ of SU(N) ($j \neq 1$ nor $N-1$ and $N \geq 4$),

where N and (m + m′) are integer solutions of the following:

$$m + m' = \frac{2N\,(j-1)!\,(N-j-1)!}{[N(N+1)-6j(N-j)]\,(N-4)!}.$$

The largest group with a solution up to SU(100) is SU(24) with j = 2. We do not know whether there exists a solution beyond SU(100).

v) $\Lambda$ = adjoint $\oplus$ (any number, any chirality)(<u>3</u> and <u>3</u>*) of SU(3).

<u>D = 10</u> (all solutions are regular except those with a $\sharp$ superscript)

i) $\Lambda$ = adjoint $\oplus$ m vectors $\oplus$ m′ spinors of SO(N),

where $m + 2^{[\frac{N-7}{2}]} m' = 32 - N$. For N = 2n, m′ is the sum of the numbers of spinors, $\lambda_n$ and $\lambda_{n-1}$, except N=8 and 12 where their numbers must be equal. For m = m′ = 0, N must be 32, which is the Green-Schwarz solution.

ii) $\Lambda^{\sharp}$ = adjoint $\ominus$ (N + 32) vectors of Sp(N) (arbitrary even N)

iii) $\Lambda$ = adjoint of $E_8$ (the Green-Schwarz solution)

iv) $\Lambda^{\sharp}$ = $\ominus$<u>78</u> (adjoint) $\ominus$<u>27</u> $\oplus$ <u>351</u> ($\lambda_2$) of $E_6$

v) $\Lambda$ = <u>78</u> $\oplus$ m <u>27</u> $\oplus$ m′ <u>27</u>* (m + m′ = 6) of $E_6$.

<u>D = 14</u> (all regular solutions)

i) $\Lambda$ = <u>153</u> $\ominus$ 26 <u>18</u> $\ominus$ <u>256</u> of SO(18)

ii) $\Lambda$ = <u>91</u> $\ominus$ 22 <u>14</u> $\ominus$ m <u>64</u> $\ominus$ m′ <u>64</u>* (m + m′ = 4) of SO(14)

iii) $\Lambda$ = <u>45</u> $\ominus$ 18 <u>10</u> $\ominus$ m <u>16</u> $\ominus$ m′ <u>16</u>* (m + m′ = 16) of SO(10)

iv) $\Lambda$ = <u>133</u> $\ominus$ 10 <u>56</u> of $E_7$ (the Schellekens solution)

v) $\Lambda$ = <u>78</u> $\ominus$ m <u>27</u> $\ominus$ m′ <u>27</u>* (m + m′ = 18) of $E_6$.

$\underline{D=18}$ (regular solution)

$\Lambda = \underline{248}$ of $E_8$ (the Thierry-Mieg solution).

We have eliminated trivial subgroup solutions from the list. Note that we must add an appropriate number of singlets to cancel the pure gravitational anomaly. We could not find any theories with just one antichiral spin 1/2 field, except those found by Green-Schwarz[3] and Thierry-Mieg.[6] Most of the solutions cannot be obtained from higher dimensional theories, although some of them may be, just like the Schellekens solution at D=14 which comes from D=18 by compactifying four dimensions as $K_3$. We have not investigated this possibility. We do not know at this stage whether these new anomaly-free configurations could be derivable from string theories or not. We have also found few solutions which do not contain an adjoint rep:

$\underline{D=6}$

$\Lambda = \lambda_2 \oplus m\,\lambda_1$ $(N+m=8)$ for Sp(N) (N:even) and SU(N)

$\underline{D=10}$

i) $\Lambda = \lambda_4$ ($\underline{495}$) of SO(12)

ii) $\Lambda = \lambda_2$ ($\underline{495}$) of Sp(32)

iii) $\Lambda = \lambda_2 \oplus \lambda_4$ ($\underline{65} + \underline{429}$) of Sp(12)

iv) $\Lambda = \lambda_2 \oplus (32-N)\lambda_1$ of Sp(N) (N:even)

v) $\Lambda = (2\lambda_1) \ominus (N+32)\lambda_1$ of SO(N)

$\underline{D=18}$

$\Lambda = \lambda_3 \ominus \lambda_1$ ($\underline{273} \ominus \underline{26}$) of $F_4$.

Note that none of these is a regular solution.

Although we regard shadow matter fields as gauge singlets, they can be gauged, provided that: 1) We assign another group *only* to the shadow world. 2) Only shadow particles have quantum numbers for the shadow gauge group. 3) The Y.M. constraint is satisfied by themselves. The simplest well-known example is the $E_8$ case at D=10. The adjoint $\underline{248}$ of $E_8$ does satisfy the D=10 Y.M. constraint, but we need $\ell = -247$ shadow particles. When we assign the shadow group $E_8$ to 248 of them, then the whole theory becomes the $E_8 \otimes E_8$ theory with the rep ($\underline{248}$, 1) $\oplus$ (1, $\underline{248}$) and only one auxiliary field.

For theories without gravitinos, we have a solution with two irreps in D=10 as follows:[1] The rep,

$$\Lambda = m \text{ spinors} \ominus m\; 2^{[\frac{N-7}{2}]} \text{ vectors of SO(N)} \tag{4.14}$$

where m is an arbitrary positive or negative integer. Note that for even N, any combinations of two kinds of spinors are acceptable, except for N=8 and 12 where the numbers of the two different kinds of spinors have to be equal. For SO(8), we do not need any singlets:

$$\Lambda = m\ (\underline{8}\ (\text{spinor}) \oplus \underline{8}'\ (\text{spinor}'))\ominus 2m\ \underline{8}\ (\text{vector}) \tag{4.15}$$

for any positive or negative m. A more complicated solution is derived from the heterotic string theory by Alvarez-Gaume et al. and by Dixon and Harvey.[5]

## V. ANOMALY-FREE SOLUTIONS IN D=4k

### A. Anomaly-free constraints

As mentioned in Section II, in D=4k only the Yang-Mills particles are constrained. Hereafter, we omit the chirality symbol $\epsilon$.

For theories in $D=4k$, we have the explicit form

$$I_{4k+2} = \hat{A}_k \frac{i\mathrm{Tr}\ F}{2\pi} + \hat{A}_{k-1}\frac{i^3\mathrm{Tr}\ F^3}{(2\pi)^3} + \dots + \frac{i^{2k+1}\mathrm{Tr}\ F^{2k+1}}{(2\pi)^{2k+1}} \tag{5.1}$$

Consequently, any theory with $\mathrm{Tr}\ F^{odd} = 0$ ($odd \leq 2k+1$) is anomaly-free in $D=4k$. That is, any group which does not contain U(1), SU(n) ($n \geq 3$), SO(4n+2) ($n \geq 2$), or $E_6$, can be used to make a theory anomaly-free, since in this case we have $F = -SF^tS^{-1}$ for some non-singular matrix S, and thus $\mathrm{Tr}\ F^{odd} = 0$. However, as we have seen in Section III, these groups cannot yield chiral fermions at four dimension, if no isometries are generated after compactification.

In order to have an anomaly-free theory at $D=4k$ *a la* Green-Schwarz, we must satisfy

$$\begin{cases} \mathrm{Tr}\ F = \mathrm{Tr}\ F^3 = \dots = \mathrm{Tr}\ F^{2k-3} = 0 \\ \mathrm{Tr}\ F^{2k+1} = c\ \mathrm{Tr}\ F^{2k-1} \end{cases} \tag{5.2}$$

where c is an arbitrary constant in proportion to $\mathrm{Tr}\ F^2$ of some representation.

Note that we have for a rep $\Lambda = \sum m_j \Lambda_j$ ($\Lambda_j$ : irreducible) of G:

$$\mathrm{Tr}F(\Lambda)^n = \left(\sum m_j Q_n(\Lambda_j)\right)\ \mathrm{Tr}F(\Box)^n + (\text{products of lower order traces of } F(\Box))$$

where indices (Casimir invariants), $Q_n(\Lambda_j)$, are normalized to a rep, $\Box$.[12] Thus, it is necessary to have

$$\sum m_j Q_n(\Lambda_j) = 0 \quad \text{for } n = 1,\ 3,\ \dots,\ 2k+1 \tag{5.3}$$

*except* $n = 2k-1$.

### B. Anomaly-free solutions in D=4k

In this subsection,[2] we will find solutions to two constraints: 1) anomaly-free property and 2) chiral four dimensional theories. For simplicity, we assume that the groups G be *simple*. Then, according to Section III, G is one of SU(N) ($N \geq 3$), SO(4N+2) ($N \geq 2$), or $E_6$.

Because G uses a complex rep and still satisfies vanishing of most of odd order traces, it is usually hard to find a solution with just one irrep. Even with two irreps, it becomes harder to find a solution as one looks at a higher dimension. Here, we are satisfied with solutions up to $D = 12$ dimension. Our strategy is: 1) try to find solutions with a single irrep; 2) if we cannot find them, then try to find solutions with two irreps. We also use the Dynkin notation for an irrep: $\Lambda = m_1\lambda_1 + m_2\lambda_2 + \cdots + m_n\lambda_n = (m_1, m_2, \cdots, m_n)$.

The general solution for any k in D=4k is given by:

Any complex rep of $SO(4\ell+2)$ with either $\ell \geq k+1$ or $\ell = k-1$ in $D = 4k$, since the only non-vanishing odd order index for SO(2n) is $Q_n$.

Now, we give solutions for each D (D=4,8,12).

<u>D = 4</u>

i) Any complex rep of $E_6$ and $SO(4n+2)$ $(n \geq 2)$.

ii) A single irrep solution:[11]

$$\Lambda = \lambda_3 + \lambda_{21} \text{ of } SU(32).$$

$$\Lambda = 5\lambda_1 + \lambda_2 + 8\lambda_3 + 4\lambda_4 \text{ of } SU(5).$$

iii) Two irrep soutions with $\Lambda = \lambda_j \oplus \lambda_k$ of SU(n) where[11]

$$1)\ j = \frac{1}{2}(n \pm 1 - \sqrt{n-1}), \quad k = \frac{1}{2}(n \pm 1 + \sqrt{n-1}),$$

$$2)\ j = \frac{1}{2}(n \pm 2 - \sqrt{n}), \quad k = \frac{1}{2}(n \pm 2 + \sqrt{n}),$$

where n, j and k must be integers. One of the solution is the famous SU(5): $5^* \oplus 10$.

<u>D = 8</u>

i) any complex irrep of SU(3) and SU(4).

ii) (010 . . .) = <u>120</u> of SU(16).

iii) (110 . . .) = <u>240</u> of SU(9); (0010 . . .) = <u>2925</u> of SU(27).

iv) (020 . . .) = <u>3185</u> of SU(14); (1010 . . .) = <u>7140</u> of SU(16).

v) (10010 . . .) = <u>263120</u> of SU(25).

vi) (030 . . .) = <u>41405</u> of SU(13); (0020 . . .) = <u>1163800</u> of SU(24).

vii) Two $E_6$ solutions are

$$\Lambda = (010000) - 11(100000) = \underline{351} - 11 \cdot \underline{27},$$

$$\Lambda = (200000) - 4(010000) = \underline{351'} - 4 \cdot \underline{351}.$$

viii) Any complex irrep of SO(4n+2) ($n \geq 3$).

D = 12

i) Any complex irrep of $E_6$ is a solution.

ii) Any complex irrep of SO(4n+2) ($n \geq 2$), except those of SO(14) which have non-vanishing $Q_7$.

## VI. HOW TO SOLVE TRACE IDENTITIES

In this section, we explain how we solved trace-identities. Hereafter, we use the notations: $d(\rho)$ = dimension of the irrep $\rho$ and $d_0$ = dimension of an adjoint rep.

We discuss only the fourth order trace identity. First, we deal with the second order trace $\mathrm{Tr}F(\rho)^2$ for an irrep $\rho$. As is well-known, we have for a generator $X_\mu(\rho)$,

$$\mathrm{Tr}X_\mu(\rho)X_\nu(\rho) = \frac{d(\rho)}{d_0} I_2(\rho) g_{\mu\nu}, \tag{6.1}$$

where $I_2(\rho)$ is the second order Casimir invariant defined by $I_2(\rho) = g^{\mu\nu}X_\mu(\rho)X_\nu(\rho)$ and $g_{\mu\nu}$ is the Killing metric ($g^{\mu\nu}$ is the inverse). For a generic element, $F(\rho) = \sum \xi^\mu X_\mu(\rho)$ with some coefficient $\xi^\mu$, we have

$$\mathrm{Tr}F(\rho)^2 = \sum \xi^\mu \xi^\nu \mathrm{Tr}X_\mu(\rho)X_\nu(\rho) = f(\xi)\frac{d(\rho)}{d_0} I_2(\rho),$$

with $f(\xi) = g_{\mu\nu}\xi^\mu\xi^\nu$, or

$$\mathrm{Tr}F(\rho)^2 = Q_2(\rho)\mathrm{Tr}F(\square)^2, \tag{6.2}$$

with

$$Q_2(\rho) = \frac{d(\rho)I_2(\rho)}{d(\square)I_2(\square)}. \tag{6.3}$$

The quantity $Q_2(\rho)$ is called the second order index.[12] The irrep $\square$ is usually taken to be the lowest dimensional irrep.

For the fourth order trace of $F(\rho)$, we have[12]

$$\mathrm{Tr}F(\rho)^4 - A_4(\rho)(\mathrm{Tr}F(\rho)^2)^2 = f'(\xi)D_4(\rho), \tag{6.4}$$

with

$$A_4(\rho) = \frac{3}{d_0+2}\left(\frac{d_0}{d(\rho)} - \frac{1}{6}\frac{Q_2^0}{Q_2(\rho)}\right). \tag{6.5}$$

The quantity $D_4(\rho)$ is the fourth order Casimir invariant which does not depend on $\xi^\mu$. By dividing Eq.(6.4) by the one for the irrep $\square$ and using Eq.(6.2), we obtain

$$\mathrm{Tr}F(\rho)^4 = Q_4(\rho)\mathrm{Tr}F(\square)^4 + \left[A_4(\rho)Q_2(\rho)^2 - A_4(\square)Q_4(\rho)\right](\mathrm{Tr}F(\square)^2)^2, \tag{6.6}$$

with $Q_4(\rho) = D_4(\rho)/D_4(\square)$.

Now, we can solve the trace identity of the form

$$\mathrm{Tr}F^4 = K(\mathrm{Tr}F^2)^2, \tag{6.7}$$

Using Eqs.(6.2) and (6.4) for a rep $\Lambda = \sum m_j\Lambda_j$, we must have

$$\begin{cases} \sum m_j Q_4(\Lambda_j) = 0 \\ \sum m_j \left[A_4(\Lambda_j)Q_2(\Lambda_j)^2 - A_4(\square)Q_4(\Lambda_j)\right] = K(\sum m_j Q_2(\Lambda_j))^2. \end{cases} \tag{6.8}$$

Therefore, the trace identity of the form , Eq.(6.6), is now reduced to two equations for $m_j$ and $\Lambda_j$, using indices $Q_2(\Lambda_j)$ and $Q_4(\Lambda_j)$. That is, once we know indices for various irrreps, we can solve trace identities. Similar things can be done for higher order trace identities.[1,2,12]

## VII. GLOBAL ANOMALIES

We realize that we have not yet constructed full physical theories like superstring theories, although we have found many anomaly-free configurations. However, we hope that finding these solutions become the impetus for finding alternatives to superstring theories.

Another thing we have not dealt with is the global anomaly. The global anomalies may still kill theories which have no local anomalies. In the case of superstring theories, Witten was able to show that no global anomalies exist.[13] Note that the well-known superstring theories do not possess the global pure gauge anomalies, since $\Pi_{10}(E_8) = \Pi_{10}(SO(32)) = 0$. However, it seems that the vanishing of homotopy groups for these gauge groups is coincidental.

Unfortunately, it is usually difficult to deal with the full set of global anomalies, although there exists a general formula.[14] Here, we discuss in which dimension the homotopy group vanishes. We have also started to investigate the global pure gauge anomalies, when homotopy groups do not vanish.[15] This is because the non-vanishing homotopy group does not necessarily imply the existence of the global anomaly.

For the case of classical groups, SU(n) and SO(n), as the gauge group, one usually does not have to worry about the global pure gauge anomaly, provided that one uses a suitably high rank group. This is because of Bott's periodicity theorem:[16] For theories in even D dimension, we have

$$\Pi_D(SO(n)) = 0 \quad \text{for } n \geq D+2$$

$$\Pi_D(SU(n)) = 0 \quad \text{for } n \geq \frac{D+1}{2}$$

For the case of symplectic groups, Sp(n) ($n \geq (D-1)/4$) is always safe for $D \equiv 0$, 2 or 6 (mod 8), but not $D \equiv 4$ (mod 8):

$$\Pi_D(Sp(n)) = \begin{cases} 0 & D \equiv 0,\ 2 \text{ or } 6 \pmod 8 \\ Z_2 & D \equiv 4 \pmod 8 \end{cases}$$

Table 3. Homotopy groups for exceptional groups up to D=22.[18] The number denotes the order of a cyclic group.

| | 4 | 6 | 8 | 10 | 12 | 14 | 16 | 18 | 20 | 22 |
|---|---|---|---|---|---|---|---|---|---|---|
| $G_2$ | 0 | 3 | 2 | 0 | 0 | 168+2 | 6+2+2 | 240 | 2 | 1386+8 |
| $F_4$ | 0 | 0 | 2 | 0 | 0 | 2 | 2+2 | 720+3 | 0 | 27 |
| $E_6$ | 0 | 0 | 0 | 0 | 12 | 0 | 0 | 720+6 | 1512 | 27+3 |
| $E_7$ | 0 | 0 | 0 | 0 | 2 | 0 | 2 | 12 | 2 | 108 |
| $E_8$ | 0 | 0 | 0 | 0 | 0 | 0 | 2 | 24 | 0 | 0 |

Interestingly, those groups with smaller ranks than those specified above usually have non-vanishing homotopy groups, except SO(n) with n = 5,6,7 at D=6.[17]

For exceptional groups, there exists no analogue of Bott's periodicity theorem. The relevant homotopy groups up to D=22 are given in Table 3.[18] Except D=4 and 10, some of exceptional groups have always non-vanishing homotopy groups and thus for those groups a careful analysis is needed.

ACKNOWLEDGMENT

This work is supported in part by the US Department of Energy Contract Nos. DE-AC02-86ER40253 (Y.T.) and DE-AC02-76ER13065 (S.O.). Y.T. would like to thank Professor H. Toda for providing a table of homotopy groups for exceptional groups.

REFERENCES

1. Y. Tosa and S. Okubo, Phys. Lett. **B188**, 81 (1987); Phys. Rev. D36, No. 8 (1987) to be published.
2. Y. Tosa and S. Okubo, University of Colorado preprint COLO-HEP-151 (1987).
3. M. Green and J.H. Schwarz, Phys. Lett. **B149**, 117 (1985); Nucl. Phys. **B255**, 93 (1985); M. Green, J.H. Schwarz, and P. West, Nucl. Phys. **B254**, 327 (1985).
4. A.N. Schellekens and N.P. Warner, Phys. Lett. **B181**, 339 (1986); Nucl. Phys. **B287**, 317 (1987).
5. K.S. Narain, Phys. Lett. **B169**, 41 (1986); L. Alvarez-Gaume, P. Ginsparg, G. Moore, and C. Vafa, ibid., **B171**, 155 (1986); L. Dixon and J. Harvey, Nucl. Phys. **B274**, 93 (1986); F. Ardalam and F. Mansouri, Phys. Rev. **D9**, 3341 (1974); Phys. Rev. Lett. 56, 2456 (1986); P.G.O. Freund, Phys. Lett. **B151**, 387 (1985); A. Casher, F. Englert, H. Nicolai, and A. Taormina, ibid. **162**, 121 (1985) ; F. Englert, H. Nicolai, and A. Schellekens, Nucl. Phys. **B274**, 315 (1986); A. Bilal and J.-L. Gervais, ibid. **B284**, 397 (1987); Phys. Lett. **B187**, 39 (1987); R. Bluhm and L. Dolan, Phys. Lett. **B169**, 347 (1986); H. Kawai, D.C. Lewellen, and S.H.H. Tye, Phys. Rev. Lett. 57, 1832 (1986); Phys. Rev. **D34**, 3794 (1986); Nucl. Phys. **B288**, 1 (1987); W. Lerche, D. Lüst, A.N. Schellekens, Nucl. Phys. **B287**, 477 (1987).
6. J. Thierry-Mieg, Phys. Lett. **B171**, 163 (1986).
7. A.N. Schellekens, Phys. Lett. **B175**, 41 (1986).
8. L. Alvarez-Gaume and E. Witten, Nucl. Phys. **B234**, 269 (1983); L. Alvarez-Gaume and P. Ginsparg, Ann. Phys. (New York) **161**, 423 (1985).
9. Y. Tosa, Univ. of Colorado preprint COLO-HEP-120 (1986). See also,

J.J. Gordon, Class. Quantum Grav. 1, 673 (1984); R. Delbourgo and T. Matsuki, J. Math. Phys. **26**, 1334 (1985); A.N. Schellekens and N.P. Warner, Phys. Lett. **B177**, 317 (1986).

10. E. Witten, in *Shelter Island II*, edited by R. Jackiw *et al.* (MIT press, Cambridge, 1983); M.B. Green, J.H. Schwarz, and E. Witten, *Superstring theory*, vol II, (Cambridge Univ. Press, New York, 1987).
11. Y. Tosa and S. Okubo, Phys. Rev. **D23**, 3058 (1981); S. Okubo, ibid., **D16**, 3528 (1977).
12. S. Okubo and J. Patera, J. Math. Phys. **24**, 2772 (1983); **25**, 219 (1984); S. Okubo, ibid., **26**, 2127 (1985). See also references therein.
13. E. Witten, Comm. Math. Phys. **100**, 197 (1985).
14. J-M. Bismut and D.S. Freed, Commun. Math. Phys. **106**, 159 (1986); ibid., **107**, 103 (1986). See also, D.S. Freed, ibid., **107**, 483 (1986); S. Della Pietra, V. Della Pietra, and L. Alvarez-Gaume, ibid., **109**, 691 (1987);ibid., **110**, 573 (1987); D.S. Freed and C. Vafa, ibid., **110**, 349 (1987).
15. S. Okubo, H. Zhang, Y. Tosa, and R.E. Marshak, Virginia Tech preprint 87/5 (1987); H. Zhang, S. Okubo, and Y. Tosa, University of Rochester preprint in prepartion.
16. For example, J. Milnor, *Morse theory* (Princeton Univ. Press, Princeton, 1973).
17. Nihon Sugakkai, *Encyclopedic dictionary of mathematic*, vol. II, Edited by S. Iyanaga and Y. Kawada (MIT press, Cambridge, 1977).
18. M. Mimura, J. Math. Kyoto Univ., **6**, 131 (1967); H. Kachi, Nagoya Math. J. **32**, 109 (1968); H. Toda, Japan. J. Math. **2**, 355 (1976); M. Mimura and H. Toda, Topology, **9**, 317 (1970); H. Toda, private communication (1987).

# Chapter VIII
# Supermembrane and Other Alternatives

# SUPERMEMBRANES

K.S. Stelle

The Blackett Laboratory
Imperial College
London, SW7, England

ABSTRACT

Supersymmetric theories of extended objects with spatial dimension two or greater may prove to be more viable than has hitherto been realized. In particular, the existence of models with a local fermionic gauge invariance reopens the question whether such theories can contain massless states. In a semiclassical approximation, we show that the quantum corrections to the spin-mass relation are not inconsistent with the possibility of massless states. The analysis is carried out in a topologically-stabilized sector of the super 2-membrane theory in 11-dimensional spacetime. We also show at the classical level that this theory has a consistent truncation to the type IIA superstring in 10 dimensions.

## 1. - INTRODUCTION

Supermembrane theories provide an enlarged framework for the study of supersymmetric extended objects, including superstrings. They also establish relations to supergravity theories that are not obtainable as field theory limits of superstring theories. The most striking instance of such a relation to date has been the discovery that consistent propagation of 2-dimensional supermembranes (i.e. 2-branes) in 11-dimensional spacetime requires that the background satisfy the field equations of 11-dimensional supergravity [1].

Supermembrane theories are characterized by spacetime supersymmetry and a fermionic gauge invariance analogous to that first described for the superparticle in [2]. The first example of such a theory for a spatial dimension $p > 1$ was given in [3], where the action for a super 3-brane was given in 6-dimensional flat spacetime. This example was also significant in that the supermembrane was actually obtained as a topologically-stabilized classical solution of $d = 6$, $N = 1$ super-Maxwell theory coupled to a hypermultiplet with a Fayet-Iliopoulos term added [4]. This example gives rise to the suspicion that the other possible supermembranes may also be viewed as classical solutions of appropriate supersymmetric field theories.

For supermembranes, there is a difficulty in constructing a Ramond-Neveu-Schwarz (RNS) formulation due to the presence of a world-volume cosmological constant in the action [5]. The Nambu-Goto form for the action of the bosonic membrane ($p$=2) is

$$I = \frac{-1}{4\pi^2\Xi}\int d^3\xi\sqrt{-det(\partial_i X^\mu \partial_j X^\nu \eta_{\mu\nu})}\ , \tag{1.1}$$

where $\Xi$ is the membrane coupling constant. The action (1.1) is classically equivalent to

$$I = \frac{-1}{8\pi^2\Xi}\int d^3\xi\sqrt{-g}\,[g^{ij}\partial_i X^\mu \partial_j X^\nu \eta_{\mu\nu} - 1]\ , \tag{1.2}$$

where $g_{ij}$ is an auxiliary world-volume metric whose field equation is

$$g_{ij} = \partial_i X^\mu \partial_j X_\mu\ . \tag{1.3}$$

The cosmological term in the action (1.2) is essential to obtain the correct form of the embedded metric (1.3) as the equation of motion for $g_{ij}$. This term also prevents a normal world-volume supersymmetrization, since supergravity acquires a cosmological term only from the combination of a term linear in an auxiliary field and a quadratic term in that auxiliary field which occurs in a supersymmetric kinetic term. Thus, it is only possible to add a cosmological constant at the price of introducing extra bosonic degrees of freedom on the world-volume beyond those already present in (1.1,1.2).

Although a straightforward RNS formulation of supermembrane theories does not seem possible, the work of references [3,1] showed that a Green-Schwarz formulation does exist. In the Green-Schwarz formulation, the fermions transform as spinors under the spacetime Lorentz group. In order to cut down the number of propagating spinor components, an analog of the fermionic gauge invariance of the Green-Schwarz superstring [6] is necessary. As in the case of the superstring [7], this fermionic gauge invariance requires the presence of a Wess-Zumino term in the action. Such Wess-Zumino terms are not manifestly spacetime supersymmetric, but achieve supersymmetry only after integration by parts and use of appropriate $\gamma$-matrix identities. The existence of the appropriate $\gamma$-matrix identity for a $p$-dimensional object propagating in $d$-dimensional spacetime places strong restrictions on the possible supermembrane theories.

From the conditions required for the existence of the Wess-Zumino terms, it results [8] that there are only four sequences of supermembrane theories, each beginning on a superstring theory in a classically allowed dimension (for superstrings, $p$ = 1, one can have $d = 3, 4, 6, 10$). The allowed theories are shown in the table.

<u>Super $p$-branes in $d$ spacetime dimensions</u>

| d \ p | 0 | 1 | 2 | 3 | 4 | 5 |
|---|---|---|---|---|---|---|
| 11 | | | √ | | | |
| 10 | | √ | | | | √ |
| 9 | | | | | √ | |
| 8 | | | | √ | | |
| 7 | | | √ | | | |
| 6 | | √ | | √ | | |
| 5 | | | √ | | | |
| 4 | | √ | √ | | | |
| 3 | | √ | | | | |
| 2 | | | | | | |

N=1 (p = 1)

N=2 (p = 1) N=1 (p = 2–5)

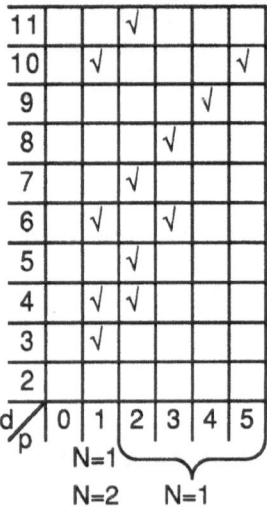

It is remarkable that the necessary $\gamma$-matrix identities hold in precisely those cases where, after fixing of the fermionic gauge symmetry and the membrane reparameterization symmetries, the numbers of propagating bosonic and fermionic components are equal [8]. Thus, for the $p$=2,$d$=11 case that we shall mainly be concerned with in the following, the original 11 bosonic components are reduced by 3 gauge symmetries ($p = 2$ spatial plus 1 time reparameterization), while the 32 fermionic components of the $d$=11 Majorana spinor are reduced to 16 by the fermionic gauge symmetry, and since these satisfy first order equations of motion, equality with the 8 propagating bosonic components holds.

The members of the four sequences are related by "double dimensional reduction" [9], which yields a ($p$-1)–brane in ($d$-1) dimensions from a $p$-brane in $d$ dimensions. This procedure was first used [9] to obtain the coupling of the type 2A superstring to $d$=10 type 2A supergravity from the super 2-brane coupled to 11-dimensional supergravity. In that case, the supermembrane target space is a supermanifold with superspace coordinates $\hat{Z}^{\hat{M}} = (\hat{X}^{\hat{m}}, \hat{\Psi}^{\hat{\mu}})$, where $\hat{m} = 1, \ldots, 11$ and $\hat{\mu} = 1, \ldots, 32$, with spacetime signature $(-, +, \ldots, +)$. The "hat" notation denotes quantities and indices in the ($d$=11, $p$+1=3) theory, and will be used to distinguish them from quantities in the ($d$=10, $p$+1=2) theory after the double dimensional reduction. One also defines $\hat{E}_{\hat{\imath}}{}^{\hat{A}} = (\partial_{\hat{\imath}} \hat{Z}^{\hat{M}}) \hat{E}_{\hat{M}}{}^{\hat{A}}(\hat{Z})$, where $\hat{E}_{\hat{M}}{}^{\hat{A}}$ is the supervielbein and $\hat{A} = (\hat{a}, \hat{\alpha})$ denotes a tangent space index ($\hat{a} = 1, \ldots, 11$ and $\hat{\alpha} = 1, \ldots, 32$). The supermembrane action is then given by [1]

$$I = \int d^3\hat{\xi} \left( \frac{1}{2}\sqrt{-\hat{g}}\hat{g}^{\hat{\imath}\hat{\jmath}} \hat{E}_{\hat{\imath}}{}^{\hat{a}} \hat{E}_{\hat{\jmath}}{}^{\hat{b}} \eta_{\hat{a}\hat{b}} - \frac{1}{6}\epsilon^{\hat{\imath}\hat{\jmath}\hat{k}} \hat{E}_{\hat{\imath}}{}^{\hat{A}} \hat{E}_{\hat{\jmath}}{}^{\hat{B}} \hat{E}_{\hat{k}}{}^{\hat{C}} \hat{A}_{\hat{C}\hat{B}\hat{A}} - \frac{1}{2}\sqrt{-\hat{g}} \right) . \tag{1.4}$$

The second term in (1.4) is the Wess-Zumino term, involving the super three-form gauge field $\hat{A}_{\hat{A}\hat{B}\hat{C}}$ of the 11-dimensional supergravity background. The third term is the world-volume cosmological term discused above.

The world-volume metric $\hat{g}_{\hat{\imath}\hat{\jmath}}$ may be eliminated from the action using its algebraic equation of motion, which is the supersymmetric generalization of (1.3):

$$\hat{g}_{\hat{\imath}\hat{\jmath}} = \hat{E}_{\hat{\imath}}{}^{\hat{a}} \hat{E}_{\hat{\jmath}}{}^{\hat{b}} \eta_{\hat{a}\hat{b}} . \tag{1.5}$$

The action then takes its Nambu-Goto form

$$I = \int d^3\hat{\xi} \left( \sqrt{-\det \hat{E}_{\hat{\imath}}{}^{\hat{a}} \hat{E}_{\hat{\jmath}}{}^{\hat{b}} \eta_{\hat{a}\hat{b}}} - \frac{1}{6}\epsilon^{\hat{\imath}\hat{\jmath}\hat{k}} \hat{E}_{\hat{\imath}}{}^{\hat{A}} \hat{E}_{\hat{\jmath}}{}^{\hat{B}} \hat{E}_{\hat{k}}{}^{\hat{C}} \hat{A}_{\hat{C}\hat{B}\hat{A}} \right) . \tag{1.6}$$

We now suppose the 11 bosonic dimensions of the target space to be $R_{10} \times S_1$ and we consider a membrane wrapped around the $S_1$ direction. We now make a two-one split of the world-volume coordinates

$$\hat{\xi}^{\hat{\imath}} = (\xi^i, \rho), \quad i = 1, 2 \tag{1.7}$$

and a ten-one split of the spacetime coordinates

$$\hat{X}^{\hat{m}} = (X^m, y), \quad m = 1, \ldots, 10. \tag{1.8}$$

We can now make the partial gauge choice

$$\rho = y , \tag{1.9}$$

which identifies the eleventh spacetime dimension with the third dimension of the world-sheet. Upon dimensional reduction by shrinking the $S_1$ direction of the target space, we will also shrink away the second spatial dimension of the membrane, giving a string in $d$=10.

The Kaluza-Klein Ansatz for the $N$=1, $d$=11 supervielbein is

$$\hat{E}_{\hat{M}}{}^{\hat{A}} = \begin{pmatrix} \hat{E}_M{}^a & \hat{E}_M{}^\alpha & \hat{E}_M^{11} \\ \hat{E}_y{}^a & \hat{E}_y{}^\alpha & \hat{E}_y{}^{11} \end{pmatrix} , \tag{1.10}$$

$$= \begin{pmatrix} E_M{}^a & E_M{}^\alpha + A_M \chi^\alpha & \Phi A_M \\ 0 & \chi^\alpha & \Phi \end{pmatrix} , \tag{1.11}$$

where $E_M{}^A = (E_M{}^a, E_M{}^\alpha)$ is the $N$=2$A$, $d$=10 supervielbein, $A_M$ is the superspace $U(1)$ gauge field, and $\Phi$ and $\chi^\alpha$ are superfields whose leading components are the dilaton and the dilatino, respectively. A partial $d$=11 Lorentz gauge choice has been made to set $\hat{E}_y{}^a = 0$. The Ansatz for the superspace three-form potential $\hat{A}_{\hat{M}\hat{N}\hat{P}}$ is simply

$$\hat{A}_{MNP} = A_{MNP} , \tag{1.12}$$

$$\hat{A}_{MNy} = A_{MN} . \tag{1.13}$$

All of the 10-dimensional superfields $E_m{}^A$, $\chi^\alpha$, $A_m$, $\Phi$, $A_{MN}$ and $A_{MNP}$ are taken to be independent of $y$. Since the 10-dimensional spinor indices run from 0 to 32 just as in 11 dimensions, $\alpha$ and $\hat{\alpha}$ can simply be identified. The Ansatz is completed by imposing

$$\partial_\rho Z^M = 0 . \tag{1.14}$$

Substituting the Ansätze (1.11-1.13) into the action (1.6) and using (1.9) and (1.14), one obtains the action for a type IIA superstring coupled to an $N$=2A supergravity background

$$I = \int d^2\xi \left( \Phi \sqrt{-\det E_i{}^a E_j{}^b \eta_{ab}} - \frac{1}{2} \epsilon^{ij} \partial_i Z^M \partial_j Z^N A_{NM} \right) . \tag{1.15}$$

The dilaton superfield factor $\Phi$ in (1.15) may be removed by a rescaling of the supervielbein.

The double dimensional reduction discussed above compactifies both the spacetime and the world volume on the same circle. The superstring in $d$=10 is then obtained after a consistent truncation that discards all the modes that are $y$-dependent (by imposing (1.14)). Classically, this is equivalent to letting the membrane coupling constant $\Xi$ tend to zero together with the radius $R$ of the compactifying $S_1$, so that the string coupling constant

$$\alpha' = \frac{2\pi\Xi}{R} \tag{1.16}$$

remains fixed. If one keeps the tower of states that are $y$-dependent, then the supermembrane may be viewed as a "tower of superstrings" in $d$=10, with the massless level being the type IIA superstring.

There is thus a remarkable classical pattern of supermembrane theories including the superstring theories. How much of this structure is preserved at the quantum level? It is of course possible that the supermembranes may have a role to play only as extended object solutions to the effective field theories of superstrings. But it is tempting to think that there may also be some meaning in quantizing the supermembrane theories themselves. One immediate objection is that only for $p$=1 (superstrings) does one have a renormalizable theory on the world-volume. This is a problem that certainly must be addressed, although we cannot give a definite resolution at this time. One can envisage a number of possible outcomes. For example, the non-renormalizability could simply be the field-theoretic expression of the impossibility of consistently decoupling the massive sector from the massless sector of the theory. Or, there may be some non-trivial ultraviolet fixed point with a finite-dimensional critical surface of attractive flows in the space of coupling constants, so that the idea of asymptotic safety [10] is realized. Or, perhaps, supermembranes are just one sector of the fundamental theory, but may nonetheless consistently describe the behaviour of the theory at some appropriate level, as quantized skyrmions are thought to represent the behaviour of QCD on the scale of hadrons, despite their non-renormalizability.

Whatever the resolution to the problem of world-volume non-renormalizability, it is the structure of the supermembrane's spectrum that will determine whether it can ever be related to the low-energy world of particle phenomenology. The primary question is whether the spectrum includes massless states.

Indeed, for the purely bosonic membrane ($p$=2), it has been argued [11] that there can be no massless states. This argument depends upon the result of a semiclassical calculation of the quantum correction to the bound on the maximal angular momentum for states of a given mass, in analogy with the familiar Regge formula for strings. If one assumes that a tree-level $p$-brane theory has a single coupling constant $\alpha_p$ (of dimensions $[\dim \hbar]^{-1}(\text{length})^{p+1}$), then dimensional analysis plus the requirement that the classical term be $\hbar$-independent yields the spin-mass formula in the vicinity of zero mass,

$$J = A(\alpha_p)^{\frac{1}{p}} c^{\frac{p+1}{p}} M^{\frac{p+1}{p}} + B\hbar \,, \tag{1.17}$$

where $c$ is the speed of light, M is the mass of the state, and $A$ and $B$ are dimensionless numbers. In eq. (1.17), $J$ should be interpreted as the maximal eigenvalue for any angular momentum generator $M^{\mu\nu}$ applied to states of mass $M$. By the assumption that there is only a single, dimensionful, coupling constant $\alpha_p$, the dimensionless nature of $B$ would imply that it is $\alpha_p$-independent. In that case, $B$ could be accurately calculated from the semiclassical approximation, where $\alpha_p \to 0$. The result of the semiclassical calculation of ref. [11] is that $B$ cannot be integral for any integral dimension $d$ of spacetime. In that case, (1.17) is inconsistent with the existence of massless states with integrally quantized angular momentum.

One may certainly argue with the assumption of a single dimensionless coupling constant in the above discussion, especially in view of the world-volume non-renormalizability of $p$-brane theories with $p \geq 2$. Nonetheless, it is something of a miracle for the bosonic string that $B$ in (1.17) takes the values 1 (2) for the open (closed) string in exactly $d$=26 dimensions. Such a miracle seems unlikely to recur for purely bosonic theories with $p \geq 2$.

Supersymmetry provides a possible mechanism to circumvent the above difficulty. In the Green-Schwarz formulation of the superstring, the contributions to $B$ cancel mode by mode as a result of the combination of spacetime supersymmetry and the fermionic gauge invariance. In the case of the superstring, this set of invariances is equivalent to the world-sheet supersymmetry in the RNS formulation of the theory. The mode-by-mode cancellation of contributions to $B$ then follows automatically from the world-sheet supersymmetry. In the remainder of this article, we examine whether such a mechanism is indeed active in the $p = 2$ supermembrane [12].

## 2. – THE CLASSICAL SUPERMEMBRANE

The action for the supermembrane in flat superspace is [1]

$$I = \frac{-1}{4\pi^2} \int d^3\xi \left[ \frac{1}{2}\sqrt{-g}\, g^{ij}\Pi_i^\mu \Pi_{j\mu} - \frac{1}{2}\sqrt{-g} + \varepsilon^{ijk}\Pi_i^A \Pi_j^B \Pi_k^C A_{CBA} \right], \tag{2.1}$$

where $\Pi_i^A = (\Pi_i^\mu, \Pi_i^\alpha)$ with

$$\Pi_i^\mu = \partial_i X^\mu - i\overline{\Psi}\Gamma^\mu \partial_i \Psi, \tag{2.2}$$

$$\Pi_i^\alpha = \partial_i \Psi^\alpha. \tag{2.3}$$

$\Psi^\alpha$ is a 32-component Majorana spinor, $\mu = 0, 1, \cdots, 10$, and we have set the membrane tension $\Xi$ equal to unity. The super 3-form $A_{CBA}$ is such that dA = H, with all components of H vanishing except $H_{\mu\nu\alpha\beta} = -\frac{1}{3}(C\Gamma_{\mu\nu})_{\alpha\beta}$. Our conventions are that $\varepsilon^{012} = -1, \{\Gamma^\mu, \Gamma^\nu\} = -2\eta^{\mu\nu}$ with $\eta^{\mu\nu} = diag(-1, 1, \cdots 1)$, and $\overline{\Psi}_\alpha = -\Psi^\beta C_{\beta\alpha}$. The charge conjugation matrix $C_{\alpha\beta}$ is antisymmetric.

Solving for $A_{CBA}$, the action may be written as

$$\begin{aligned} I = \frac{-1}{8\pi^2} \int d^3\xi\, [\sqrt{-g}\, g^{ij}\Pi_i^\mu \Pi_{j\mu} - \sqrt{-g} + \varepsilon^{ijk}\Pi_i^\mu \Pi_j^\nu \overline{\Psi}\Gamma_{\mu\nu}\partial_k\Psi \\ - i\varepsilon^{ijk}\Pi_i^\mu \overline{\Psi}\Gamma_{\mu\nu}\partial_j\Psi\overline{\Psi}\Gamma^\nu\partial_k\Psi \\ - \frac{1}{3}\, \varepsilon^{ijk}\overline{\Psi}\Gamma_{\mu\nu}\partial_i\Psi\overline{\Psi}\Gamma^\mu\partial_j\Psi\overline{\Psi}\Gamma^\nu\partial_k\Psi]. \end{aligned} \tag{2.4}$$

In addition to world-volume general coordinate invariance and Poincaré invariance, the action is also invariant under rigid spacetime supersymmetry transformations

$$\delta\Psi = \varepsilon, \tag{2.5}$$

$$\delta X^\mu = i\, \overline{\varepsilon}\, \Gamma^\mu \Psi, \tag{2.6}$$

and local Siegel transformations [1-3]

$$\delta\Psi = (1 + \Gamma)\kappa, \tag{2.7}$$

$$\delta X^\mu = i\overline{\Psi}\Gamma^\mu \delta\Psi, \tag{2.8}$$

where $\kappa$ is a world-volume scalar, spacetime Majorana spinor function of $\tau, \sigma$ and $\rho$, and

$$\Gamma = \frac{i}{6\sqrt{-g}}\, \varepsilon^{ijk}\Pi_i^\mu \Pi_j^\nu \Pi_k^\rho \Gamma_{\mu\nu\rho}. \tag{2.9}$$

For the action (2.4), Siegel invariance requires that $g_{ij}$ also transform. The variation of $g_{ij}$ is complicated, and may be found in Ref.[1]. In practice, it is more convenient to exhibit the Siegel invariance by going on "half-shell", i.e. substituting the algebraic equation of motion for $g_{ij}$,

$$g_{ij} = \Pi_i^\mu \Pi_{j\mu}, \tag{2.10}$$

into the action (2.4), and working in 1.5 order formalism.

Varying (2.4) with respect to $X^\mu$ and $\Psi$ yields

$$\partial_i(\sqrt{-g}\, g^{ij}\Pi_j^\mu) + \varepsilon^{ijk}\Pi_i^\nu \partial_j \overline{\Psi}\Gamma^\mu{}_\nu \partial_k \Psi = 0, \tag{2.11}$$

$$(1 - \Gamma)g^{ij}\Pi_i^\mu \Gamma_\mu \partial_j \Psi = 0, \tag{2.12}$$

with $\Gamma$ given by (2.9). In deriving (2.12) we have used the identity

$$\Gamma\Pi_k^\mu\Gamma_\mu = \Pi_k^\mu\Gamma_\mu\Gamma = \frac{-i}{2\sqrt{-g}}\, g_{k\ell}\varepsilon^{ij\ell}\Pi_i^\mu\Pi_j^\nu\Gamma_{\mu\nu}. \tag{2.13}$$

Note that after using the half-shell condition (2.10), we have $\Gamma^2 = 1$, and hence $\frac{1}{2}(1 \pm \Gamma)$ act as projection operators. Therefore (2.12) gives equations of motion for only 16 of the 32 components of $\Psi$. This is a consequence of the Siegel invariance of the action; equation (2.7) may be used to gauge away the other 16 unphysical components of $\Psi$. A convenient gauge to choose is

$$\Gamma\Psi = -\Psi. \tag{2.14}$$

The canonical momentum $K^\mu = \frac{\partial L}{\partial \dot{X}^\mu}$ is given by

$$K^\mu = -\frac{1}{4\pi^2}\sqrt{-g}\; g^{0i}\Pi_i^\mu - \frac{1}{4\pi^2}\,\varepsilon^{0ij}(\Pi_i^\nu + \frac{i}{2}\,\overline{\Psi}\,\Gamma^\nu\partial_i\Psi)\overline{\Psi}\Gamma^\mu{}_\nu\partial_j\Psi. \tag{2.15}$$

The total conserved momentum is then given by

$$P^\mu = \int d\sigma d\rho K^\mu, \tag{2.16}$$

One can easily verify that $P^\mu$ is invariant under supersymmetry transformations as the supersymmetry algebra requires it to be. The formula for the total angular momentum $M^{\mu\nu}$ is

$$\begin{aligned} M^{\mu\nu} = \int d\sigma d\rho [X^\mu K^\nu - X^\nu K^\mu \\ - \frac{i}{8\pi^2}\sqrt{-g}g^{0i}\Pi_i^\rho\overline{\Psi}\Gamma_\rho\Gamma^{\mu\nu}\Psi + \frac{1}{16\pi^2}\varepsilon^{0ij}\Pi_i^\rho\Pi_j^\sigma\overline{\Psi}\Gamma_{\rho\sigma}\Gamma_{\mu\nu}\Psi \\ + \frac{i}{16\pi^2}\varepsilon^{0ij}(\Pi_i^\rho + \frac{i}{3}\overline{\Psi}\Gamma^\rho\partial_i\Psi)(\overline{\Psi}\Gamma^\sigma\Gamma^{\mu\nu}\Psi\overline{\Psi}\Gamma_{\rho\sigma}\partial_j\Psi + \overline{\Psi}\Gamma_{\rho\sigma}\Gamma^{\mu\nu}\Psi\overline{\Psi}\Gamma^\sigma\partial_j\Psi)]. \end{aligned} \tag{2.17}$$

The remaining generator in the eleven-dimensional super-Poincaré algebra is the supercharge $Q^\alpha$, which we find to be

$$\begin{aligned} Q = \frac{1}{4\pi^2}\int d\sigma d\rho[-2i\sqrt{-g}g^{0i}\Pi_i^\mu\Gamma_\mu\Psi + \varepsilon^{0ij}\Pi_i^\mu\Pi_j^\nu\Gamma_{\mu\nu}\Psi \\ + \frac{4i}{3}\varepsilon^{0ij}(\Pi_i^\mu + \frac{2i}{5}\overline{\Psi}\Gamma^\mu\partial_i\Psi)(\Gamma_{\mu\nu}\Psi\overline{\Psi}\Gamma^\nu\partial_j\Psi + \Gamma^\nu\Psi\overline{\Psi}\Gamma_{\mu\nu}\partial_j\Psi)]. \end{aligned} \tag{2.18}$$

## 3. – SEMICLASSICAL QUANTIZATION

Owing to the intrinsically non-linear structure of the membrane theory, we are forced to adopt a semiclassical approach to quantization. This involves choosing a stable classical solution and quantizing the fluctuations about it. From this procedure we may investigate the relative contributions of the bosons and fermions to the vacuum energy in the sector of the theory defined by the classical solution.

In order to simplify the problem as much as possible, we consider a membrane propagating in a spacetime with topology $R^9 \times S^1 \times S^1$. With such a topology, there is a classical solution with a toroidal membrane stretched around the $S^1 \times S^1$. Choosing the $S^1 \times S^1$ to be in the $X^1$ and $X^2$ directions, the classical solution takes the form of a purely bosonic background, with

$$X^1 = \ell_1 R_1 \sigma \quad , \quad X^2 = \ell_2 R_2 \rho \quad , \quad X^I = 0, \quad I = 3, \cdots, 9, \tag{3.1}$$

$$\Psi = 0, \tag{3.2}$$

where $0 \leq \sigma \leq 2\pi$, $0 \leq \rho \leq 2\pi$, $R_1$ and $R_2$ are the radii of the two circles, and $\ell_1$ and $\ell_2$ are integers characterizing the winding numbers of the membrane around the two circles. The most general flat 2-torus is characterized by 3 parameters, but for simplicity we take the two circles to be orthogonal.

For the bosonic gauge choices, we work in an analogue [13] of the light-cone gauge for the string. We first use the three general coordinate invariances to set

$$X^+ = p^+\tau, \tag{3.3}$$

$$g_{0a} = 0 \quad , \quad (a = 1,2) \tag{3.4}$$

where $X^\pm = \frac{1}{\sqrt{2}}(X^0 \pm X^{10})$.

Although these conditions fix the gauge-freedoms that are functions of all three coordinates $(\tau, \sigma, \rho)$, there remains a residual $\tau$-independent reparametrization invariance of (3.3) and (3.4),

$$\delta\xi^0 = 0, \quad \delta\xi^a = f^a(\sigma, \rho). \tag{3.5}$$

Consequently, at a fixed time $\tau^*$, one may impose further gauge conditions. Thus, we use one of the two gauge freedoms (3.5) to impose

$$g_{00} = -\gamma \tag{3.6}$$

at $\tau = \tau^*$, where $\gamma \equiv det\ \gamma_{ab}$. Eq.(3.6) is in fact maintained for all $\tau$, by virtue of the $\mu = +$ component of the field equation (2.11) and the $\Psi$ field equation (2.12). This may be seen by substituting (3.3) and (3.4) into (2.11,2.12), to obtain

$$\frac{\partial}{\partial\tau}\left(\frac{\gamma}{g_{00}}\right)^{\frac{1}{2}} = 0. \tag{3.7}$$

There still remains one reparametrization freedom in (3.7) that leaves the gauge condition (3.6) invariant, namely diffeomorphisms of the form (3.5) for which $\partial_a(\delta\xi^a) = 0$, i.e. [13]

$$\delta\xi^0 = 0, \quad \delta\xi^a = \varepsilon^{ab}\partial_b f(\sigma, \rho), \tag{3.8}$$

where $\varepsilon^{12} = -\varepsilon^{21} = 1$. The Jacobian of this transformation is unity. It is the generalization for membranes of the residual freedom in the case of the string to perform a shift of $\sigma$ by a constant. For the time being, we shall leave this residual gauge freedom unfixed.

In the background (3.1,3.2), the world-volume metric is flat,

$$g_{ij} = diag(-(\ell_1\ell_2R_1R_2)^2, (\ell_1R_1)^2, (\ell_2R_2)^2), \tag{3.9}$$

and solving for $X^-$ we find

$$X^- = \frac{1}{2p^+}(\ell_2\ell_2R_1R_2)^2\tau. \tag{3.10}$$

The classical $(\text{mass})^2$ of the solution is given by $-P^\mu P_\mu$, where $P^\mu$ is given by (2.16), and is

$$(\text{mass})^2 = (\ell_1\ell_2R_1R_2)^2. \tag{3.11}$$

For simplicity we shall set $\ell_1 = \ell_2 = R_1 = R_2 = 1$ for the rest of the calculations in this section, restoring them in the final results. Thus $X^+ = p^+\tau, X^- = \frac{1}{2p^+}\tau, X^1 = \sigma, X^2 = \rho$ and $g_{ij} = diag(-1,1,1)$ in the background.

We now consider fluctuations $\underline{Z}$ of the transverse coordinate around the classical solution, and write $\underline{X} = \underline{X}_{classical} + \underline{Z}$. Thus

$$X^1 = \sigma + Z^1,$$

$$X^2 = \rho + Z^2,$$

$$X^I = Z^I. \tag{3.12}$$

The fermions, being zero in the background, are pure fluctuations $\Psi$. Substituting (3.12) into (2.11) and (2.12), and keeping only the terms of linear order in $\underline{Z}$ and $\Psi$, we find, in the gauges chosen $(X^+ = p^+\tau, g_{0a} = 0, g_{00} = -\gamma, \Psi = -\Gamma\Psi)$ ,

$$\ddot{Z}^1 = \partial_\sigma\partial_\sigma Z^1 + \partial_\sigma\partial_\rho Z^2, \tag{3.13}$$

$$\ddot{Z}^2 = \partial_\rho\partial_\rho Z^2 + \partial_\sigma\partial_\rho Z^1, \tag{3.14}$$

$$\ddot{Z}^I = \partial_\sigma\partial_\sigma Z^I + \partial_\rho\partial_\rho Z^I, \tag{3.15}$$

$$\dot{\Psi} = i\Gamma_2\partial_\sigma\Psi - i\Gamma_1\partial_\rho\Psi. \tag{3.16}$$

At this stage, we must fix the remaining gauge invariance of Eq.(3.8). The gauge choice $g_{0a} = 0$ yields an equation that can be solved for $\partial_a X^-$. Taking the curl, we find an integrability condition for this equation. Upon linearization about our background, this condition is

$$\partial_\rho \dot{Z}^1 = \partial_\sigma \dot{Z}^2. \tag{3.17}$$

This can be integrated to give $\partial_\rho Z^1 = \partial_\sigma Z^2 + h(\sigma, \rho)$, which leads us to exploit the residual gauge symmetry (3.8) to set the undetermined function $h(\sigma, \rho)$ to zero, thus obtaining for all $\tau$

$$\partial_\rho Z^1 = \partial_\sigma Z^2. \tag{3.18}$$

This allows us to cast (3.13) and (3.14) into the form of standard wave equations,

$$\ddot{Z}^1 = \partial_\sigma\partial_\sigma Z^1 + \partial_\rho\partial_\rho Z^1, \tag{3.19}$$

$$\ddot{Z}^2 = \partial_\sigma\partial_\sigma Z^2 + \partial_\rho\partial_\rho Z^2. \tag{3.20}$$

The relation (3.18) reduces the number of independent functions of $\tau, \sigma$ and $\rho$ by one, yielding 8 bosonic degrees of freedom. Since the fermion equation of motion (3.16) is of first order, the 16 Siegel-gauge-fixed components of $\Psi$ satisfying (2.14) also give rise to 8 fermionic degrees of freedom.

For the eleven-dimensional $\Gamma$-matrices, we use a representation appropriate to the 11=9+2 split,

$$\Gamma^I = \gamma^I \otimes \sigma_3 \otimes \sigma_3, \quad I = 3, \cdots 9,$$

$$\Gamma^\pm = 1 \otimes \sigma_\pm \otimes 1, \quad \sigma_\pm = \frac{1}{\sqrt{2}}(\sigma_1 \pm i\sigma_2),$$

$$\Gamma^a = 1 \otimes \sigma_3 \otimes i\sigma_a, \quad a = 1, 2. \tag{3.21}$$

Using this representation, we can express $\Psi$ in the Siegel gauge (2.14) as

$$\Psi = (16\sqrt{2}\, p^+)^{-\frac{1}{2}} \begin{pmatrix} \chi \\ -i\chi^* \\ -\sqrt{2}p^+\chi \\ -i\sqrt{2}p^+\chi^* \end{pmatrix} \tag{3.22}$$

where $\chi$ is a complex 8-component spinor of $SO(7)\times U(1) \subset SO(9)$. Substituting (3.22) into (3.16) casts the fermion equation into the form

$$\dot{\chi} = \partial_\sigma \chi^* - i\partial_\rho \chi^*. \tag{3.23}$$

Substituting (3.12) and (3.22) into the gauge conditions $g_{oo} = -\gamma$ and $g_{oa} = 0$, and keeping terms to quadratic order in fluctuations,

$$\begin{aligned}\dot{X}^- =& \frac{1}{2p^+}\Big[1 + \dot{\underline{Z}}^2 + (\partial_\sigma \underline{Z})^2 + (\partial_\rho \underline{Z})^2 + G(\underline{Z}),\\ &+\frac{i}{2}\left[(\chi^\dagger \partial_\sigma \chi^* + \chi^T \partial_\sigma \chi) - i(\chi^\dagger \partial_\rho \chi^* - \chi^T \partial_\rho \chi)\right]\Big],\end{aligned} \tag{3.24}$$

$$\partial_\sigma X^- = \frac{1}{p^+}\left[\dot{Z}^1 + \partial_\sigma \underline{Z}\cdot \dot{\underline{Z}} + \frac{i}{8}\left[(\chi^T \dot{\chi} + \chi^\dagger \dot{\chi}^*) + (\chi^\dagger \partial_\sigma \chi + \chi^T \partial_\sigma \chi^*)\right]\right], \tag{3.25}$$

$$\partial_\rho X^- = \frac{1}{p^+}\left[\dot{Z}^2 + \partial_\rho \underline{Z}\cdot \dot{\underline{Z}} + \frac{i}{8}\left[i(\chi^T \dot{\chi} - \chi^\dagger \dot{\chi}^*) + (\chi^\dagger \partial_\rho \chi^* + \chi^T \partial_\rho \chi^*)\right]\right], \tag{3.26}$$

where

$$G(\underline{Z}) = 2(\partial_\sigma Z^1 + \partial_\rho Z^2) + 4(\partial_\sigma Z^1 \partial_\rho Z^2 - \partial_\rho Z^1 \partial_\sigma Z^2). \tag{3.27}$$

When we come to construct the mass formula, G($\underline{Z}$) will drop out in the $\int d\sigma d\rho$ integral.

As a preliminary to quantization, we write the general solutions of (3.19), (3.20), (3.15) and (3.23) as

$$\underline{Z} = \underline{z}_o + \underline{p}\tau + \frac{1}{\sqrt{2}} \sum_{m^2+n^2\neq 0} \frac{1}{\omega_{mn}} e^{i(m\sigma+n\rho)}[\underline{\alpha}^\dagger_{mn} e^{i\omega_{mn}\tau} + \underline{\alpha}_{-m-n} e^{-i\omega_{mn}\tau}], \tag{3.28}$$

$$\chi = \sqrt{2} S_{oo} + \sum_{m^2+n^2\neq 0} e^{i(m\sigma+n\rho)}[\frac{m-in}{\omega_{mn}} S^\dagger_{mn} e^{i\omega_{mn}\tau} + S_{-m-n} e^{-i\omega_{mn}\tau}], \tag{3.29}$$

and

$$\omega_{mn} = (m^2+n^2)^{\frac{1}{2}}, \tag{3.30}$$

where, in (3.28), the Fourier coefficients have been chosen so that $\underline{Z}$ is real.

Proceeding to the semiclassical quantization, we must distinguish between the unconstrained variables $Z^I$ ($I = 3,\cdots,9$) and the two variables $Z^1$ and $Z^2$ which satisfy the constraint (3.17) and the gauge condition (3.18). For the unconstrained variables we have the canonical commutation relation $[K^I, Z^J] = -i\delta^{IJ}\delta(\sigma-\sigma')\delta(\rho-\rho')$ which implies that

$$[\dot{Z}^I, Z^J] = -(2\pi)^2 i\delta^{IJ}\delta(\sigma-\sigma')\delta(\rho-\rho'). \tag{3.31}$$

For the constrained variables $Z^1$ and $Z^2$, we must proceed differently, owing to the need to incorporate (3.17) and (3.18). For this, we adopt Dirac's procedure for quantizing constrained variables [14]. Eqs.(3.17) and (3.18) constitute a non-commuting (i.e. second class) system of constraints. The canonical momenta conjugate to $Z^1$ and $Z^2$ are $k^1 = (2\pi)^{-2}\dot{Z}^1$, $k^2 = (2\pi)^{-2}\dot{Z}^2$. Denoting the constraints (3.17) and (3.18) by $\phi_s = 0, s = 1, 2$, with

$$\begin{aligned}\phi_1 &= \partial_\rho k^1 - \partial_\sigma k^2,\\ \phi_2 &= \partial_\rho Z^1 - \partial_\sigma Z^2,\end{aligned} \tag{3.32}$$

we have the Poisson bracket algebra

$$[\phi_1(\sigma,\rho), \phi_2(\sigma',\rho')]_P = \nabla^2\delta(\sigma-\sigma')\delta(\rho-\rho') \equiv C_{12}, \tag{3.33}$$

where $\nabla^2 = \partial_\sigma^2 + \partial_\rho^2$.

The Dirac brackets are now defined by

$$[A, B]_D = [A, B]_P - [A, \phi_s]_P (C^{-1})^{st} [\phi_t, B]_P, \tag{3.34}$$

for arbitrary operators A and B. The Dirac brackets have the property that $\phi_1$ and $\phi_2$ now commute with each other and with the Hamiltonian. The Dirac brackets of the conjugate pairs $(Z^1, k^1)$ and $(Z^2, k^2)$ are:

$$\begin{aligned} [k^1(\sigma, \rho), Z^1(\sigma', \rho')]_D &= -\left(1 - \frac{\partial_\rho^2}{\nabla^2}\right)\delta(\sigma - \sigma')\delta(\rho - \rho'), \\ [k^2(\sigma, \rho), Z^2(\sigma', \rho')]_D &= -\left(1 - \frac{\partial_\sigma^2}{\nabla^2}\right)\delta(\sigma - \sigma')\delta(\rho - \rho'). \end{aligned} \tag{3.35}$$

We now pass from the Dirac brackets to the quantum commutators

$$\begin{aligned} [\dot{Z}^1(\sigma, \rho), Z^1(\sigma', \rho')] &= -(2\pi)^2 i\left(1 - \frac{\partial_\rho^2}{\nabla^2}\right)\delta(\sigma - \sigma')\delta(\rho - \rho'), \\ [\dot{Z}^2(\sigma, \rho), Z^2(\sigma', \rho')] &= -(2\pi)^2 i\left(1 - \frac{\partial_\sigma^2}{\nabla^2}\right)\delta(\sigma - \sigma')\delta(\rho - \rho'). \end{aligned} \tag{3.36}$$

For the fermions, one could proceed as above using the Dirac formalism for the eleven-dimensional spinor satisfying the Majorana and Siegel conditions. In practice, we find it more convenient to work with the unconstrained spinor $\chi$. Substituting (3.22) into the action (2.1), one can quantize $\chi$ canonically, yielding

$$\{\chi^{*A}, \chi^B\} = 2(2\pi)^2 \delta^{AB} \delta(\sigma - \sigma')\delta(\rho - \rho'), \tag{3.37}$$

where $A, B = 1, \cdots 8$ are SO(7) spinor indices.

Substituting (3.28) and (3.29) into (3.31), (3.36) and (3.37), we find the following commutation relations for the $\alpha$ and S oscillators:

$$[\alpha^1_{mn}, \alpha^{1\dagger}_{m'n'}] = \frac{m^2}{\omega_{mn}} \delta_{mm'} \delta_{nn'}, \tag{3.38}$$

$$[\alpha^2_{mn}, \alpha^{2\dagger}_{m'n'}] = \frac{n^2}{\omega_{mn}} \delta_{mm'} \delta_{nn'} \tag{3.39}$$

$$[\alpha^I_{mn}, \alpha^{J\dagger}_{m'n'}] = \omega_{mn} \delta^{IJ} \delta_{mm'} \delta_{nn'}, \tag{3.40}$$

$$\{S^A_{mn}, S^{B\dagger}_{m'n'}\} = \delta^{AB} \delta_{mm'} \delta_{nn'}, \tag{3.41}$$

$$[p^1, z^1_0] = [p^2, z^2_0] = -i \quad , \quad [p^I, z^J_0] = -i\delta^{IJ}, \tag{3.42}$$

$$\{S^A_{00}, S^{B\dagger}_{00}\} = \delta^{AB}, \tag{3.43}$$

with all other independent commutators vanishing. Note that the constraint (3.18) implies that

$$n\alpha^1_{mn} = m\alpha^2_{mn} \tag{3.44}$$

and that this is consistent with (3.38) and (3.39). It also implies that $[\alpha^1_{mn}, \alpha^{2\dagger}_{m'n'}] = \frac{mn}{\omega_{mn}} \delta_{mm'} \delta_{nn'}$, etc.

Substituting (3.28) and (3.29) into (3.24), and then the result into (2.15) and (2.16), we find

$$P^- = \frac{1}{2p^+}[1 + \underline{p}^2 + \sum_{m^2+n^2\neq 0} (\underline{\alpha}_{mn} \cdot \underline{\alpha}^\dagger_{mn} + \underline{\alpha}^\dagger_{mn} \cdot \underline{\alpha}_{mn}) + \sum_{m^2+n^2\neq 0} \omega_{mn} \, (-S^A_{mn} S^{A\dagger}_{mn} + S^{A\dagger}_{mn} S^A_{mn})]. \tag{3.45}$$

Using the fact that $P^+ = p^+$ and $\underline{P} = \underline{p}$, we therefore have the eleven-dimensional mass formula

$$(\text{mass})^2 = (\ell_1 \ell_2 R_1 R_2)^2 + H, \tag{3.46}$$

where

$$H = 2 \sum_{m^2+n^2\neq 0} (\underline{\alpha}^\dagger_{mn} \cdot \underline{\alpha}_{mn} + \omega_{mn} S^{A\dagger}_{mn} S^A_{mn}). \tag{3.47}$$

In obtaining the result (3.46) from (3.45), we have restored the winding numbers $\ell_1$ and $\ell_2$, and radii $R_1$ and $R_2$ of the two circles, which we had set equal to unity earlier; now

$$\omega_{mn} = [(m\ell_2 R_2)^2 + (n\ell_1 R_1)^2]^{\frac{1}{2}}. \tag{3.48}$$

The key result is that the mass formula (3.46, 3.47) is correct as it stands, without a vacuum energy term. This is because the bosonic contribution, which takes the form of an Epstein zeta function,

$$\Delta_B(\text{mass})^2 = \sum_{m^2+n^2\neq 0} [(m\ell_2 R_2)^2 + (n\ell_1 R_1)^2]^{\frac{1}{2}}, \tag{3.49}$$

coming from normal-ordering (3.45), is cancelled mode by mode by an equal but opposite fermionic contribution. Note that $\alpha^\dagger_{mn}$ and $S^\dagger_{mn}$ are creation operators for all m and n, while $\alpha_{mn}$ and $S_{mn}$ are annihilation operators. The difference between this convention and the usual string convention stems from the form of our oscillator expansions (3.28) and (3.29).

The cancellation of the vacuum energy contributions (3.48) between bosons and fermions depends crucially on the Siegel symmetry (2.7, 2.8), which caused the number of fermionic physical degrees of freedom to be halved, (eq.(3.22)). This cancellation indicates a striking difference in the quantum behaviour of the supermembrane as compared to the bosonic membrane.

## 4. - CONCLUSION

The cancellation that we have observed between semi-classical boson and fermion contributions to the vacuum energy gives support to our expectation that this cancellation will take place also in the exact theory. Although the supermembrane has only a Green-Schwarz formulation, the combination of spacetime supersymmetry and the fermionic gauge invariance appears to have the same consequence for the cancellation of the vacuum energy correction as would a world-volume supersymmetry.

One can make the roles of spacetime supersymmetry and the fermionic gauge invariance clearer by considering the gauge-fixed supersymmetry of the theory [15]. If one imposes the gauge condition (2.14), then the supersymmetry transformations (2.5,2.6) acquire compensating fermionic gauge transformation terms so that (2.14) continues to be valid. This requires

$$(1+\Gamma)\kappa = -\frac{1}{2}(1+\Gamma)\epsilon \tag{4.1}$$

and the gauge-fixed supersymmetry transformation of $\Psi$ is now

$$\delta\Psi = \frac{1}{2}(1-\Gamma)\epsilon. \tag{4.2}$$

Thus, in the background (3.1,3.2), with the bosonic gauge conditions (3.3,3.4,3.6), there is an unbroken part of the gauge-fixed supersymmetry, with $\epsilon$ satisfying $\Gamma_{background}\epsilon = \epsilon$. The unbroken part of the spacetime supersymmetry causes the fluctuations (3.12) about the background (3.1,3.2) to have a spectrum symmetric under the unbroken supersymmetry, with cancellation of the vacuum energy contributions between bosons and fermions.

The cancellation of the vacuum energy corrections is a necessary but not sufficient condition for the existence of massless particles in the supermembrane spectrum. It remains to be determined whether the spectrum is supersymmetric in the sector of the theory with $l_1$ or $l_2$ equal to zero, where the massless states would be found. In these sectors, the full non-linearity of the supermembrane theory will have to be faced, as there seems little basis for trust in the semiclassical approximation there. Clearly, the consequences of spacetime supersymmetry and the fermionic gauge symmetry require more careful study.

The present work has shown how the supermembrane is likely to evade the difficulties with the vacuum energy term $B$ in eq. (1.1) that apparently rule out massless states for the purely bosonic membrane. It is worth noting that in our topologically-stabilized sector of the theory, the mass formula (3.46-3.48) does not give a correction to $M^{\frac{3}{2}}$ as expected from (1.1), but rather gives a correction to $M^2$, just like the case of the string. The difference from the expected relation is due to the presence of the dimensional compactification scale $R$. Thus, in the compactified theory, the dimensional arguments of ref. [10] are not in fact applicable. For small fluctuations about the topologically-stabilized vacuum, the $(\text{mass})^2$-spin relation is linear, like for superstrings.

ACKNOWLEDGMENTS

This work reviewed in this article was performed in collaboration with M.J. Duff, P.S. Howe, T. Inami, C.N. Pope and E. Sezgin. I would like to thank Dr. P.K. Townsend for stimulating discussions.

## REFERENCES

[1] E. Bergshoeff, E. Sezgin, and P.K. Townsend, *Phys.Lett.* **189B** (1987) 75.

[2] W. Siegel, *Phys.Lett.* **128B** (1983) 397.

[3] J. Hughes, J. Liu, and J. Polchinski, *Phys.Lett.* **180B** (1986) 370.

[4] P. Fayet, *Nucl. Phys.* **B263** (1986), 649.

[5] P.S. Howe, private communication. This point is implicit in P.S. Howe and R.W. Tucker, *J. Phys.* **A10** (1977) L155.

[6] M.B. Green and J.H. Schwarz, *Phys. Lett.* **136B** (1984) 367.

[7] A. Achúcarro, J.M. Evans, P.K. Townsend and D.L. Wiltshire, University of Cambridge D.A.M.T.P. preprint (1987).

[8] M. Henneaux and L. Mezincescu, *Phys. Lett.* **152B** (1985) 340.

[9] M.J. Duff, P.S. Howe, T. Inami and K.S. Stelle, *Phys. Lett.* **191B** (1987) 70.

[10] S. Weinberg, in "General Relativity; An Einstein Centenary Survey", eds. S.W. Hawking and W.Israel, Cambridge University Press (1979).

[11] K. Kikkawa and M. Yamasaki, *Prog.Theor.Phys.* **76** (1986) 1379.

[12] M.J. Duff, T. Inami, C.N. Pope, E. Sezgin and K.S. Stelle, *Nucl. Phys.* **B** (in press).

[13] J. Hoppe, Aachen preprint PITHA 86/24; and Ph.D. Thesis, MIT (1982).

[14] P.A.M. Dirac "Lectures on Quantum Mechanics", Belfer Graduate School of Science, Yeshiva University, New York (1964).

[15] E. Bergshoeff, E. Sezgin and P.K. Townsend, ICTP preprint IC/87/255.

# ON THE MASSLESS EXCITATIONS OF THE 11d SUPERMEMBRANE

L. Mezincescu

Department of Physics

University of Miami, Coral Gables, Florida 33124

## INTRODUCTION

In this talk I will describe some work done in collaboration with Rafael Nepomechie and Peter van Nieuwenhuizen.[1] As K. Stelle at this conference outlined some general features of the supermembrane theory I will focus mostly on the special topic which we investigated – the existence of massless modes in the supermembrane.

As it is well known there are two actions which do describe the superstrings – The RNS- Ramond and Neveau Schwarz[2] and the Green–Schwarz (G.S).[3] RNS model has world sheet local supersymmetry but may not be appropriate to describe higher dimensional objects.[4] The Green–Schwarz action has manifest space time supersymmetry and along with it also has an additional gauge invariance called Siegel[5] gauge invariance which is necessary for the space time supersymmetry of the system. This gauge invariance also plays an important role in the coupling of the Green–Schwarz superstring to its massless modes.[6]

Recently a generalization of Siegel invariance to the supermembrane was accomplished by Hughes, Liu and Polchinski[7] who wrote down an action for a 4 dimensional object moving in a superspace with 6 dimensional bose components. Later their approach was generalized for other high dimensional objects, by Bergshoeff, Sezgin and Townsend.[8]

An exciting feature which emerges from ref. 8 is that along with the eleven dimensional supergravity one can formulate a supermembrane. In many supergravity theories in order to accomplish a superspace formulation one is led to introduce along with the torsion and curvature forms, also some additional closed superspace forms. These closed superspace forms can be used to construct Wess-Zumino terms– for example for the G-S superstring action one needs to have a closed superspace – 3 form.[9]

For the supermembranes the situation is similar. The closed 4–form which appears in the superspace formulation of 11-d supergravity can be used to formulate an action for a three dimensional object the supermembrane, and also to couple the corresponding action to the 11-d supergravity background.[8] By analogy with the superstrings this suggests that the supermembrane may contain massless particles.

In another development Kikkawa and Yamasaki[9] (K.Y.), within a semiclassical approximation, investigated the question of the existence of spin 2 and spin 1 massless particles for the open bosonic membrane. In that framework it was concluded that the bosonic membrane cannot have massless excitations.

It should be stressed that with the present understanding of the higher dimensional objects one can make sense only of those extended objects which do have massless excitations in their spectrum.

The K.Y approach relates the spin of the massless particle to the vacuum expectation value – the Casimir energy of a certain hamiltonian to be outlined below. Therefore it is natural to ask whether by supersymmetrizing the membrane action one can improve the situation and this will be the subject of this talk.

I should stress however that even if K-Y approach may be plausible it is not rigorous. To our understanding there are two main flaws. First, one deals with a theory which by power counting is not renormalizable and it is not clear to what extent one can apply semiclassical methods. Second, partly for computational facility, a very special background is selected, from the class of solutions to the classical membrane equations proposed by Kikkawa and Yamasaki. Finally, the use of the generalized $\zeta$-function regularization (see ref.[11] for the string), for the summation of the vacuum energies, cannot at the present be justified by the use of other methods.

These questions still do apply to our computation even if for the supermembrane the Siegel symmetry may be of help for the question of renormalizability.

In this talk I will present the computation of the Casimir energy for the supermembrane. I will start with a brief description of the generalization of the Green Schwarz action to the supermembrane. Then I will outline the semiclassical approach for the bosonic string and subsequently I will present the main ideas of our computation of the massless spectrum of the supermembrane. I will end up with conclusions.

## GREEN-SCHWARZ ACTION FOR THE SUPERMEMBRANE

The Green-Schwarz action describes a string moving in a superspace of coordinates $(x\ ,\theta)$. It can be written[7] as a sum of the invariant volume of the world sheet and the Wess-Zumino (W-Z) term.

The induced metric is

$$g_{\alpha\beta} = \pi^{\mu}_{\alpha}\eta_{\mu\nu}\pi^{\nu}_{\beta} \tag{2.1}$$

where

$$\pi^{\mu}_{\alpha} = \partial_{\alpha} x^{\mu} - i\bar{\theta}\Gamma^{\mu}\theta_{,\alpha} \tag{2.2}$$

The W-Z term is the pullback to the world sheet of a superspace closed three form

$$W = d\Omega \tag{2.3}$$

The action is given by

$$S = \int \left(\sqrt{-detg} + b\Omega\right) dv \tag{2.4}$$

and for a particular value of the coefficient $b$, exhibits an invariance under the Siegel transformations:

$$\begin{aligned} \delta x^{\mu} &\sim \bar{\theta}\Gamma^{\mu}\delta\theta \\ \delta\theta &\sim \left(g^{\alpha\beta} + \frac{\epsilon^{\alpha\beta}}{\sqrt{-detg}}\right)\hat{\pi}_{\alpha}\kappa_{\beta} \end{aligned} \tag{2.5}$$

where the local parameter $\kappa_{\beta}$ is a spacetime spinor.

As it stands the transformations (2.5) look very much dependent on the fact that the world sheet has 2 dimensions.

Hughes, Liu and Polchinski[6] rewrote these transformations in a form which can be easily generalized to higher dimensional objects. Using the identity

$$\frac{\epsilon^{\alpha\beta}}{2!}\hat{\pi}_{\alpha}\hat{\pi}_{\beta}g^{\sigma\delta}\hat{\pi}_{\delta} \sim \frac{\epsilon^{\sigma\delta}}{\sqrt{-detg}}\hat{\pi}_{\delta} \quad , \tag{2.6}$$

the second term in (2.5) can be written (with a change in the definition of the local parameter $\kappa_{\alpha}$) in the form

$$\delta\theta = (1+\Gamma)\kappa \tag{2.7}$$

where

$$\Gamma = \frac{\epsilon^{\alpha\beta}}{2!}\hat{\pi}_{\alpha}\hat{\pi}_{\beta} \qquad \kappa = g^{\alpha\beta}\hat{\pi}_{\alpha}\kappa_{\beta} \tag{2.8}$$

It is now straight forward to extend the $\kappa$ transformation rules to higher dimensions of the base manifold. The action for the supermembrane is the sum of the invariant volume plus the Wess-Zumino term.

$$L = \sqrt{-detg} - \frac{1}{2}\epsilon^{\alpha\beta\gamma}\pi^{\mu}_{\alpha}\pi^{\nu}_{\beta}\bar{\theta}\Gamma_{\mu\nu}\partial_{\gamma}\theta + O(\theta^4) \tag{2.9}$$

where $g_{\alpha\beta}$ and $\pi^{\mu}_{\alpha}$ are given respectivelly by (2.1) and (2.2) with $\alpha$ runnimg now from 0 to 2. The $0(\theta^4)$ stands for the fact that only the bilinear term in $\theta$ are shown, exactly.

The action for the supermembrane looks extremely nonlinear. It does not appear that these nonlinearities may be overcome by some sort of gauge fixing. Therefore in order to get some information concerning the spectra associated with it one must rely on some approximations. Kikkawa and Yamasaki proposed, as will be outlined in the next section, that for the study of the massless states one can apply a semiclassical treatment.

An analysis of the coupling constants appearing in the theory reveals that this theory is of a nonrenormalizable type. However the question of the nonrenormalizability of this action is not yet settled as no one was yet able to construct counterterms which lead to a theory which is invariant under the Siegel transformations. For an attempt to construct higher derivative invariants under Siegel transformations within the string theory see ref.[12].

## OUTLINE OF THE SEMICLASSICAL APPROACH

Barring the questions raised before we will assume that a meaningful quantum theory can be constructed and that no new parameters are forced upon us in constructing such a theory. Following K-Y we try to get information about the spectrum of this theory (we restrict our discussion at the moment to a $n$-dimensional bosonic membrane)

$$S = \frac{\chi_n}{4\pi} \int d^n\sigma d\tau \left(-det g_{\alpha\beta}\right)^{1/2} \tag{3.1}$$

$\chi_n$ is the membrane tension and has the dimension $[\chi] = l^{n+1}$ By dimensional arguments and by the lack of any dimensional parameters other than masses and the membrane tension, any state of a given spin in the theory must be a function of the dimensionless ratio $\left(\frac{m^{n+1}}{\chi_n}\right)$ where $m$ is the mass of the state, (provided such a relation exists)[13]

$$J = f\left(\frac{m^{n+1}}{\chi_n}\right) \tag{3.2}$$

If we demand analyticity in $1/\chi_n$, which amounts to assuming that the semiclassical approximation is correct, then the spin of the massless state does not depend on $\chi_n$ and correspondingly we can let $\chi_n \to \infty$ i.e. get to the semiclassical regime. One concludes that the massless spectrum should come out exact in the semiclassical approximation. For the bosonic open string for example as we will see one gets

$$J = f(0) = -\frac{D-2}{2}\left(\sum n\right) = \frac{D-2}{24} \tag{3.3}$$

the spin of the massless state is

$$J = 1$$

This heuristic argument justifies the application of semiclassical treatment. In order to obtain the function $f(0)$, Kikkawa and Yamasaki compute the propagator of the bosonic membrane

$$\langle J \mid \frac{1}{E-H} \mid J\rangle \tag{3.4}$$

in the semiclassical approximation. The poles at $E = o$ will then be the massless states in the theory. The states $| J >$ are states of a given angular momentum. To obtain the propagator one needs the transition function

$$T_{fi} = \langle f \mid e^{-i\,TH} \mid i \rangle = \int_i^f DX^\mu e^{iS} \tag{3.5}$$

evaluated over periodic paths. In the rest of this section we will restrict to the computation of (3.4) for the case of a bosonic string. In the semiclassical regime one expands the full action $S$ in (3.5) around a periodic classical solution

$$X^\mu(\sigma, \tau) = X^\mu_{cl} + Z^\mu \tag{3.6}$$

which is taken to be a rigidly rotating string

$$\begin{aligned} X^1_{cl} &= \sigma \cos \omega\tau \\ X^2_{cl} &= \sigma \sin \omega\tau \end{aligned} \tag{3.7}$$

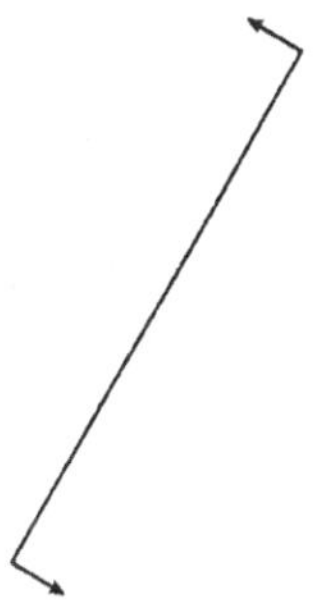

One performs a partial gauge fixing $X_o = \tau$ and does not take a fluctuation in that direction. It can easily be verified that this particular gauge fixing does not lead to the introduction of ghosts.

The induced metric corresponding to the above classical solution is

$$g_{\alpha\beta} = (-A,\ 1) \tag{3.8}$$

where $A = (1 - \omega^2\sigma^2)$. This metric is singular at $\sigma = \pm\frac{1}{\omega}$ which are therefore the end points of the string moving with the velocity of light.

The relation between the frequency $\omega$ and the classical mass $E_{cl}$ of the string is

$$\omega = \frac{\pi\chi}{E_{cl}} \tag{3.9}$$

which shows that for fixed $E_{cl}$ and $\chi \to \infty$, $\omega \to \infty$. Correspondingly, as it is well known, in the semiclassical limit $\chi \to \infty$ the string shrinks to a point.

We must develop the $\sqrt{-g}$ around the classical solution and keep only terms quadratic in the quantum fluctuations

$$L = \sqrt{-g_{cl}} = -\frac{1}{2}\sqrt{-g_{cl}}\partial_\alpha Z^\mu \left[g_{cl}^{\alpha\beta}\eta_{\mu\nu} - P^{\alpha\gamma\beta\delta}\partial_\gamma X_{cl\mu}\partial_\delta X_{cl\nu}\right]\partial_\beta Z \tag{3.10}$$

The above lagrangian depends explicitly on time. However by going to a frame rigidly rotating with the string (this rotating frame is noninertial much like the light cone for the string)

$$\hat{Z}^1 + i\hat{Z}^2 = e^{i\omega\tau}(Z^1 + iZ^2) \tag{3.11}$$

one of the degrees of freedom drops out and the explicit time dependence in the action disappears. We can then write the action as:

$$S^{(2)} = S_{\parallel}^{(2)} + S_{\perp}^{(2)}(D-3 \text{ coord}) \tag{3.12}$$

$S_{\perp}^{(2)}$ is the action for the $D-3$ transversal degrees of freedom with the induced metric (3.8), while $S_{\parallel}^{(2)}$ corresponds to the fluctuations along the string and is given by

$$S_{\parallel}^{(2)} = \int d\tau df \sqrt{A}\{-\frac{1}{A^2}\dot{Z}_2^2 - \frac{2\omega^2 f}{A}Z_2 Z_{2,f} + \frac{1}{A}Z_{2,f}^2\} \tag{3.13}$$

In order to find the normal modes of the above system one solves the equations of motion with the ansatz

$$Z = e^{iE\tau}R(f) \tag{3.14}$$

Then one finds

$$E^{\parallel} = n\omega \qquad E^{i\perp} = n\omega \tag{3.15}$$

Performing the computation of (3.4) in the semiclassical limit K.Y. find that the propagator of the string is of the form

$$\langle J \mid \frac{1}{E-H} \mid J\rangle \sim \frac{1}{\alpha(E) - J} \tag{3.16}$$

where

$$\alpha(E)\mid_{E=0} = -\langle 0 \mid H \mid 0\rangle = J(0) \tag{3.17}$$

where $H$ is the Hamiltonian corresponding to the system (3.12). The computations for the membrane and for the supermembrane go along the same lines and will be outlined in the next section.

## CASIMIR ENERGY FOR THE SUPERMEMBRANE

Because supersymmetry is difficult to implement at the ends of the membrane we are going to compute the normal modes for the closed supermembrane. The closed membrane solution at the classical level is taken to be a superposition of two rigidly rotating discs.

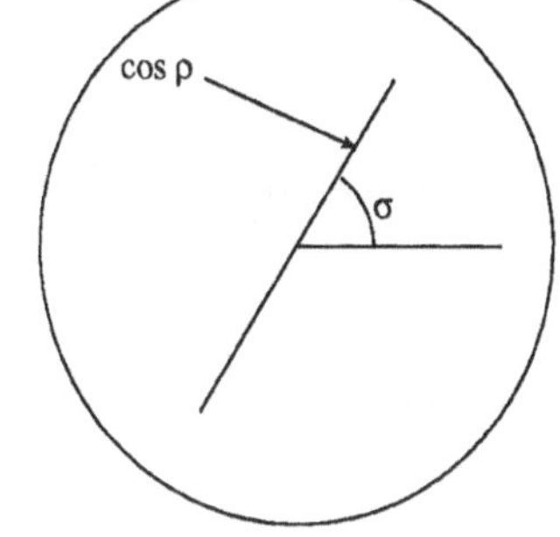


A convenient parameterization for the membrane rotating in the planes 1-2 and 3-4 is

$$X_0 = \tau \qquad X^1_{cl} + iX^2_{cl} = fe^{i\omega_1\tau}$$
$$X^3_{cl} + iX^4_{cl} = ge^{i\omega_2\tau} \tag{4.1}$$

In order to describe a closed membrane $f$ and $g$ are of the form:

$$f = \frac{1}{\omega_1}\cos\rho\cos\sigma \quad g = \frac{1}{\omega_2}\cos\rho\sin\sigma \quad 0 \le \rho \le 2\pi \quad \text{and} \quad 0 \le \sigma \le \pi \tag{4.2}$$

The fermions at the classical level are taken to be zero.

The induced metric is

$$\big(-A = -(1 - \omega_1^2 f^2 - \omega_2^2 g^2),\ \ 1,\ \ 1\big) = g_{\alpha\beta} \tag{4.3}$$

and is singular on the contour of the disc. The boundary of the disc rotates therefore with the velocity of light. Because in computing the propagator of the supermembrane one has to expand around periodic solutions, $\omega_1$ and $\omega_2$ must be comensurate i.e.

$$\omega_1/\omega_2 = p/q \tag{4.4}$$

By looking at the classical mass one obtains the relation

$$\sqrt{2\omega_1\omega_2} = \left(\frac{\chi_2}{E_{cl}}\right)^{1/2} \tag{4.5}$$

Correspondingly in the semiclassical regime $\chi_2 \to \infty$ (for fixed $E_{cl}$) the square root of the product of the two frequencies goes to infinity. The membrane therefore will shrink in at least one direction. We will however assume that the membrane, shrinks in both directions in the same way. i.e. $\omega_1 = \omega_2$. There does not exist a good reason[14] for this assumption other than that the computations are simpler in such a regime. In order to compute the spectra of fluctuation one makes an expansion analogous to (3.6) introducing also the fermionic fluctuations around zero. Then the lagrangian for the fluctuations contains a bosonic piece similar to (3.10) plus additional terms for the fermionic fluctuations

$$L^{(2)} = L^B + i\sqrt{-g_{cl}}g_{cl}^{\alpha\beta}\partial_\alpha X^\mu_{cl}\bar{\theta}\Gamma_\mu\partial_\beta\theta - \frac{1}{2}\epsilon^{\alpha\beta\gamma}\partial_\alpha X^\mu_{cl}\partial_\beta X^\nu_{cl}\bar{\theta}\Gamma_{\mu\nu}\partial_\gamma\theta \qquad (4.6)$$

In order to find the normal modes we go again to the noninertial frame rotating also the fermions and through this rotation two bosonic and half of the fermionic degrees of freedom decouple.

The action can again be written as

$$S = S^B_{\parallel} + S^B_{\perp}(D - 5 \text{ degree}) + S_F \qquad (4.7)$$

$S^B_\perp$ is the action for $D - 5 = 6$ degrees of freedom with the induced metric (3.5) while $S^B_\parallel$ and $S_F$ are more complicated actions which lead to a set of coupled equations among $Z_2$ and $Z_4$, respectively $(1 - \Gamma)\theta's$. In order to find the normal modes we look for solutions of the form

$$\hat{Z} = e^{iE\tau}R(f,g) \text{ and } \theta = e^{iE\tau}P(f,g) \qquad (4.8)$$

For the $D-5$ transverse degree of freedom getting the normal modes is easy. There are two sets of eigenvalues, which are obtained by using the regularity conditions at the boundary of the membrane and at the center of the membrane. One set corresponds to the eigenfunctions which are analytic at the boundary while the other set to the eigenfunctions which are finite but are not analitic on the boundary of the supermembrane. One finds the eigenvalues

$$E^i_I = 2\omega\left[\left(n+\mid\ell\mid+\frac{1}{4}\right)\left(n+\frac{1}{4}\right) - \frac{1}{16}\right]^{1/2}$$

$$E^i_{II} = 2\omega\left[\left(n+\mid\ell\mid+\frac{3}{4}\right)\left(n+\frac{3}{4}\right) - \frac{1}{16}\right]^{1/2} \qquad (4.9)$$

$n = 0,\ 1,\ 2, \dots,\ \ell = 0, \pm1, \pm2 \dots,\ (i = 1, ..D - 5)$

For the other degrees of freedom the corresponding equations are

$$\left\{\left(-\Box+\omega^2\right)\begin{pmatrix}1 & 0\\ 0 & 1\end{pmatrix} - 2\omega^2\begin{pmatrix}f\partial_f & g\partial_f\\ f\partial_g & g\partial_g\end{pmatrix}\right\}\begin{pmatrix}\hat{Z}^2\\ \hat{Z}^4\end{pmatrix} = 0 \qquad (4.10)$$

$$(1-\Gamma)\left[\partial_\tau + (A)^{\frac{1}{4}}\omega\left(\Gamma^3\partial_f - \Gamma^1\partial_g\right) - \frac{\omega}{2}\left(\Gamma^1\Gamma^2 + \Gamma^3\Gamma^4\right)\right]\theta = 0 \qquad (4.11)$$

In order to integrate these equations one must find convenient radial equations, by separating the variables. This is accomplished in a similar manner with the way one solves the Coulomb potential Dirac equation in the spherical coordinates (using spinors which are eigenfunctions of the total angular momentum.) The corresponding conserved quantity in this case is a combined reparametrization and space time rotation. The form of this operator is

$$J_\phi = \frac{\partial}{\partial \phi} + \begin{cases} i\sigma^2 & \text{-bosons} \\ \frac{1}{2}\left(\Gamma_1\Gamma_3 + \Gamma_2\Gamma_4\right) & \text{-fermions} \end{cases} \tag{4.12}$$

It is convenient to use a representation for the Dirac matrices of the form $SO(4) \times SO(6.1)$ in which

$$\begin{aligned} &\Gamma_\mu = \gamma_\mu \otimes I(8 \times 8) &&\Gamma_A = \gamma_5 \times \gamma_A \\ &\mu = 1, \ldots 4 &&A = 5 \ldots 10 \end{aligned} \tag{4.13}$$

so that the equation to integrate becomes a 4 dimensional Dirac equation. Imposing regularity conditions at the boundary and at the center of the membrane we get the system of eigenvalues

$$E = 2\omega\left[(n + \mid \ell \mid + c)(n + c) + b\right]^{1/2} \tag{4.14}$$

with $n = 0,\ 1, \ldots; \ell = 0, \pm 1,\ \pm 2 \ldots$

For the bosons one has the following set of values for $b$ and $c$

$$(b,\ c) = (-1/16, 1/4), (7/16, 3/4), (-1/16, 3/4), (7/16, 1/4) \tag{4.15}$$

For the fermions we get a set of zero modes and a doubly degenerate set of eigenvalues:

$$(b,\ c) = (0,\ 1) \quad \text{and} \quad (0,\ 1/2) \tag{4.16}$$

The eigenvalues for fermions are always different from the ones for bosons. For the fermions $b = 0$ and $c$ takes different values corresponding to different sets of eigenvalues while for the bosons both $b$ and $c$ are different from zero for the different sets of eigenvalues.

A feature of the above set is that we have both bosonic and fermionic zero modes. The presence of the fermionic zero modes does indicate a certain supersymmetry of the system but their number does not match the number of supersymmetries in the original system. (as it happens when one quantizes by the same method the Green-Schwarz superstring action) We wish to stress that had we looked at the full nonlinear equations (without expanding around the given background) we would have found that we have $D - 3$ bosonic zero modes and $\frac{N}{2}$ fermion zero modes. In the present situation the number of zero modes is reduced as compared with the above case which presumably happens because our system has less symmetry. If this is true (that the reduction of symmetries is related to a reduction in the # of zero modes) then one can easily convince oneself that the choice $\omega_1 = \omega_2$ in fact corresponds to the maximum allowed symmetry within the class of solutions which we are considering.

The difference between the fermionic and bosonic eigenvalues makes the cancellations we are seeking improbable. It is on these cancellations that the hope of getting massless particles at the quantum level is based. We did also perform a generalized $\zeta$ function computation of the Casimir energy (even if such a computation lacks the justification one gives to it in the string case)

$$Z_\nu(b,\ c,\ a) = \sum_{n=0}^{\infty} \sum_{\ell=-\infty}^{+\infty} [(n + g \mid \ell \mid + c)(n + c) + b]^{-\frac{\nu}{2}} = Z(\text{analytical}) + Z(\text{numeric}) \tag{4.17}$$

and obtained

$$E(\text{vac}) = -1.7488\omega \tag{4.18}$$

By generalizing the connection between the Casimir energy and the spin of massless states one is led to the conjecture

$$J = \frac{E(\text{vac})}{\omega} + n, \tag{4.19}$$

where $n$ is an integer In order to have zero mass particles $E(\text{vac})/\omega$ must be therefore an integer, which it is not.

CONCLUSIONS

I have presented the computation of the Casimir energy for the supermembrane within the approach of Kikkawa and Yamasaki. Within this approach in order to be able to have massless particles one must obtain a Casimir energy which is zero or an integer number. The real hope in doing this computation was that (as it happens for many supersymmetric theories) the vacuum energy could have been zero. This did not happen. It may be argued that this is because the corresponding solution around which we expand does break the supersymmetry. The problem is however that this is basically the only type of solution which is avaiable. An improvement of this result within the class of solutions of the Kikkawa Yamasaki type does not seem very probable to us in view of the argument about the zero modes which we gave. The other types of solutions[15] which were considered and which lead to more symmetry for the system are valid only in the sector which definitely does not contain the graviton. This is because the stability of the system is related to nontrivial number of windings in the compactified sector.

Then one may contemplate the supermembrane action in the light cone gauge where the hamiltonian becomes

$$H \sim \frac{\dot{x}^2}{2} + \text{fourth order terms}$$

It is plausible that the quantization of the zero modes (together with the corresponding part for the fermions) will give good counting results[16] which will include a massless gravition. However the semiclassical approximation for this type of expansions (that is neglecting the interactions) is known to fail. We must therefore conclude that within the present understanding of the supermembranes there are no indications that they may contain massless excitations in their spectrum.

ACKNOWLEDGEMENTS

I wish to thank P.G.O. Freund and K.T. Mahanthapa for giving me the oportunity to speak at the Workshop on Superstrings in Boulder. I am also indebted to my collaborators R.I. Nepomechie and P. van Nieuwenhuizen as well as to I. Bars, E. Bergshoeff, M.T. Grisaru, I. Khlebanov, J. Nearing, J. Polchinski, M. Peskin, E. Sezgin, and L. Susskind. This work was supported in part by a NSF-Grant No.Phy-87-03-390.

REFERENCES

1. L. Mezincescu, R.I. Nepomechie, and P. van Nieuwenhuizen, preprint UMTG-139/ ITP-SB-87-43, to be published.
2. P. Ramond, Phys Rev **D3** ,2415( 1971) A. Neveu,and J. Schwarz, Nucl Phys **B31** ,86(1971)
3. M. Green, and J. Schwarz, Phys Lett **136B** ,367 (1984)
4. P. Howe, and R.W. Tucker, J. Phys **A10**, L155 (1977) T. Higashijima, T. Uematsu,and Y-Z Yu, Tokyo preprint UT-423 (1984)
5. W. Siegel, Phys Lett **128B**, 397 (1983)
6. E. Witten, Nucl Phys. **B266**, 245 (1986) M.T. Grisaru, P. Howe, L. Mezincescu, B.E.W. Nilsson, and P.K. Townsend Phys Lett **162B**, 116 (1985)
7. J. Hughes, J. Liu, and J. Polchinski Phys Lett **180B** ,370 (1986)
8. E.Bergshoeff,E. Sezgin, and P.K. Townsend Phys Lett **189B** , 75 (1987)
9. M. Henneaux, and L. Mezincescu Phys Lett **152B** ,340 (1985)
10. K. Kikkawa and M. Yamasaki Prog. Theor. Phys. **76** , 1379 (1986)
11. L. Brink, and H.B. Nielsen, Phys Lett **45B** , 332(1973)
12. T. Curtright and P. van Nieuwenhuizen preprint ITP-SB-87-20
13. I am indebted to L. Susskind and I. Khlebanov for a discussion on this point
14. I wish to thank J. Polchinski for stressing this point
15. M.J. Duff, T. Inami, C.N. Pope, E. Sezgin, and K. Stelle preprint CERN- Th 4731/87, IC/87/74, to be published in Nucl Phys
16. I. Bars, C.N. Pope, and E. Sezgin, preprint to be published

HIGHLIGHTS OF PARASTRING THEORIES

Freydoon Mansouri

Physics Department
University of Cincinnati
Cincinnati, Ohio 45221

## I. INTRODUCTION

Most of the attempts at the construction of consistent string theories in four-dimensional space-time begin with the construction of the corresponding string theories in space-times which have more than four dimensions and then proceed to the compactification of the extra dimensions.[1] The schemes which have been developed to this end so far are elegant and promising but not completely satisfactory. For one thing, the non-existence of a compactification scheme which is uniquely dictated by the string theory itself runs counter to notion that we are dealing with an ultimate theory. From the physical point of view, it is also important to keep in mind that the current enthusiasm for ten-dimensional superstring theories is based on the plausibility that these theories make contact with phenomenology through a grand unified model such as $E_6$ or SO(10). If it turns out that in the next generation of experiments quarks or leptons begin to exhibit structure, then a fundamental theory of elementary particles and its unification with gravity would have to be related to a superstring theory at a deeper level.

An alternative to this approach is to construct consistent string theories directly in four-dimensional space-time and to make use of the extra dimensions only in so far as they provide a rationale for the existence of internal symmetries.[2] To this end, let us consider the two physical requirements which lead to the critical dimensions in these theories and which we are unwilling to give up. These are Poincaré invariance and quantum mechanics. The compatibility between these principles in a string theory can be ensured in one of two ways: (a) We can fix the statistics of the particles (fields) to be the standard fermi and bose and vary the Poincaré group or, equivalently, the dimension, D, of space-time. Then one arrives at D = 26 for bosonic strings and D = 10 for the sypersymmetric varieties. (b) We can fix the Poincaré group or, equivalently, D, and search for statistics for which the corresponding quantum theory would be consistent. There is no a priori reason to expect that such statistics exist. But it turns out that for parastatistics of order Q, it is possible to obtain a number of other critical dimensions at which the string theories are again consistent. For bosonic parastrings the critical dimension and the order Q are related by the expression[3]

$$D = 2 + \frac{24}{Q} \qquad \text{(I.1)}$$

For supersymmetric parastrings, the critical dimensions are given by[3]

$$D = 2 + \frac{8}{Q} \quad , \tag{I.2}$$

Thus, in both cases, D = 4 is a distinct possibility. We note that the critical dimensions (I.2) are the same as those in which sypersymmetric Yang-Mills theories can be constructed. Given the intimate relation between the zero mass sector of string theories and the supersymmetric Yang-Mills theories, this is of course not surprising. These considerations provide sufficient justification for further study of these models.

The remaining sections of this work are organized as follows: In section II, parastatistics is briefly reviewed with emphasis on those features which are relevant to string theories. In section III, the quantization of open bosonic parastring is presented. In section IV, I'll discuss the spectra of parastring theories and their zero point fluctuations. In section V, a vertex operator is constructed and the explicit expression for the simplest 4-point amplitude is given. In section VI, the planar loop correction for the open bosonic parastring is carried out and is shown to be consistent with Lovelace analyticity. In this section, the modular invariance of closed bosonic string is also demonstrated. Finally, in section VII, the results are summarized and are commented on.

## II. A BRIEF REVIEW OF PARASTATISTICS

Parafield theories differ from fermion theories in that the dynamical variables satisfy not bilinear but trilinear relations[4]. In the Hilbert space, $\underline{A}$, associated with a parafield theory the observables of the theory are determined by the requirement of locality. Although it is, in principle, possible to study various features of these theories within the Hilbert space, $\underline{A}$, it is often convenient to put this space in correspondence with a larger Hilbert space, $\underline{B'}$, in which the operators satisfy bilinear relations. Traditionally, for Fock-type irreducible representation of para theories with a unique vacuum,

$$a_k \, |0\rangle = 0, \tag{II.1}$$

this is done by means of the Green ansatz[5]:

$$a_k = \sum_{\alpha=1}^{Q} a_k^{(\alpha)} \tag{II.2}$$

where Q is the order of the parafield, $\alpha$ is the Green index, and $a_k^{(\alpha)}$ are Green components satisfying bilinear but <u>anomalous</u> (anti) commutation relations:

$$\left[a_k^{(\alpha)}, a_\ell^{(\alpha)\dagger}\right]_{\pm} = \delta_{k\ell} \; ; \quad \left[a_k^{(\alpha)}, a_\ell^{(\beta)}\right]_{\mp} = 0 \; ; \; \alpha \neq \beta \tag{II.3}$$

One can remove this anomaly by means of a Klein transformation. Then for real fields, the Green components $a_k^{(\alpha)}$ will transform as a representation of the group SO(Q). The main disadvantage of Klein transformations is that they are non-linear and non-local. Although it is possible to show[4] that for a large class of parafield theories the locality requirement makes the observables independent of these non-local operators, some ambiguities remain. For one thing, normal ordering in space $\underline{B'}$ does not necessarily imply normal ordering in space $\underline{A}$, and vice versa.

It is important to note that a parafield theory is formulated in the Hilbert space $\underline{A}$. One can put the states and the observables of this space into correspondence with a <u>subset</u> of the states and observables of the Hilbert space $\underline{B'}$ which has a larger number of states and a larger class of observables. But not every one of the operators and states in space $\underline{B'}$ can be represented in space A. In particular, observables in space B' which carry uncontracted SO(Q) indices have no equivalents in the original space A. Since the parafield theory is defined in the Hilbert space A, the Hilbert space B' is useful, but not necessary, to the extent that it facilitates making deductions about the space A.

An alternative to the Green ansatz and the corresponding Klein transformations is an ansatz suggested by Greenberg and Macrae[6]:

$$\Psi(x) = \sum_{\alpha=1}^{Q} e^{\alpha}\, \psi^{\alpha}(x) \quad ; \quad \alpha = 1,\ldots,Q \tag{II.4}$$

where $\psi^{\alpha}(x)$ are ordinary fermions, if $\Psi(x)$ is a parafermion field and ordinary boson if $\Psi(x)$ is a paraboson field, and $e^{\alpha}$ are elements of the real Clifford algebra:

$$\left\{e^{\alpha},\, e^{\beta}\right\} = 2I\delta^{\alpha\beta} \quad ; \alpha,\ \beta = 1,\ldots,Q \tag{II.5}$$

Moreover, $[e^{\alpha},\, \psi^{\beta}] = 0$, $\alpha,\ \beta = 1,\ldots,Q$. For example, the Majorana parafermions satisfy the trilinear relations

$$\Big[[\Psi(x),\, \Psi(y)]_{+} - \langle[\Psi(x),\, \Psi(y)]_{+}\rangle_{0}\ ,\ \Psi(z)\Big]_{-} = -2\delta^{3}(\vec{x} - \vec{z})\ \Psi(y) \tag{II.6}$$

where the symbol $\langle\ \rangle_0$ stands for the vacuum expectation value. The ansatz (II.4) relates the parafield operators and states to a set of field operators and states in a fermionic Hilbert space, B, say. The linearity of this relation, and the fact that $\psi^{\alpha}(x)$ are standard fermions with an SO(Q) internal symmetry, facilitate a comparison of parafermion theory with a fermion theory. The linearity also allows us to invert (II.4): $\psi^{\alpha} = \{\psi,\, e^{\alpha}\}$.

As mentioned above, the structure of the observable in parafermion theories are restricted by the requirement of locality. For a parafermion theory of odd order, the observables are limited to the functionals of the commutators $:[\Psi(x_1),\, \Psi(x_2)]:$, and for parafermion theories of even order, they may at most be functionals of this commutator and the symmetric product $:[\Psi(x_1)\ldots\Psi(x_Q)]_{+}:$, where Q is the order of parafermion theory. We note that the restrictions imposed by locality on the structure of observables is not limited to parafermion theories: It is precisely this condition which restricts the structure of the observables of standard fermions to their even products. The above observables can be further restricted by additional symmetries which the theory might carry. For example, in most cases of interest, chiral invariance or conformal invariance rule out the symmetric product as a possible term in the action.

Let us next consider the structure of states in the para space A and their relation to the states in the Hilbert space B. Since the para operators do not simply commute or anticommute, the corresponding Fock space is more complicated than those of fermion or boson operators. Using very general arguments of quantum field theory,[4] it is possible to construct a set of linearly independent basis states in terms of which an arbitrary state in space A can be expressed. For example, an n-parafermion basis state can be written in the form

$$|n,\, b,\, c,\, Q\rangle = \left[\Psi_1,\, \Psi_2\right]_{-} \cdots \left[\Psi_{2b-1},\, \Psi_{2b}\right]_{-} \left[\Psi_{2b+1} \cdots \Psi_{2b+c}\right]_{+} |0\rangle \tag{II.7}$$

where

$$n = 2b+c \quad ; \quad c \leq Q \tag{II.8}$$

These states can be classified according to the irreducible representations of the symmetric group $S_n$.

The ansatz (II.4) provides the link between the operators and states in the Hilbert space $\underline{A}$ and a subset of states and operators in the Hilbert space $\underline{B}$. Let $\underline{H}$ be the restriction of $\underline{B}$ which is in one-to-one correspondence with $\underline{A}$. That $\underline{H}$ cannot be all of $\underline{B}$ can be seen by noting that in $\underline{A}$, and therefore in $\underline{H}$, the operators do not carry an SO(Q) index, so that any operator in space $\underline{B}$ which is not an invariant under SO(Q) transformations has no equivalent in space $\underline{A}$. Such operators acting on a state in $\underline{H}$ will connect it to states in $\underline{B}$ which are not in $\underline{H}$. In general, from the set of states in $\underline{B}$ which belong to an irreducible representation of SO(Q), one can construct one, and only one, state which is equivalent, via the ansatz (II.4), to a state in $\underline{A}$. It is interesting to note that such a state still carries a Q-fold degree of freedom. This can be seen from the form of the ansatz (II.4). It can also be seen by applying the number operator to a one paraparticle state:

$$N|1\ \text{para}\rangle = Q|1\ \text{para}\rangle \tag{II.9}$$

As an illustration, consider[7] a free two-dimensional Majorana parafermi theory of order Q with a chiral SO(N) symmetry. Using a light cone basis, both for the coordinates and for the fields, the action can be written as

$$S = \frac{i}{4}\int dx_+ dx_- \left( \left[\psi_-^{\,i},\ \partial_+\,\psi_-^{i}\right] + \left[\psi_+^{i}\ ,\ \partial_-\,\psi_+^{i}\right] \right) \tag{II.10}$$

It conforms to the general form of the observables in space $\underline{A}$ stated above. Using the ansatz (II.4), we can find the equivalent action in the subspace $\underline{H}$:

$$S = \frac{i}{2}\int dx_+ dx_- \sum_{\alpha=1}^{Q} \left( \psi_-^{\alpha i}\partial_+\,\psi_-^{\alpha i} + \psi_+^{\alpha i}\partial_-\,\psi_+^{\alpha i} \right) \tag{II.11}$$

where $\psi_\pm^{\alpha i}$ are ordinary fermions. This action is indistinguishable from a free fermion action with SO(Q) x SO(N) symmetry. It can therefore be extended to the entire space $\underline{B}$. This is a general feature of the subclass of observables in space $\underline{B}$ which have equivalents in space $\underline{A}$. Their structure does not change in restriction from $\underline{B}$ to $\underline{H}$. The underlying reason is not hard to see: The Hilbert space $\underline{B}$ is a carrier of the irreducible representations of SO(Q). In each irreducible representation, there is one state (or ray) which belongs to the subspace $\underline{H}$. Since the eigenvalues of the SO(Q) invariant operators are the same for all states of a given irreducible representation, their structure is independent of which SO(Q) states they act on. This independence can be used to make deductions about a parafield theory defined in $\underline{A}$ while working in the more convenient Hilbert space $\underline{B}$. For example, consider the SO(N) currents, $J^{ij}$, of the theory given by (II.10):

$$J_\pm^{ij} = \frac{1}{2} : \left[\psi_\pm^{i}\ ,\ \psi_\pm^{j}\right] : \tag{II.11}$$

where the normal ordering is defined with respect to the parafields $\psi_\pm^{i}$. Note that this expression conforms to the general structure of the observables deduced from locality. To rewrite $J_\pm^{ij}$ in terms of the fermion field operators in space $\underline{B}$, we make use of the ansatz (10). Since in that ansatz $[e^\alpha,\ \psi^\beta] = 0$, the normal ordering of the parafields in $\underline{A}$-space does not affect the ordering of the elements of the clifford algebra $\{e^\alpha\}$, and it reduces to the normal ordering of the fermion field operators in the $\underline{B}$-space. We thus get:

$$J_\pm^{ij}(+) = :\ \psi_\pm^{i\alpha}\ \psi_\pm^{j\alpha}\ :$$

Not surprisingly, these are precisely the expressions as those for the O(N)xO(Q) free fermion theory. Therefore, the algebraic structure of these currents is the same as the SO(N) currents of the SO(N) x SO(Q) fermion theory. In particular, the central charge is the same for both. One can show that the virasoro algebra of the parafermion theory is also identical to that of the corresponding free fermion theory.

It is worth noting again that these equivalences involve only the SO(Q) invariant operators of the Hilbert space B which leave the subspace H invariant. An example of an operator in the space B which has no equivalent in space A is provided by the SO(Q) currents of the fermion theory:

$$J_{\pm}^{\alpha\beta} = :\psi_{\pm}^{\alpha i}\,\psi_{\pm}^{\beta i}:$$

These currents are not SO(Q) invariants and do not leave the subspace H invariant.[7]

## III. LIGHT CONE QUANTIZATION OF PARASTRINGS

Consider the paraquantization of the bosonic open string in the light-cone gauge.[3] According to (II.4), we can write the transverse string variable as

$$y^{i}(\tau,\sigma) = \sum_{\beta=1}^{Q} e^{\beta} y^{i\beta}(\sigma,\tau)\ , \tag{III.1}$$

where Q is the order of parastatistics and

$$y^{i\beta}(\tau,\sigma) = x^{i\beta} + 2\alpha' p^{i\beta}\tau + \sqrt{2\alpha'}\sum_{n=1}^{\infty}\frac{1}{\sqrt{n}}\left[a_n^{i\beta}\exp(-in\tau) + a_n^{i\beta\dagger}\exp(in\tau)\right]\cos n\sigma . \tag{III.2}$$

The canonical variable conjugate to $y^{i\beta}(\tau,\sigma)$ is

$$\Pi^{i\beta}(\tau,\sigma) = (1/2\pi\alpha')\dot{y}^{i\beta}(\tau,\sigma)\ , \tag{III.3}$$

so that, at equal times

$$\left[y^{i\alpha}(\sigma),\ \Pi^{j\beta}(\sigma')\right] = i\delta^{ij}\delta_{\alpha\beta}\delta(\sigma - \sigma)\ . \tag{III.4}$$

A "dot" on top of a quantity means it is $\tau$-derivative, and a prime means it is $\sigma$-derivative. With these commutation relations, the variable $y^{i}(\tau,\sigma)$ and its conjugate $\Pi^{i}(\tau,\sigma)$ satisfy the trilinear relations given by (II.6).

The elements of the Virasoro algebra can be constructed in terms of the parastring oscillators. They are

$$T_m = \alpha'\vec{p}^{\,2}\delta_{m,0} - i\sqrt{2\alpha'}m\sum_{\beta=1}^{Q}\vec{p}^{\beta}\cdot\vec{a}_m^{\beta} + \sum_{\beta=1}^{Q}\sum_{n=1}^{\infty}\sqrt{n(m+n)}\,\vec{a}_n^{\beta\dagger}\cdot\vec{a}_{m-n}^{\beta}$$

$$-\frac{1}{2}\sum_{\beta=1}^{Q}\sum_{n=1}^{m-1}\sqrt{n(m-n)}\,\vec{a}_n^{\beta}\cdot\vec{a}_{m-n}^{\beta}\ , \tag{III.5}$$

where the arrows refer to the transverse vectors. Their commutator is given by

$$\left[T_{m},\ T_n\right] = (n-m)T_{n+m} + \tfrac{1}{12}Q\,(D-2)(n^3 - n)\delta_{m+n,0}\ . \tag{III.6}$$

Note the dependence of the central charge on Q. It is precisely this feature of the bosonic as well as the supersymmetric parastrings which allows for the possibility of dimensions other than D = 26 or D = 10, respectively. Also note that the Virasoro operator, $T_m$, is an SO(Q) invariant, so that it is consistent with the notion of an observable in space B which has an equivalent in space A.

In the light-cone gauge quantization of the string theories, one of the

D components, say $y^+$, of the dynamical variable $y^\mu(\tau,\sigma)$ is taken to be proportional to the parameter $\tau$. The paraquantization in this gauge is carried out similarly: of a total of D components of $y^\mu(\tau,\sigma)$, we identify one to be proportional to $\tau$. That is, we set

$$y^+ = 2\alpha' p^+ \tau = \sum_{\beta=1}^{Q} e^\beta y^{+\beta} , \tag{III.7}$$

where

$$y^{+\beta} = p^{+\beta}\tau \quad , \beta = 1, \ . \ . \ . \ , Q \tag{III.8}$$

This implies that $p^+$ is given by

$$p^+ = \sum_{\alpha=1}^{Q} e^\alpha p^{+\alpha} \tag{III.9}$$

One can then solve the geometrical constraint equations to obtain

$$\Pi^-(\sigma, \tau) = \left(\pi/(p^+)^{-1}\right) \left[\vec{\Pi}^2 + (2\pi\alpha')^{-2}\vec{y}'^2\right] \tag{III.10}$$

$$y'^- = (\pi/p^+)(\vec{\Pi}\cdot\vec{y}') \ . \tag{III.11}$$

The null plane hamiltonian is given by

$$p^- = \sum_{\alpha=1}^{Q} e^\alpha H^\alpha \tag{III.12}$$

Here,

$$H^\alpha = (2\alpha')^{-1} \ (p+)^{-2} \ p^{+\alpha} \left(T_o - \alpha_o\right) \tag{III.13}$$

From this, it follows that the mass-squared operator is given by

$$M^2 = 2p^+p^- - \vec{p}^2 = \frac{1}{\alpha'} \sum_{\beta=1}^{Q} \sum_{n=1}^{\infty} \vec{a}^{\,\beta+} \cdot \vec{a}_n^{\,\beta} - \alpha_o . \tag{III.14}$$

Of these operators, $M^2$ is a bosonic operator and an observable, and its eigenvalues can label the states in the para space $\underline{A}$. But $p^+$ and $p^-$ are not, in general. Since $p^-$ plays the role of the Hamiltonian and is responsible for the time development of the theory, it is natural to require that the states in the Hilbert space $\underline{A}$ be eigenstates of $p^-$. As can be seen from (III.12) and (III.13), this requires that the states also be eigenstates of $p^+$. The operator $p^+$ plays a special role in the light cone quantization of string theories and appears in the expressions for other observables such as the Lorentz generators as well. In order for $p^+$ to have definite eigenvalues in the Hilbert space $\underline{A}$, it is necessary and sufficient that in the Hilbert space $\underline{B}$ the eigenvalues of the operators $p^{+\alpha}$, $\alpha = 1, \ldots , Q$, be the same for all $\alpha$.[8] In more physical terms, if we identify the index $\alpha$ as some sort of color index and $p^{+\alpha}$ as the component of momentum along the length of the $\alpha^{th}$ constituent, then this amounts to requiring the density of momentum along the string be the same for all constituents.

The consistency of this paraquantization with Lorentz invariance can be checked by explicitly constructing the Lorentz algebra. In the new formulation these are given by the following expressions ($\alpha' = 1/2$):

$$M^{ij} = \sum_{\alpha=1}^{Q} \left(x^{i\alpha} p^{j\alpha} - x^{j\alpha} p^{i\alpha}\right) + i \sum_{\alpha=1}^{Q} \sum_{n=1}^{\infty} \left[a_n^{i\alpha} a_n^{j\alpha\dagger} - a_n^{j\beta} a_n^{i\alpha\dagger}\right]$$

$$M^{i-} = \frac{1}{2} \sum_{\alpha=1}^{Q} \left(p^+\right)^{-2} p^{+\alpha} \left[x^{i\alpha} \left(T_o - \alpha_o\right) + \left(T_o - \alpha_o\right) x^{i\alpha} - 2\, x^{-\alpha} p^{i\alpha}\right]$$

$$+ \sum_{\alpha=1}^{Q} \left(p^+\right)^{-2} p^{+\alpha} \sum_{n=1}^{\infty} \frac{1}{\sqrt{n}} \left[a_n^{i\alpha} T_n^{\dagger} + T_n a_n^{i\alpha\dagger}\right]$$

$$M^{i+} = \sum_{\alpha=1}^{Q} x^{i\alpha} p^{+\alpha}$$

$$M^{+-} = -\frac{1}{2} \sum_{\alpha=1}^{Q} \left(p^{+\alpha} x^{-\alpha} + x^{-\alpha} p^{+\alpha}\right)$$

The vanishing of the commutator $[M^{i-}, M^{j-}]$ for $i \neq j$ will then lead to the critical dimensions given by equation (1) and the condition $\alpha_o = 1$.

The light cone quantization of closed bosonic parastring as well as supersymmetric parastrings can be carried out in a similar way.[3]

## IV. THE SPECTRA AND ZERO-POINT ENERGIES

The spectrum of a parastring model is similar to the corresponding string model in that for every mass point in the string model, there is a mass point in the parastring model. As can be seen from the example (II.7), in bosonic parastrings the states can be either bosons or parabosons. Similarly, in supersymmetric parastrings, the states can be bosons, fermions, parabosons, or parafermions. Below, I'll give a few examples of these.

For open bosonic parastring the ground state is a tachyon. A simple way to see this is to consider the zero-point fluctuations.[9] The expression (III.8) for $M^2$ can be rewritten as

$$M^2 = \frac{1}{\alpha'} \sum_{i=1}^{D-2} \sum_{\alpha=1}^{Q} \sum_{n=1}^{\infty} n a_n^{i\alpha\dagger} a^{i\alpha} + M_o^2 , \qquad \text{(IV.1)}$$

where, in appropriate units, $M_o^2$ is the zero point energy of the system and is given by

$$\alpha' M_o^2 = \frac{1}{2}(D-2)Q \sum_{n=1}^{\infty} n . \qquad \text{(IV.2)}$$

This expression can be regularlized by means of the Riemann $\xi$-function:

$$\alpha' M_o^2 = \frac{1}{2}(D-2)Q\xi(-1) = \frac{-Q(D-2)}{24} \qquad \text{(IV.3)}$$

Thus, the (mass)$^2$ of the ground state is fixed at

$$\alpha' M^2 |0\rangle = -\frac{Q(D-2)}{24} |0\rangle . \qquad \text{(IV.4)}$$

The first excited state is constructed by the application of the transverse oscillator $a_1^{i\dagger}$ on the vacuum:

$$|v^i\rangle = a_1^{i\dagger} |0\rangle \qquad \text{(IV.5)}$$

This para-vector forms a representation of the group SO(D -2). Since it is the only state at this mass level, to be a representation of the Poincare group in D dimensions, it must be massless. On the other hand, the application of $M^2$ - operator on this state gives

$$\alpha' M^2 |v^i\rangle = \left[-\frac{Q(D-2)}{24} + 1\right] |v^i\rangle . \qquad \text{(IV.6)}$$

To ensure a vanishing eigenvalue for $M^2$, it is necessary that

$$D = 2 + \frac{24}{Q} \,. \tag{IV.7}$$

Similar considerations apply to the spectrum of closed bosonic parastrings. The ground state is again a tachyon. Since there are now two sets of para oscillator modes, $a^i$ and $\hat{a}^i$, say, the physical states are constructed by symmetric and antisymmetric products of these as discussed in section II. The symmetric combination of the first excited states is given by

$$|S^{ij}\rangle = \tfrac{1}{2}\left(a_1^{i\dagger}\,\hat{a}_1^{j\dagger} + \hat{a}_1^{j\dagger}\,a_1^{i\dagger}\right)|0\rangle \tag{IV.8}$$

This is a massless bosonic spin 2 state. The simplest way to see that it is indeed a boson and not a paraboson is to make use of the ansatz (II.4) to rewrite (IV.8) in the form

$$|S^{ij}\rangle = \frac{1}{2}\sum_{\alpha=1}^{Q}\left(a_1^{i\alpha\dagger}\,\hat{a}_1^{j\alpha\dagger} + \hat{a}_1^{j\alpha\dagger}\,a_1^{i\alpha\dagger}\right)|0\rangle \tag{IV.9}$$

With the trace removed, we are left with a unique massless spin 2 state which can be identified as the graviton. It follows that the most important selling point of the standard string theories is retained in parastring theories. Moreover, such a gravitational theory is not in ten but in four dimensions.

The main feature of the spectra of parastring theories is that the graviton as well as other bosonic and fermionic states are all composite. Since their massless sectors contain parabosons and parafermions, it is clear that the contact with the phenomenology of particle physics cannot be made at the level of the standard model or of grand unified theories. Their most natural interpretation seems to be in the context of preon theories.

The spectra and the zero point energies of the supersymmetric parastrings can be studied along the lines sketched above. The details are discussed elsewhere.[9]

## V. INTERACTIONS AND TREE AMPLITUDES

Following the analogy with conventional string theories, we write down a ground state vertex operator for the bosonic parastring of order Q in the form

$$V(k) = :\exp[\tfrac{1}{2}\, i\, \{\vec{k}\cdot\vec{y}(0)\}]: \,, \tag{V.1}$$

where the momenta are taken to be transverse and

$$\{\vec{k}\cdot\vec{y}(0)\} = \vec{k}\cdot\vec{Y}(0) + \vec{Y}(0)\cdot\vec{k} \,.$$

Note that although both $\vec{k}$ and $\vec{Y}(0)$ are paraquantities, $V(k)$ is an $SO(Q)$ singlet bosonic operator. Then, with $\Delta = (P^2 - M^2)^{-1}$, dual tree amplitudes can be constructed in the same way as for conventional models. A typical n-point term is

$$B_N = \langle 0, K_N|V(k_{N-1})\,\Delta \,.\,.\,.\, V(k_2)|0, k_1\rangle \,. \tag{V.2}$$

As an example, consider the four-point amplitude in more detail. Following the standard method of evaluating these expressions and using the vertex and the propagator given above, $B_4$ takes the form

$$B_4 = \int_0^1 dx\,x^{-\alpha(s)-1} M \,, \tag{V.3}$$

where

$$M = (1 - x)^{-2 \sum_{\alpha=1}^{Q} \vec{k}_2^{\alpha} \cdot \vec{k}_3^{\alpha}} = (1 - x)^{-\alpha(t)-1} \tag{V.4}$$

For consistency, it is necessary to set $m^2 = \Sigma_{\alpha=1}^{Q} m_\alpha^2 = -1$, so that the ground state is again a tachyon.

Two important conclusions can be drawn from this result: (i) the kinematical relations which emerge are consistent with the interpretation given above that these amplitudes correspond to the scattering of composite objects consisting of Q constituents, (ii) the dual tree amplitudes cannot tell whether they arise from a bosonic theory with D = 26 or from a parastring theory with critical dimensions given by (I.1).

Since the vertex operator V(k) is bosonic, the emitted or absorbed particle is a boson. The vertex operators for excited states such as spin-1 and spin-1/2 particles can be constructed by the same method as in standard string and superstring theories. That is, the expression for V(k) is multiplied by an appropriate spin factor. In all of these cases the emitted and absorbed particles are not para but standard bosons and fermions. This is because in each case the vertices are constructed from a symmetric product of para operators. As we have seen in section II, there are also states involving antisymmetric products of para operators. Therefore, in parastring theories, it is possible to construct amplitudes which have no analogs in standard string theories.

## VI. LOOP CORRECTIONS AND MODULAR INVARIANCE

It will be recalled that in dual models, critical dimensions first arose in connection with the analytic behavior of the one-loop corrections to the dual tree amplitudes.[10] To check this in parastring theories, let us consider a typical planar one-loop amplitude in the open bosonic parastring. Using the notation of reference 11, we have for M external lines,

$$E = \int dP \; \mathrm{tr}\, [\Delta V(k_1)\Delta V(k_2)\Delta \; . \; . \; . \; V(k_M)]$$

$$= (\tfrac{1}{2})^M \int dP \prod_{I=1}^{M} \left[\int_0^1 d\,x_I \; x_I^{1/2P^2 - 2}\right] T \tag{VI.1}$$

where

$$T = \mathrm{tr}\left[x_1^N \, V_o(k_1) \; x_2^N \; V_o(k_2) \; . \; . \; . \; x_M^N \; V_o(k_M)\right] \tag{VI.2}$$

In these expressions, tr = trace, and

$$V_o(k) = g\!: \exp[\tfrac{i}{2}\{\bar{k} \cdot \bar{Y}(0)\}] = g\!: \exp\left[i \sum_{\alpha} \vec{k}^{\alpha} \cdot \vec{y}^{\alpha}(0)\right] ; \tag{VI.3}$$

$$N = \sum_{n=1}^{\infty} \sum_{\alpha=1}^{Q} n \, a_n^{\alpha\dagger} \, a_n^{\alpha} \tag{VI.4}$$

In a first quantized parastring theory, the momenta, and hence their differentials are para quantities, so that the integration over loop momentum requires some explanation. The most physical approach for dealing with this question is to translate it to the Hilbert space $\underline{B}$ where we can regard $p^{\mu\alpha}$, $\alpha = 1, . . ., Q$, as the momenta associated with the Q constituents of a composite particle. Then, it becomes clear that integrals over loop momenta will involve Q-fold integrations:

$$dp = \prod_{\alpha=1}^{Q} d^D p^{\alpha} \tag{VI.5}$$

where D is the dimension of space-time. In applying this expression to our parastring theory, we note from section III that $p^{+\alpha}$ have the same eigenvalues for all $\alpha$. Since off-shell momenta are analytic continuations of their respective on-shell values, this must also hold off-shell. Therefore, we must insert (Q - 1) appropriate $\delta$-functions to ensure that there is not Q but one independent degree of freedom in $p^+$ direction. A similar argument also hold for $p^{-\alpha}$.

Taking these results into account, the loop correction (VI.I) can be readily evaluated. The result is

$$E = \frac{1}{\pi} g^M \int_0^1 \left[ \prod_{I=1}^{M-1} \theta(\nu_{I+1} - \nu_I) d\nu_I \right] \int_0^1 dq\, q^{-1-Q(2-D)/12} W^{-1-Q(2-d)/24}$$

$$\times \left[ \frac{-2\pi^2}{\ln q} \right]^M [f(q^2)]^{(2-D)Q} \prod_{I<J} (\psi_{IJ})^{k_I \cdot k_J} \qquad \text{(VI.6)}$$

Here $\theta$ is a step function, $\psi_{IJ}$ is related to a Jacobi function, and W is related to q by the relation $\ln q \ln W = 2\pi^2$. In this expression, for $D \leq 26$, the integrand is meromorphic at $q = 0$ only at critical dimensions given by (I.1). Similar statements hold for other loop amplitudes. For example, for closed bosonic parastring, the planar one-loop amplitude with M external closed parastring tachyon states is given by

$$A = \int d^2\tau (\mathrm{Im}\tau)^{-2} C(\tau) F_\nu(\tau) \;, \qquad \text{(VI.7)}$$

where we use the notation of reference 11. For example,

$$C(\tau) = 4(\tfrac{1}{2}\mathrm{Im}\tau)^{-Q(D - 2)/2} e^{4\pi \mathrm{Im}\tau} |f(e^{2\pi i\tau})|^{-2Q(D - 2)} \qquad \text{(VI.8)}$$

We can apply modular transformations

$$\tau \rightarrow \frac{a\tau + b}{c\tau + d}$$

to the expression (VI.7) for A and explicitly verify that it is modular invariant only for the critical dimensions given by (I.1). These results lend further support to the non-triviality of these critical dimensions.

## VII. CONCLUDING REMARKS

We have seen in the preceding sections that it is possible to enlarge the number of critical space-time dimensions in which consistent string theories can be constructed. These critical dimensions are given by (I.1) and (II.2) and allow D = 4 as a possibility. To achieve this, it was necessary to generalize the statistics of the particles in the spectrum to include parafermions and parabosons. This is certainly compatible with principles of quantum mechanics and Poincaré invariance. Parastring theories have many features in common with standard string theories. Among these are Lorentz invariance, modular invariance, and the Lovelace criterion for analyticity. They also offer the hope for a consistent quantum theory of gravity, a feature which is an important rationale for interest in any string theory. In parastring theories, the graviton is a composite spin 2 particle in 4-dimensional space-time, so that complications related to compactification do not cloud the picture here.

A number of problems remain to be explored. On the phenomenological front, since the spectra of parastrings contain paraparticles (or an exact internal symmetry), and since the graviton turns out to be a composite particle, it seems natural to interpret these theories as preon theories. On the theoretical front, it is of interest to determine the gauge groups compatible with supersymmetric parastrings. Since the structure of anomolies in, say, four space-time dimensions is different from those in ten dimensions, they are not as restrictive in determining the gauge groups. Our preliminary results[8] indicate that a wide variety of gauge groups are allowed in four dimensions. In connection with supersymmetric parastrings, we also note that the critical dimensions (I.2) as well as other properties of the supersymmetric parastrings were first studied within the framework of the original Ramond-Neveu-Schwarz formalism. It would therefore be of interest to see if these parastring theories can be reformulated in terms of the superstring formalism of Green and Schwarz. This turns out to be the case.[3] Various properties of these theories will be reported elsewhere.[8]

This work was supported in part by the Department of Energy under the contract No. DOE-FG02-84ER40153.

REFERENCES

1. P. Candelas, G. Horowitz, A. Strominger, and E. Witten, Nucl. Phys., B258, 46 (1985).

2. For attempts to understand these symmetries as arising from the compactification of the bosonic string, see P.G.O. Freund, Phys. Lett., 151B, 387 (1985); A. Casher, F. Englert, H. Nicolai, and A. Taormina, Phys. Lett., 162B, 121 (1985).

3. F. Ardalan and F. Mansouri, Phys. Rev., D9, 3341 (1974); Phys. Lett., 176B, 99 (1986); Phys. Rev. Lett., 56, 2456 (1986). F. Mansouri, "Proceedings of XIV International Colloquium on Group Theoretical Methods in Physics," ed. R. Gilmore, World Scientific, in press.

4. K. Druhl, R. Haag, and J. E. Roberts, Commun. Math. Phys., 18, 204 (1970); Y. Ohnuki and S. Kamefuchi: Quantum Field Theory and Parastatistics, Springer-Verlag, New York (1982).

5. H. S. Green, Phys. Rev., 90, 270 (1953); O. W. Greenberg and A.M.L. Messiah, Phys. Rev., 138, B1155 (1965).

6. O. W. Greenberg and K. I. Macrae, Nucl. Phys., B219, 358 (1983).

7. F. Mansouri and X. Wu, Univ. of Cincinnati preprint, UCTP/87.

8. F. Mansouri and X. Wu, in preparation.

9. F. Mansouri and X. Wu, Mod. Phys. Lett., A2, 215 (1987).

10. C. Lovelace, Phys. Lett., 34B, 500 (1971).

11. J. H. Schwarz, Phys. Rep, 89, 223 (1982).

12. M. B. Green and J. H. Schwarz, Phys. Lett., 109B, 444 (1982), and 149B, 117 (1984).

LINE FUNCTIONAL THEORY FOR STRINGS

Yutaka Hosotani

School of Physics and Astronomy
University of Minnesota
Minneapolis, MN 55455

## 1. Introduction

Despite recent enormous developments in string field theory two of the important issues, namely those concerning field content and symmetry principles, do not seem to be satisfactorily settled. More precisely, one might ask (a) what fundamental mathematical entities we should take to represent string fields, and (b) what symmetry principles we should impose on the string field action. Of course, these two issues are not independent, but rather go hand in hand. In gauge theory and general relativity vector potentials ($A_\mu(x)$) and the metric of spacetime ($g_{\mu\nu}(x)$) are the fundamental entities, in terms of which the gauge invariance and the general coordinate invariance are most naturally and conveniently implemented. What we have to first know in string field theory is an analogue of $A_\mu(x)$ and $g_{\mu\nu}(x)$ and associated invariance principles.

Classical dynamics of Nambu-Goto strings are dictated by the requirement of two dimensional world-sheet reparametrization invariance. In view of this it is reasonable to ask, first of all, in terms of what fundamental entities the defining symmetry of string theory, reparametrization invariance, is most naturally implemented in the context of field theory. In our view these must be line functionals (functions of geometric lines), whose values depend only on the location and shape of lines (loops in the case of closed strings), but not on their parametrization[1-3]. In this particular respect the standard string field theory as defined in the literature is not quite satisfactory. There string fields are functions of parametrized lines, but not of geometric lines.

A theory of line functionals is not necessarily a theory of string. The next task is to implement string dynamics. If one insists on making a connection with classical geometric strings in the $x_0 = \tau$ gauge for the reason explained in Sec. 2, then one is naturally led to new generalizations of Dirac algebra. It implies that string fields nontrivially Lorentz-transform. We remind the reader that in the standard string field theory it is an implicit assumption that a string field itself is a Lorentz scalar, although a single string field, expanded in eigen-modes, contains various spin states. This is also obvious in the path integral formalism of quantum strings. Recall that basic quantities in QED are spinors and vectors, and tensors in general relativity, all of which non-trivially Lorentz-transform.

As explained at the beginning, it is desirable to find symmetry principles, in addition to reparametrization invariance, to be imposed on the string field action. A partial answer is provided by analyzing our equations in three dimensions. One finds that the equations have an infinite number of additional symmetries. There seem huge hidden symmetries underlying the line functional theory.

Some properties of line functionals were first investigated by Volterra many years ago[4]. In string theory Marshall and Raymond proposed to formulate a string field theory on the basis of line functionals in 1975[5]. Their action is quite different from ours. We determine the action by requiring the correspondence to classical string in the $x_0 = \tau$ gauge. Further advances in line functional theory took place not in string theory, but in QCD. Nambu, Polyakov, and others considered differential equations obeyed by matrix elements of Wilson loop integrals in QCD, which may be viewed as functions of integration paths (loops) [6-8] Recently, with the upsurge of interests in string theory, Bowick and Rajeev have proposed to formulate string theory on the basis of the complex geometry of loop (line) space [9]. It might be pointed out, however, that their procedure of introducing the complex structure in loop space is not manifestly reparametrization invariant. It is our philosophy to keep manifest $\sigma$- reparametrization invariance at all stages to construct a line functional theory for strings.

## 2. String Dynamics in the Dirac Form

Consider the Nambu-Goto action in the $x_0 = \tau$ gauge.

$$I = -\frac{1}{2\pi\alpha'}\int d\tau d\sigma\, L, \quad L = \{(\dot{\vec{x}}\vec{x}')^2 + (1 - \dot{\vec{x}}^2)\, \vec{x}'^2\}^{1/2}. \tag{1}$$

The conjugate momenta $\vec{p}(\sigma) = \delta I/\dot{\vec{x}}(\sigma)$ are not completely independent, but obey one set of constraints

$$\phi(\sigma) = \vec{x}'(\sigma) \cdot \vec{p}(\sigma) = 0. \tag{2}$$

The canonical Hamiltonian is given by

$$H_0 = \int d\sigma \,\{ \vec{p}^2 + \frac{\vec{x}'^2}{(2\pi\alpha')^2} \}^{1/2}. \tag{3}$$

The corresponding Schrödinger equation is

$$i \frac{\partial}{\partial\tau} \Psi[\tau;\vec{x}(\sigma)] = H_0 \Psi[\tau;\vec{x}(\sigma)], \tag{4}$$

where $\vec{p}(\sigma)$ in $H_0$ is replaced by $-i\delta/\delta\vec{x}(\sigma)$. The wave function $\Psi$ must satisfy, in addition, a constraint equation following from (2);

$$\vec{x}'(\sigma) \frac{\delta}{\delta\vec{x}(\sigma)} \Psi = 0. \tag{5}$$

Now one can see why the $x_0 = \tau$ gauge is ideally suited for a line functional theory. Eq. (5) implies that the string field $\Psi$ is $\sigma$-reparametrization invariant, i.e., $\Psi$ is a line functional. Eqs. (4) and (5) are compatible, since $H_0$ is $\sigma$-reparametrization invariant. In the covariant orthonormal gauge one obtains, instead, two sets of constraints

$$x'_\mu(\sigma)p^\mu(\sigma) = 0 \qquad p(\sigma)^2 + \frac{1}{(2\pi\alpha')^2} x'(\sigma)^2 = 0 \tag{6}$$

with a vanishing Hamiltonian $H_0 = 0$. At the quantum level these two constraints are incompatible. One is forced to require that only a half of the constraints are obeyed by $\Psi$, i.e., in terms of Virasoro generators $\{L_n\}$, $(L_0 - 1)\Psi = 0$ and $L_n \Psi = 0$ $(n>0)$. Then the first condition in (6) for $\sigma$-reparametrization invariance, which is implies $L_n = L_{-n}$, is not satisfied by $\Psi$. Therefore, in the standard string field theory, string fields are not line functionals, but functions of parametrized lines.

Eq. (4), however, is not quite what we are looking for. First, the Hamiltonian $H_0$ in Eq. (4) is ill-defined because of the presence of the square root in the integral. This difficulty is resolved by Dirac's trick. In the point particle case the relation $H_0 = (\vec{p}^2 + m^2)^{1/2}$ leads to a wave equation $i\partial\Psi/\partial t = (\vec{p}^2 + m^2)^{1/2}\Psi$. Dirac interpreted $(\vec{p}^2 + m^2)^{1/2}$ as a linear differential operator whose square is $\vec{p}^2 + m^2$, and replaced it by $\vec{\alpha}\cdot\vec{p} + \beta m$ to arrive at a perfectly well-defined equation. We employ the same trick to replace (4) by

$$i\frac{\partial}{\partial\tau}\Psi = \int d\sigma \sqrt{\vec{x}'^2}\,\mathcal{H}(\sigma)\,\Psi, \quad \mathcal{H}(\sigma) = \vec{\alpha}(\sigma)\cdot\frac{\vec{p}(\sigma)}{\sqrt{\vec{x}'(\sigma)^2}} + \beta(\sigma)\frac{1}{2\pi\alpha'} . \tag{7}$$

Because of the constraint equation (5) $\mathcal{H}(\sigma)$ only needs to satisfy

$$\mathcal{H}(\sigma)^2 = \frac{\vec{P}^2}{\vec{x}'^2} + \frac{1}{(2\pi\alpha')^2} + (\vec{x}'\cdot\vec{p}\ \text{terms}). \tag{8}$$

The arbitrainess represented in the last term in (8) is important to find non-trivial $\vec{\alpha}(\sigma)$ and $\beta(\sigma)$ as will be seen below.

## 3. Covariant Equations and Generalizations of the Dirac Algebra

Eq. (7) is not yet the final one. It lacks manifest Lorentz covariance. We shall make one more jump to write down

$$\int_0^1 d\sigma \sqrt{-x'(\sigma)^2}\{\Gamma^\mu(\sigma)\frac{P_\mu(\sigma)}{\sqrt{-x'^2}} - \frac{1}{2\pi\alpha'}\Lambda(\sigma)\}\Psi = 0, \tag{9}$$

$$x'^\mu(\sigma)\frac{\delta}{\delta x^\mu(\sigma)}\Psi = 0. \tag{10}$$

Here $\Psi$ is a function of spacetime curves. Eq. (10), generalizing Eq. (5), ensures that $\Psi$ is a line functional . $\Gamma^\mu(\sigma)$ and $\Lambda(\sigma)$ are determined by the following conditions: i) Eq. (9) is manifestly Lorentz covariant and translation invariant. ii) It also is manifestly $\sigma$-reparametrization invariant. iii) $\Gamma^\mu(\sigma)$ and $\Lambda(\sigma)$ depend on $\sigma$ only through $x^\mu(\sigma)$ and its derivatives. (This requirement is for the sake of economy.) iv) Eq. (9) reduces to Eq. (7) with Eq. (8) in the $x_0 = \tau$ subspace of lines so that the equation describes string dynamics in the Dirac form. v) $\vec{\alpha}(\sigma)$ and $\beta(\sigma)$ in Eq. (7) so obtained by condition iv) are local in $\sigma$-space. Finally vi) $\Gamma^\mu(\sigma)$ and $\Lambda(\sigma)$ commute with $p_\mu(\sigma)$.

Conditions i), ii), iii), and vi) imply that $\Gamma^{\mu}(\sigma)$ and $\Lambda(\sigma)$ can be expanded in a Taylor series in $t_{\mu}(\sigma) = x'_{\mu}(\sigma)/(-x'^{2})^{1/2}$ with constant matrix coefficients. Conditions iv) and v), then, determine an algebra to be satisfied by these matrix coefficients. The result for closed strings is

$$\Gamma^{\mu}(\sigma) = a_1\Gamma^{\mu} + a_2 i\Gamma^{\mu\nu}_{(-)}\, t_{\nu}(\sigma) + a_3\, i\, \Gamma^{\mu\nu}_{(+)}\, t_{\nu}(\sigma),$$

$$\Lambda(\sigma) = b_1 + b_1'\, i\, \Lambda_o + b_2\, i\Lambda^{\mu\nu}\, t_{\mu}(\sigma)\, t_{\nu}(\sigma)\ (+\ldots). \tag{11}$$

Here $\Gamma^{\mu\nu}_{(-)}$ ($\Gamma^{\mu\nu}_{(+)}$ or $\Lambda^{\mu\nu}$) is anti-symmetric (symmetric and traceless) in the Lorentz indices $\mu$ and $\nu$. Notice that only three distinct terms are allowed in the expansion of $\Gamma^{\mu}(\sigma)$ because of condition v). Though $\Lambda(\sigma)$ could contain any powers of $t_{\mu}(\sigma)$ in principle, the analysis of the equation in three dimensions indicates that higher-order terms in $\Lambda(\sigma)$ give rise to inconsistency, unless higher-order terms are included in $\Gamma^{\mu}(\sigma)$.

Condition iv), namely the condition for Eq. (9) to reduce to Eq. (7) in the $x_o = \tau$ subspace of lines, is satisfied, provided (a) the $\Gamma$'s and $\Lambda$'s satisfy the algebras

$$\{\Gamma^{\mu}, \Gamma^{\nu}\} = 2\eta^{\mu\nu}, \qquad \{\Gamma^{\mu\nu}_{(-)}, \Gamma^{\rho\sigma}_{(-)}\} = 2(\eta^{\mu\rho}\eta^{\nu\sigma} - \eta^{\mu\sigma}\eta^{\nu\rho})$$

$$\{\Gamma^{\mu\nu}_{(+)}, \Gamma^{\rho\sigma}_{(+)}\} = \{\Lambda^{\mu\nu}, \Lambda^{\rho\sigma}\} = 2(\eta^{\mu\rho}\eta^{\nu\sigma} + \eta^{\mu\sigma}\eta^{\nu\rho} - \frac{2}{d}\eta^{\mu\nu}\eta^{\rho\sigma}), \tag{12}$$

and $\Lambda_o^2 = 1$ with all other anticommutators zero, and (b) the c-number coefficients $a_i$ and $b_i$ obey the relations

$$a_1^2 + a_2^2 + a_3^2 = b_1^2 + b_1'^2 + 2(1-\frac{1}{d})\, b_2^2 = 1. \tag{13}$$

Indeed, by making use of (12) and (13) the square of $\mathcal{H}(\sigma)$ in (7) is shown to be

$$a_1^2 \mathcal{H}(\sigma)^2 = \frac{\vec{P}^2}{\vec{x}'^2} + \frac{1}{(2\pi\alpha')^2} - \{a_2^2 - (1-\frac{2}{d})\, a_3^2\}\, \frac{(\vec{x}'\cdot\vec{P})^2}{(\vec{x}'^2)^2}. \tag{14}$$

Therefore the requirement (8) is satisfied after rescaling $\tau$. Substituting (11) into (9) we finally arrive at the equation,

$$D\Psi = [a_1\, \Gamma^{\mu}\, P_{\mu} + a_2\, \frac{i}{2}\, \Gamma^{\mu\nu}_{(-)}\, M^{(-)}_{\mu\nu} + a_3\, \frac{i}{2}\, \Gamma^{\mu\nu}_{(+)}\, M^{(+)}_{\mu\nu}$$

$$-\frac{1}{2\pi\alpha'}\{(b_1 + b_1'\, i\Lambda_o)\ell + b_2\, i\Lambda^{\mu\nu}\, N_{\mu\nu}\}]\, \Psi = 0 \tag{15}$$

$$P_{\mu} = i\int_o^1 d\sigma\, \frac{\delta}{\delta x^{\mu}(\sigma)}, \qquad M^{(\pm)}_{\mu\nu} = i\int_o^1 d\sigma\, \{t_{\mu}(\sigma)\, \frac{\delta}{\delta x^{\nu}(\sigma)} \pm t_{\nu}(\sigma)\, \frac{\delta}{\delta x^{\mu}(\sigma)}\},$$

$$\ell = \int_o^1 d\sigma\, (-x'^2)^{1/2}, \qquad N_{\mu\nu} = \int_o^1 d\sigma\, (-x'^2)^{1/2}\, t_{\mu} t_{\nu}.$$

Eq. (15) corresponds to the Dirac equation in the point particle case. It is a linear differential equation for the line functional (string field) $\Psi$. Clearly $\Psi$ cannot be a Lorentz scalar. The generates of Lorentz transformations are

$$J^{\mu\nu} = i\int_0^1 d\sigma \{x^\mu(\sigma) \frac{\delta}{\delta x_\nu(\sigma)} - x^\nu(\sigma) \frac{\delta}{\delta x_\mu(\sigma)}\}$$

$$+\frac{i}{4} [\Gamma^\mu, \Gamma^\nu] + \frac{i}{4} \{\Gamma^{\mu\alpha}_{(-)}, \Gamma^\nu_{(-)\alpha}] + \frac{i}{4} [\Gamma^{\mu\alpha}_{(+)}, \Gamma^\nu_{(+)\alpha}] + \frac{i}{4} [\Lambda^{\mu\alpha}, \Lambda^\nu{}_\alpha]. \quad (16)$$

## 4. The $\Gamma^{\mu\nu}_{(\pm)}$ and $\Lambda^{\mu\nu}$ algebras

The algebras (12), taken separately or in any combination, specify Clifford algebras. The algebras for $\Gamma^\mu$, $\Gamma^{\mu\nu}_{(-)}$, and $\Gamma^{\mu\nu}_{(+)}$ $(\Lambda^{\mu\nu})$ are equivalent, in d spacetime dimensions, to d, d(d-1)/2, and (d-1) (d+2)/2 dimensional Clifford algebras, respectively. It is of great interest to find how $\Psi$ transforms under four dimensional proper Lorentz transformations and to determine the spin content of $\Psi$ in reduced four-dimensional spacetime. It is sufficient for that purpose to consider each $\Gamma$- or $\Lambda$-algebra separately, since the $\Gamma$- and $\Lambda$- algebras are orthogonal to each other in the algebraic sense and each piece in the intrinsic spin components of the Lorentz generators $J^{\mu\nu}$, (16), commutes with the others. If we keep more than one of the $\Gamma$- and $\Lambda$- terms in Eq. (15), the spin content of $\Psi$ is just the direct product of the spin content determined by each $\Gamma$ or $\Lambda$.

Four-dimensional Lorentz transformation properties are characterized by values of a pair of spins, namely by a pair of integers or half-odd-integers, $(j_+, j_-)$. The results are

$$\begin{aligned} &\Gamma^\mu: && (0,1/2) \oplus (1/2, 0) \\ &\Gamma^{\mu\nu}_{(-)}: && \{(1/2, 1/2) \oplus (1/2, 1/2)\} \otimes \{(0, 1/2) \oplus (1/2, 0)\}^{d-4} \qquad (17) \\ &\Gamma^{\mu\nu}_{(+)}(\Lambda^{\mu\nu}): && \{(1/2, 3/2) \oplus (3/2, 1/2)\} \otimes \{(0, 1/2) \oplus (1/2, 0)\}^{d-4}. \end{aligned}$$

$\{\Gamma^\mu\}$ always defines spinors, irrespectively of spacetime dimensionality d, while the spin content defined by $\{\Gamma^{\mu\nu}_{(\pm)}\}$ or $\{\Lambda^{\mu\nu}\}$ depends on d. It follows from (17) that $\{\Gamma^{\mu\nu}_{(\pm)}\}$ or $\{\Lambda^{\mu\nu}\}$ defines bosonic (fermionic) objects in even (odd) dimensions.

Another problem associated with the new $\Gamma$- and $\Lambda$- algebras is the definition of the Dirac conjugate, which is necessary to construct a string field action. The Dirac conjugate $\bar{\Psi} = \Psi^\dagger\Omega$ is defined such that $\bar{\Psi}\Psi(\bar{\Psi}\Lambda_o\Psi)$, $\bar{\Psi}\Gamma^\mu\Psi$, and $\bar{\Psi}\Gamma^{\mu\nu}_{(\pm)}\Psi$ $(\bar{\Psi}\Lambda^{\mu\nu}\Psi)$ transform under Lorentz transformations as a scalar, a vector, and a tensor, respectively. Contrary to the point particle case there turn out to be several possibilities for $\Omega$ in a particular theory which give a Lorentz invariant action.

The general construction of $\Omega$ goes as follows. First we note that the Clifford subalgebra of the diagonal components $\Gamma^{\mu\nu}_{(+)}$ $(\Lambda^{\mu\nu})$ in (12) is not of canonical form owing to the tracelessness condition. It is an easy exercise to show that these d diagonal components can be expressed in terms of d - 1 hermitian matrices $\gamma_m$ obeying $\{\gamma_m, \gamma_n\} = 2\delta_{mn}$ (m, n = 1,..., d-1). Secondly we define

$$A_1 = \Gamma^o, \qquad \bar{A}_1 = \Gamma^1\Gamma^2 \ldots \Gamma^{d-1},$$

$$A_2 = \prod_{i<j}^{d-1} \Gamma^{ij}_{(-)}, \qquad \bar{A}_2 = \Gamma^{01}_{(-)} \ldots \Gamma^{0,\, d-1}_{(+)},$$

$$A_3 = \prod_{k=1}^{d-1} \gamma_k \prod_{i<j}^{d-1} \Gamma^{ij}_{(+)}, \qquad \bar{A}_3 = \Gamma^{01}_{(+)} \ldots \Gamma^{0,\, d-1}_{(+)}, \tag{18}$$

$$A_4 = A_3(\Gamma^{\mu\nu}_{(+)} \to \Lambda^{\mu\nu}), \qquad \bar{A}_4 = \bar{A}_3(\Gamma^{\mu\nu}_{(+)} \to \Lambda^{\mu\nu}),$$

$$A_5 = \Lambda_o.$$

Each $A_i$ ($\bar{A}_i$) is composed entirely of (anti-)hermitian matrices. It can be shown that in a theory which involves some collection C of the algebras (12) the products

$$\Omega = e^{i\delta} \prod_{i \in C} (A_i \text{ or } \bar{A}_i) \tag{19}$$

provide suitable candidates for $\Omega$. We shall see in Sec. 7 that the requirement of discrete symmetries (P,C,T, etc.) severely restricts possible forms of $\Omega$.

## 5. Line Space Measure and the String Field Action

The action for string fields must have the form

$$I = \int D(\text{line})\ \bar{\Psi} D \Psi, \tag{20}$$

where D is defined in Eq. (15). In line functional theory line space measure D(line) cannot be

$$DX = \Pi dx^\mu_n, \quad x^\mu(\sigma) = \sum_n x^\mu_n\, e^{2\pi i n\sigma},$$

which is commonly used in the literature. For in terms of the coordinates $x^\mu_n$, each line is counted an infinite number of times. Since the Lagrangian density $\bar{\Psi}D\Psi$ is reparametrization invariant, the action with the measure DX shall always diverge, containing the infinite volume factor of the reparametrization group.

The infinite volume is removed a la Faddeev and Popov. We select as gauge conditions the functionals

$$h_n[x] = \frac{1}{2\pi i n} \int_o^1 d\sigma e^{-2\pi i n\sigma} \left(\frac{\sqrt{-x'(\sigma)^2}}{\ell} - 1\right), \quad (n \neq 0), \tag{21}$$

corresponding to the arclength parametrization. It is straight forward to check that the corresponding Faddeev-Popov determinant is exactly unity. Thus the line space measure is determined to be

$$D(\text{line}) = DX \prod_{n \neq 0} \delta[h_n(x)]. \tag{22}$$

The above measure (22) has a distinct property. It satisfies

$$\int D(\text{line})\ M_{\nu_1 \ldots \nu_m, \mu}\ \Psi(\text{line}) = 0, \tag{23}$$

where $\Psi$ is an arbitrary line functional and

$$M_{\nu_1 \ldots \nu_m, \mu} = \int_o^1 d\sigma\ t_{\nu_1}(\sigma) \ldots t_{\nu_m}(\sigma) \frac{\delta}{\delta x^\mu(\sigma)}.$$

The differential operators $P_\mu$ and $M^{(\pm)}_{\mu\nu}$ in (15) are special cases of the above M-operators. Owing to the property (23) one can integrate in parts in the action (20), bringing differential operators from $\Psi$ to $\bar{\Psi}$. This property is absolutely necessary for the consistency of the theory.

Otherwise two equations obtained by varying $\Psi$ and $\bar{\Psi}$ in (20) would be incompatible with each other. We note that if an additional factor is multiplied in the definition of D(line), (22), then the measure so defined would not satisfy (23) in general.

## 6. Solutions in Three Dimensions and the Excitation Spectrum

Eq. (15) is not easy to solve in general. Tremendous simplification is achieved in $d = 3$ in the $x_o = \tau$ subspace. The equation reads $D\Psi = 0$ where

$$D = a_1(i\Gamma^o\frac{\partial}{\partial\tau} + \Gamma^k P_k) + a_2 \frac{i}{2} \Gamma^{jk}_{(-)} M^{(-)}_{jk} + a_3 \frac{i}{2} \Gamma^{jk}_{(+)} M^{(+)}_{jk} - \frac{1}{2\pi\alpha'} (b_1\ell + b_2\, i\Lambda^{jk}N_{jk}). \tag{24}$$

Here $\Psi$ is a function of $\tau$ and spacelike plane curves $\{\vec{x}(\sigma)\}$, satisfying (5). (The $b_1'$-term in (15) has been suppressed, since it can be transformed away in most cases.)

Eq. (24) can be solved completely in the subspace spanned by reparametrization invariant coordinates $\tau$, $\vec{y}$, $\ell$, $\kappa_n (n = \pm 3, \pm 4, \ldots)$, and $\lambda_\pm$ :

$$\vec{y} = \frac{1}{\ell}\int_o^1 d\sigma\sqrt{\vec{x}'(\sigma)^2}\,\vec{x}(\sigma), \qquad \kappa_n = \int_o^1 d\sigma\sqrt{\vec{x}'(\sigma)^2}\{t_1(\sigma) + it_2(\sigma)\}^n,$$

$$\lambda_\pm = \ell\int_o^1 d\sigma\{x_\pm(\sigma) - y_\pm\}\{t_1'(\sigma)\, t_2(\sigma) - t_1(\sigma)\, t_2'(\sigma)\}, \tag{25}$$

where $x_\pm(\sigma) = x_1(\sigma) \pm ix_2(\sigma)$ etc.. It is important to recognize that Eq. (24) has an infinite number of symmetries. In particular,

$$[D, \frac{\partial}{\partial\lambda_\pm} + \frac{iP_\mp}{4\pi q\ell}] = 0 \quad , \qquad [D, \frac{\partial}{\partial\kappa_{\pm 3}}] \propto a_3(\frac{\partial}{\partial\lambda_\pm} + \frac{iP_\mp}{4\pi q\ell})$$

$$[D, \frac{\partial}{\partial\kappa_n}] = 0 \quad (n = \pm 4, \pm 5, \ldots). \tag{26}$$

Here $P_\pm$ and q are momentum eigenvalues and the rotation index (= an integer) of a closed plane curve, respectively. Corresponding to the symmetries (26), one might impose the following set of physical state conditions:

$$(\frac{\partial}{\partial\lambda_\pm} + \frac{iP_\mp}{4\pi q\ell})\,\Psi_{phys} = 0, \quad \frac{\partial}{\partial\kappa_n}\Psi_{phys} = 0 \ (n = \pm 3, \pm 4, \ldots). \tag{27}$$

Of course, the $(\text{mass})^2$-spectrum remains unchanged even if the conditions (27) are dropped, though the multiplicity is increased by a factor $\infty$. One physical change would result in possible values of the orbital angular momentum L, which, in the rest frame ($p_\pm = 0$), is given by

$$L = \lambda_+\frac{\partial}{\partial\lambda_+} - \lambda_-\frac{\partial}{\partial\lambda_-} + \Sigma n\kappa_n\frac{\partial}{\partial\kappa_n}. \tag{28}$$

The conditions (27) avoid a bizarre spectrum in which L is totally unconstrained with respect to $(\text{mass})^2$.

The resulting $(\text{mass})^2$- spectrum is

$$(\text{mass})^2 = n_- m_-^{\,2} + n_+ m_+^{\,2}, \quad (n_\pm = 0, 1, 2, \ldots), \tag{29}$$

$$m_-^2 = \frac{2a_2 bq}{a_1^2\, \alpha'}, \quad m_+^2 = \frac{3a_3 b_2 q}{a_1^2\, \alpha'}$$

where $b = (b_1^2 + 1/3\, b_2^2)^{1/2}$. Notice that ground states are always massless, irrespective of the values of $a_i$ and $b_i$. The first and second terms in (29) correspond to radial ($\ell$-) and orbital ($\kappa_{\pm 2}$-) excitations, respectively. In particular, eigen functions of massless states depend only on $\tau$, $\vec{y}$, $\ell$, and $\lambda_\pm$, but not on $\kappa_{\pm 2}$. They are maximally symmetric in line space. (Recall that the $\lambda_\pm$-dependence enters in a trivial fashion dictated by (27).) In view of this tachyonic states are expected to be absent not only in the subspace we defined above, but also in the full space. Non-trivial $\kappa_{\pm 2}$ dependence increases both L, (28), and $(\text{mass})^2$, (29). The line functional theory yields an equally spaced $(\text{mass})^2$-spectrum and linearly rising trajectories, although detail of the spectrum differs from that in the standard string theory. In the case at hand ground states are not necessarily Lorentz scalars, since the string field $\Psi$ carries an intrinsic spin. (In the most general case $a_i \neq 0$, $b_i \neq 0$, their total angular momenta, $J_{12}$, are 0, ±1, or ±2.)

## 7. Discrete Symmetries

As was shown in Sec. 4, there are several possibilities for $\Omega$ in defining the Dirac conjugate $\bar{\Psi} = \Psi^\dagger \Omega$. The requirement of the invariance of the action (20) under discrete symmetry operations, space inversion P, charge conjugation C, time reversal T, and strong reflection SR, partially removes the ambiguity in selecting $\Omega$ from (19).

We quote some of the results in ref. 3, which are valid in the minimal representations of the $\Gamma$- or $\Lambda$- algebras.

i) $\{\Gamma^\mu\}$, $\{\Gamma^{\mu\nu}_{(-)}\}$, or $\{\Gamma^{\mu\nu}_{(+)}\}$

Let's first consider a single algebra represented by $\Gamma^\mu$, $\Gamma^{\mu\nu}_{(-)}$, or $\Gamma^{\mu\nu}_{(+)}$. There are two possible $\Omega$'s in each case, namely $A_i$ or $\bar{A}_i$ (i = 1,2,3), up to a phase factor. (See (19)). The results are

| | $\Omega$ | $A_1$ | | | | $\bar{A}_1$ | | | |
|---|---|---|---|---|---|---|---|---|---|
| $\{\Gamma^\mu\}$ | d(mod 4) | P | C | T | SR | P | C | T | SR |
| | 2 | o | o | o | o | x | x | x | o |
| | 4 | o | o | o | o | x | x | o | o |

| | $\Omega$ | $A_2$ | | | | $\bar{A}_2$ | | | | |
|---|---|---|---|---|---|---|---|---|---|---|
| $\{\Gamma^{\mu\nu}_{(-)}\}$ | d(mod 8) | P | C | T | SR | P | C | T | SR | |
| | $2^*$ | x | x | o | o | — | — | — | — | o: yes |
| | 4 | o | o | o | o | x | o | x | o | x: no |

| | $\Omega$ | $A_3$ | | | | $\bar{A}_3$ | | | |
|---|---|---|---|---|---|---|---|---|---|
| $\{\Gamma^{\mu\nu}_{(+)}\}$ | d(mod 8) | P | C | T | SR | P | C | T | SR |
| | 2 | o | x | x | o | x | x | o | o |
| | $4^*$ | x | x | x | o | — | — | — | — |

Here the asterisk * indicates that the dimensionarity of the Clifford algebra is odd so that $A_i$ and $\bar{A}_i$ are essentially the same.

ii) $\{\Gamma^{\mu}, \Gamma^{\mu\nu}_{(-)}\}$ or $\{\Gamma^{\mu}, \Gamma^{\mu\nu}_{(+)}\}$

The next examples are combinations of two Γ's.

$\{\Gamma^{\mu}, \Gamma^{\mu\nu}_{(-)}\}$

| $\Omega$ | $A_1A_2$ | | | | $A_1\bar{A}_2$ | | | | $\bar{A}_1A_2$ | | | | $\bar{A}_1\bar{A}_2$ | | | |
|---|---|---|---|---|---|---|---|---|---|---|---|---|---|---|---|---|
| d(mod 8) | P | C | T | SR | P | C | T | SR | P | C | T | SR | P | C | T | SR |
| $2^*$ | o | o | o | o | x | o | x | o | — | — | — | — | — | — | — | — |
| 4 | o | x | x | o | x | x | o | o | x | o | x | o | o | o | o | o |

$\{\Gamma^{\mu}, \Gamma^{\mu\nu}_{(+)}\}$

| $\Omega$ | $A_1A_3$ | | | | $A_1\bar{A}_3$ | | | | $\bar{A}_1A_3$ | | | | $\bar{A}_1\bar{A}_3$ | | | |
|---|---|---|---|---|---|---|---|---|---|---|---|---|---|---|---|---|
| d(mod 8) | P | C | T | SR | P | C | T | SR | P | C | T | SR | P | C | T | SR |
| 2 | o | o | o | o | x | o | x | o | x | o | x | o | o | o | o | o |
| $4^*$ | o | x | x | o | x | x | x | o | — | — | — | — | — | — | — | — |

Again * indicates the odd dimensionality of the Clifford algebras. In the first example, for instance, $A_1A_2 \simeq \bar{A}_1\bar{A}_2$ in d = 2 (mod 8).

iii) P, C, T invariant theories.

From the above examples we see that a theory is not invariant under P, C, and T in general. (Although symmetries can be restored, for instance, by doubling the number of components of $\Psi$, we stick to minimal representations of the Γ- or Λ- algebras.) In d = 2 (mod 8) theories defined by $\{\Gamma^{\mu\nu}_{(-)}\}$, $\{\Gamma^{\mu\nu}_{(+)}\}$, $\{\Gamma^{\mu\nu}_{(-)}, \Gamma^{\mu\nu}_{(+)}\}$, and $\{\Gamma^{\mu}, \Gamma^{\mu\nu}_{(-)}, \Gamma^{\mu\nu}_{(+)}\}$ are not invariant under either P, or C, or T for any choice of $\Omega$. In d = 4 (mod 8) theories defined by $\{\Gamma^{\mu\nu}_{(+)}\}$, $\{\Gamma^{\mu}, \Gamma^{\mu\nu}_{(-)}, \Lambda_o\}$, $\{\Gamma^{\mu}, \Gamma^{\mu\nu}_{(+)}\}$. $\{\Gamma^{\mu\nu}_{(-)}, \Gamma^{\mu\nu}_{(+)}\}$, and $\{\Gamma^{\mu}, \Gamma^{\mu\nu}_{(-)}, \Gamma^{\mu\nu}_{(+)}\}$ are not invariant.

## 8. The Second Quantization of String Fields

We defined string field theory as a line functional theory which incorporates string dynamics in the Dirac form, and were led to Eq. (15) and the action (20). In the three dimensional case we found, in addition to the equally spaced $(\text{mass})^2$-spectrum and linearly rising trajectories, that massless states always exist irrespective of the values of the five parameters defining the theory. Moreover, we observed the indication of the non-existence of tachyons and the existence of huge hidden symmetries.

We solved the string field equation which corresponds to the Dirac (or Schrodinger) equation in point particle theory. Note that solving the classical string field equation is equivalent to solving the Schrodinger wave equation in the first quantizied theory in the one-string sector. This is how we could get the discrete $(mass)^2$ spectrum.

If one finds the complete set of solutions to our linear field equation, and knows all local symmetries underlying the theory, then the second quantization of string fields can be carried out by imposing canonical commutation relations on them after eliminating gauge degrees of freedom. Although we have only limited knowledge in either aspect mentioned above, still we can make important observations.

First, since the field equation is linear (although the differential operator D in Eq. (15) is highly non-linear with respect to $x^\mu(\sigma)$), the quantization would not change the $(mass)^2$ spectrum. In particular, massless states remain massless in quantum theory. One might wonder why the tuning of the parameters is unnecessary in our case to guarantee the existence of massless states. In the standard field theory the zero intercept $\alpha_0$, which appears in field equations, must be set to be one to have massless states. There is no such parameter in Eq. (15) as $\alpha_0$. The reason is that our equation is the first order differential equation, the standard normal ordering procedure being unnecessary. As for the Lorentz invariance and the unitarity, the problem is more subtle. The action (20) is Lorentz invariant at the classical level. If one keeps the manifest Lorentz invariance in the quantizaition, then there would appear negative norm states corresponding to excitations in the $x_0$-direction. Accordingly, consistent physical state conditions ought to be found to guarantee that all physical states have positive or zero norms. If one takes the $x_0 = \tau$ subspace as we did in Sec. 6, then the closure of the Lorentz algebra has to be checked. We have not done these analyses yet. Finally, as we found in the three-dimensional case, the line functional theory is expected to have infinite-dimensional symmetries, which means that the gauge fixing procedure is necessary in the quantization.

## References


1. L. Carson and Y. Hosotani, Phys. Rev. Lett. 56, 2144 (1986).
2. L. Carson and Y. Hosotani, Univ. Minn. Preprint, UMN-TH-613/87.
3. C.L. Ho, L. Carson, and Y. Hosotani, Univ. of Minn. Preprint UMN-TH-614/87.
4. W. Volterra, Theory of Functionals (Blackie and Son Limited, Glasgow, 1930).
5. C. Marshall and P. Ramond, Nucl. Phys. B85, 375 (1975).
6. Y. Nambu, Phys. Lett. 80B, 372 (1979).
7. A.M. Polyakov, Nucl. Phys. B164, 171 (1979).
8. J.L. Gervais and A. Neveu, Phys. Lett. 80B, 255 (1979).
9. M.J. Bowick and S.G. Rajeev, Phys. Rev. Lett. 58, 535 (1987), MIT preprint MIT-CIP-1450.

# INTERNAL SYMMETRY FOR TYPE I STRINGS IN OSCILLATOR FORMALISM

L. Clavelli

Dept. of Physics and Astronomy
University of Alabama
Tuscaloosa AL 35487

Since the late 1960's, string theory has profited from the existence of an operator formalism for constructing scattering amplitudes. The idea of a vertex operator, generalizing the concept of a source current for point particles to an operator describing the emission of an extended object, was itself one of the greater contributions of string theory to physics and mathematics. However, up until recently, internal symmetry in type I string theories was incorporated by merely appending to the amplitudes certain traces of products of generators of the internal symmetry group (Chan-Paton factors.) Although this procedure was fully consistent mathematically, it always seemed a somewhat artificial treatment of what should have been a dynamical degree of freedom connected with the emission process. Within the last year a proposal(1) has been made to remedy this by associating to the vertex operators an internal symmetry piece which provides the Chan-Paton factors. The group at Alabama(2) has constructed a particular Fock space realization of this proposal including explicit construction of the twist operator in the internal symmetry space. The present report reviews these developments and discusses the status of the $\xi$ parameter relating the internal symmetry factors of the Mobius loop to those of the planar loop. We also discuss some preliminary attempts to generalize the Chan-Paton ansatz using the oscillator formalism.

In the operatorial scheme, D fermionic degrees of freedom are assigned to the string endpoints corresponding to the canonical anti-commuting oscillators

$$b^I, b^{I\dagger} \qquad I=1,2,\ldots D \qquad \left\{b^I, b^{J\dagger}\right\} = \delta^{IJ} \quad . \tag{1}$$

With these we can construct D hermitian fermionic fields at each end of the string.

$$\Psi_0{}^I = b^I + b^{I\dagger} \quad , \quad \Psi_\pi{}^I = i(b^{I\dagger} - b^I) \tag{2}$$

At this stage in our understanding, D is any even integer, not neccessarily related to the dimension of space-time. Requiring that the type I string theory be finite at one loop order relates the dimension of the internal symmetry space to the dimension of the Minkowski space, thus identifying the D of eq. 1 to the space-time dimension. With a $2^{D/2} \times 2^{D/2}$ representation of the Dirac gamma matrices we can construct the generators of the Clifford algebra

$$\sigma^{\{I\}} = i^{n(n-1)/2} \prod_{\substack{j=1 \\ I_1<I_2<\ldots<I_n}}^{n} \gamma^{I_j} \quad . \tag{3}$$

These satisfy

$$\mathrm{tr}\, \sigma^{\{I\}}\sigma^{\{J\}} = 2^{D/2}\, \delta^{\{I\}\{J\}} \tag{4}$$

and

$$\sum_{\{I\}} \sigma^{\{I\}}_{\alpha\beta}\, \sigma^{\{I\}}_{\rho\tau} = 2^{D/2}\, \delta_{\beta\rho}\delta_{\alpha\tau} \quad . \tag{5}$$

The $\sigma^{\{I\}}$ are in one-to-one correspondence with the independent products of $\Psi$'s.

$$\Psi_0^{\{I\}} = i^{n(n-1)/2} \prod_{\substack{j=1 \\ I_1<I_2<\ldots<I_n}}^{n} \Psi_0^{I_j} \tag{6}$$

To emit a string with internal symmetry corresponding to the group generator $\lambda^i$, we multiply the vertex operator by an operator $J(\lambda^i)$

$$J(\lambda^i) = 2^{-D/2}\, \lambda^i{}_{\alpha\beta} \sum_{\{I\}} \sigma^{\{I\}}_{\beta\alpha}\, \Psi_0^{\{I\}} \quad . \tag{7}$$

The operator J has the composition property

$$J(A)J(B) = J(AB) \tag{8}$$

and therefore

$$\langle 0|\prod_{i=1}^{N} J(\lambda^i)|0\rangle = \langle 0|J(\prod_{i=1}^{N}\lambda^i)|0\rangle = 2^{-D/2}\,\mathrm{tr}(\prod \lambda^i) \quad . \tag{9}$$

This provides the usual Chan-Paton factors for string tree graphs. Loop amplitudes are given by Fock space traces over vertex operators so that one must consider expressions such as

$$\mathrm{Tr} \prod_{i=1}^{N} J(\lambda^i) = \mathrm{Tr}\, J(\prod_{i=1}^{N}\lambda^i) \quad . \tag{10}$$

We denote Fock space traces by Tr to distinguish them from matrix traces denoted by tr. The basic identity that will enable us to calculate such traces is

$$\mathrm{Tr}\ x^{N_b}\ \mathcal{O}(b,b^\dagger) = (1+x)^D \langle 0|\mathcal{O}(\frac{b}{1+x} + b'^\dagger,\ b^\dagger + \frac{b'x}{1+x})|0\rangle \qquad (11)$$

where $b'$ and $b'^\dagger$ are a new set of fermionic oscillators parallel to b and $b^\dagger$. By means of this identity one can easily see that

$$\mathrm{Tr}\ J(A) = 2^{D/2}\ \mathrm{tr}\ A = \mathrm{tr}\ (\lambda^0)\ \mathrm{tr}(A) \qquad (12)$$

where $\lambda^0$ is the unit matrix in the internal symmetry space. Eq. 12 is the usual Chan-Paton factor for the planar loop.

To compute the twisted loops it is necessary to construct the twist operator, $\Omega$, which we write

$$\Omega = (-1)^R\ \Omega_J \qquad (13)$$

where

$$R = L_0 - p^2/2 \qquad (14)$$

and $\Omega_J$ is the internal symmetry piece of the twist operator. At every internal line in a loop we will put a factor $1+\Omega$ so that the physical states with R odd will have $\Omega_J$ eigenvalue -1 while the physical states with R even will have $\Omega_J$ eigenvalue +1. In the bosonic theory and the covariant formulation of the superstring the gauge bosons are in the adjoint representation of the symmetry group and have R=1 while the massless fermions are also in the adjoint but have R=0. Therefore if $\lambda$ is a group generator we must have

$$\Omega_J J(\lambda)|0\rangle = -\ J(\lambda)|0\rangle \qquad \text{(boson)} \qquad (15)$$

$$\Omega_J J(\lambda)|0\rangle = +\ J(\lambda)|0\rangle \qquad \text{(fermion)}\ . \qquad (16)$$

At alternate mass levels the states fall into singlet and symmetric tensor representations and the corresponding equations are reversed in sign. Thus in the covariant formulation the twist operator in the internal space differs in sign for a boson line and a fermion line. In the manifestly supersymmetric (light-cone) approach both gauge bosons and massless fermions have R=0 and therefore positive eigenvalues of $\Omega_J$.

The generators of the orthogonal groups are antisymmetric

$$\lambda^T = -\ \lambda \qquad SO(N) \qquad (17)$$

whereas the generators of the symplectic group have a property we call "asymplectic"

$$\lambda^T = -\ A_2\ \lambda\ A_2 \qquad USp(N) \qquad (18)$$

with

$$A_2 = \begin{pmatrix} 0 & -i\mathbb{1} \\ i\mathbb{1} & 0 \end{pmatrix} \qquad (19)$$

In the case of the orthogonal group, the required operator is

$$\Omega_J = \frac{i^{N_b-1/2}+i^{-N_b+1/2}}{\sqrt{2}} \; e^{i\pi N_a} \, v \quad . \tag{20}$$

Here, v is +1 for the Bosonic string or the superstring in manifestly Lorentz covariant formalism with bosons circulating. It is -1 for the manifestly supersymmetric formalism and for the manifestly Lorentz covariant formalism with fermions circulating.

Table 1. Values of the v parameter

| | BOSONIC THEORY | TYPE I SUPERSTRING | | | |
|---|---|---|---|---|---|
| | | COVARIANT FORM. | | SUPERSYM. FORM. | |
| | | BOSON LINE | FERMION LINE | BOSON LINE | FERMION LINE |
| v | +1 | +1 | -1 | -1 | -1 |

The operator $N_b$ is the total number operator of the b oscillators while $N_a$ is the number operator of those fermionic modes which correspond to antisymmetric gamma matrices. For the unitary symplectic group, the twist operator is as in 20 except that one replaces $N_a$ with $N_{asy}$, the number operator of those fermionic modes that correspond to "asymplectic" gamma matrices. Using eq. 11 it is shown in ref. 2 that

$$\mathrm{Tr}\, \Omega_J\, J(A) = \xi \;\mathrm{tr}\, A \tag{21}$$

with

$$\xi = v\,\sqrt{2}\,\cos\left\{\frac{\pi}{4}(\eta_s-\eta_a-1)\right\} \quad . \tag{22}$$

$\eta_s-\eta_a$ is the total number of symmetric gamma matrices minus the total number of asymmetric gammas if the internal symmetry is $SO(2^{D/2})$. If the internal symmetry is $USp(2^{D/2})$ one replaces $\eta_s-\eta_a$ in eq. 22 by $\eta_{sy}-\eta_{asy}$, the total number of symplectic gammas minus the total number of asymplectic gammas. One can show(2) that, although the individual $\eta$ are representation dependent and D dependent, the differences are (independent of D) always given by

$$\eta_s-\eta_a = 0 \bmod 8 \text{ or } 2 \bmod 8 \tag{23}$$

$$\eta_{sy}-\eta_{asy} = 4 \bmod 8 \text{ or } 6 \bmod 8 \tag{24}$$

The proof depends on the fact that the $\sigma^{\{I\}}$ form a complete basis for matrices of dimension $2^{D/2}$. This gives us the $\xi$ parameters for the various loops as given in table 2. With these $\xi$ parameters the non-internal symmetry part of the twist operator is always $(-1)^R$ as in eq. 13. One can, if one is careful, assign a common $\xi$ parameter to the SO case and the opposite parameter to the USp case if compensating signs are multiplied into the stringy piece, $(-1)^R$, depending on which string theory or formulation one is dealing with. For instance, one can move the phase factor, v, from eq. 20 and 24 to eq. 13, writing $\Omega = v(-1)^R \Omega_J$. Then $\xi$ defined by eq. 21 is always +1 for SO(N) and -1 for USp(N). On the other hand, a convention in which $\Omega = -v\,(-1)^R\,\Omega_J$, leading to a universal $\xi$ of -1 for SO(N) and +1 for USp(N) would correspond to the Green-Schwarz choice.

Table 2. The internal symmetry $\xi$ parameter in string theory

| | BOSONIC THEORY | TYPE I SUPERSTRING | | | |
|---|---|---|---|---|---|
| | | COVARIANT FORM. | | SUPERSYM. FORM. | |
| | | BOSON LINE | FERMION LINE | BOSON LINE | FERMION LINE |
| SO(N) | +1 | +1 | -1 | -1 | -1 |
| USp(N) | -1 | -1 | +1 | +1 | +1 |

$$\Omega = (-1)^R \Omega_J \quad , \quad \mathrm{Tr}\; \Omega_J \prod_{i=1}^{M} J(\lambda^i) = \xi \; \mathrm{tr} \prod_{i=1}^{M} \lambda^i$$

The situation has led to some confusion in the literature. Some have assumed that the twist operator and the $\xi$ parameter should be universal for given internal symmetry group. Some errors have appeared in print and some published results are correct only due to a compensating error. The situation with respect to the bosonic theory is mistated in the Green, Schwarz, Witten book[(3)], volume 2, although the correct $\xi$ parameter could be deduced from considerations in volume 1, Ch 7. With the $\xi$ parameter of SO(8192), it has been shown that the dilaton contribution to the divergence of the bosonic string theory cancels at one-loop order and the contribution of the tachyon can be avoided by an appropriate analytic continuation of the amplitudes so that the theory is totally finite to this order.[(4)] The wrong result would lead to a bosonic theory finite for USp(8192) instead of SO(8192). On the other hand, it has been pointed out[(5)] that the tachyon divergence cancels for USp(8192)in a manner analogous to the cancellation of the massless contributions in the SO(32) superstring. These authors propose that cancellation of the leading divergence in multi-loop graphs of the bosonic theory for SO(8192) might presage finiteness of the SO(32) superstring to the same order. With respect to the extension of the finiteness proofs to all the parity even N point functions, the basic result[6] is that the planar and Mobius loop amplitudes with N external gauge bosons take the form

$$A^{N,P} = \int_0^1 dw' \ f(w')/w' \tag{25}$$

$$A^{N,M} = \int_0^1 dw' \ f(-w')/w' \tag{26}$$

The relative sign between these two amplitudes can be fixed by consistency with the zero slope limit to be -1 in the case of $SO(2^{D/2})$. If the full string amplitudes are finite at all, they are finite in the case of this gauge group. The full understanding of how this sign arises in the string theory requires a complete treatment of the $\xi$ parameter and a complete treatment of the projection operator that eliminates time-like and longitudinal modes in both the planar and Mobius loops. In ref. 6 it was assumed that the $\xi$ parameter would be the same as that given by Green and Schwarz[7] in the light cone gauge and that the effect of the Brink Olive projection operator in the Mobius loop would supply a minus sign to make the partition function, after absorbtion of a factor $1/|w|$ from the volume element, an analytic function of w. In the modern approach in which the time-like and longitudinal modes are cancelled by Fadeev-Popov ghosts[3], it is easy to see that the elimination of negative norm states provides no odd relative sign between the planar and Mobius loops. Taking this into account together with the $\xi$ parameters of table 2 , a proper understanding of the interference between these topologies would involve, in the covariant formulation with bosons circulating:

twist operator: $\Omega = (-1)^R$ , (27)

Chan Paton factors: $\mathrm{tr} \prod_{i=1}^{N} \lambda^i$ (for both planar and Mobius),

and

$$\text{modified partition function:} \quad F_{NS} = \frac{w^{1/2}}{|w|} \prod_{n=1}^{\infty} \left(\frac{1+w^{n-1/2}}{1-w^n}\right)^{D-2} . \tag{28}$$

In ref. 6, eq. 27 appears with a minus sign on the right hand side and eq. 28 appears without the modulus sign on w. This is obviously mathematically equivalent to the present eqs. 27 and 28 since the sign of $\Omega$ is only relevant to the Mobius loop and only the Mobius loop depends on $F_{NS}(-w)$. For practical calculations it may be preferable to keep the forms given in ref. 6 although the present eqs. 27 and 28 are logically more correct.

It is interesting to study the effect of the twist operator on individual vertices. The $\Psi_0$ and $\Psi_\pi$ of eq. 2 are related by

$$i^{N_b}\,\Psi_0{}^I = \Psi_\pi{}^I\, i^{N_b} \tag{29}$$

and

$$i^{-N_b}\,\Psi_0{}^I = -\,\Psi_\pi{}^I\, i^{-N_b} \quad . \tag{30}$$

Thus

$$\Omega_J\,\Psi_0{}^I = e^{i\pi N_a}\,\Psi_\pi{}^I\, e^{-i\pi N_a}\,\Omega_J' \tag{31}$$

and

$$\Omega_J'\,\Psi_0{}^I = e^{i\pi N_a}\,\Psi_\pi{}^I\, e^{-i\pi N_a}\,\Omega_J \quad , \tag{32}$$

where we have put

$$\Omega_J' \equiv \frac{i^{N_b-1/2} - i^{-N_b+1/2}}{\sqrt{2}}\, e^{i\pi N_a}\, v \quad . \tag{33}$$

In a way not totally parallel to eq. 6, we can define

$$\Psi_\pi{}^{\{I\}} \equiv \Big\{ i^{n(n-1)/2} \prod_{\substack{i=1\\ I_1<I_2<..I_n}}^{n} \Psi_\pi{}^{I_i} \Big\} \times \begin{cases} (-1)^{n_a} & ,\ n \text{ even} \\ (-1)^{n_a} e^{i\pi(N_b-1/2)} & ,\ n \text{ odd} \end{cases} \tag{34}$$

As before, n is the total number of elements in the set {I} and $n_a$ is the number of elements associated with antisymmetric $\gamma$ matrices. For a symplectic symmetry group one repaces $N_a$ and $n_a$ with $N_{asy}$ and $n_{asy}$ respectively. With this definition we have

$$\Omega_J\,\Psi_0{}^{\{I\}}\,\Omega_J = \Psi_\pi{}^{\{I\}} \quad . \tag{35}$$

The seemingly strange definition of $\Psi_\pi{}^{\{I\}}$ relative to that of $\Psi_0{}^{\{I\}}$ has the property that, although individual $\Psi_0{}^I$ anticommute with any $\Psi_\pi{}^J$, the composite operators $\Psi_0{}^{\{I\}}$ commute with arbitrary $\Psi_\pi{}^{\{J\}}$. One has then

$$\Omega_J\, J_0(\lambda)\, \Omega_J = J_\pi(\lambda) \tag{36}$$

where $J_0$ is the internal symmetry vertex operator of eq. 7 depending on the fermionic fields at the $\sigma$=0 end of the string and

$$J_\pi(\lambda) \equiv 2^{-D/2}\, \lambda_{\beta\alpha} \sum_{\{I\}} \sigma_{\beta\alpha}^{\{I\}}\, \Psi_\pi{}^{\{I\}} \quad . \tag{37}$$

Internal symmetry vertex factors at $\sigma=\pi$ commute then with factors at $\sigma$=0.

$$\big[J_0(A), J_\pi(B)\big] = 0 \tag{38}$$

Their composition law is as at $\sigma$=0:

$$J_\pi(A) J_\pi(B) = J_\pi(AB) \tag{39}$$

The operator of eq. 20 gives the desired result that a twisted tree graph and a loop with an odd number of twists are each proportional to a single Chan Paton trace while a loop graph with an even number of twists is proportional to a product of traces:

$$\langle 0|J(A)\Omega_J J(B)\Omega_J J(C)\Omega_J J(D)\Omega_J J(E)|0\rangle = \langle 0|J_0(ACE)J_\pi(BD)|0\rangle$$
$$= 2^{-D/2}\ tr(ACED^T B^T) \qquad (40)$$

$$Tr\ J(A)\Omega_J J(B)\Omega_J J(C)\Omega_J J(D) = Tr\ \Omega_J J_0(BDA)J_\pi(C)$$
$$= \xi\ tr(BDAC) \qquad (41)$$

$$Tr\ J(A)\Omega_J J(B)\Omega_J J(C)\Omega_J J(D)\Omega_J J(E) = Tr\ J_0(ACE)J_\pi(BD)$$
$$= tr(ACE)\ tr(BD) \quad . \qquad (42)$$

We suspect that the operator formalism for Chan-Paton factors may facilitate the resolution of sign ambiguities in the internal symmetry factors of multi-loop graphs but we have not, as yet, investigated this issue; nor have we solved the problem of constructing an operatorial Chan-Paton formalism for an internal symmetry such as SO(N) where $N \neq 2^{D/2}$.

Finally, one might wonder whether the operatorial chan paton formalism might suggest some generalizations of the Chan Paton scheme. For instance one might try to implement a direct product internal symmetry by associating to each particle two matrices $\lambda^i$ and $\omega^i$ transforming acccording to two representations of internal symmetry groups $G_J$ and $G_K$ respectively. The vertex for emission of such a particle would be

$$V(k^i,\lambda^i,\omega^i,z) = V(k^i,z)J(\lambda^i)K(\omega^i) \quad . \qquad (43)$$

Here, $J(\lambda)$ is the operator of eq. 7 while $K(\omega)$ is an analogous operator constructed out of independent fermionic degrees of freedom. In the operator formalism such a proposal would obviously lead to a factorizable theory with Chan Paton factors:

$$\left[\prod \lambda^i\right]\left[\prod \omega^i\right] \quad . \qquad (44)$$

The planar loop would contain two factors of eq. 12:

$$Tr\ J(\prod\lambda^i)K(\prod\omega^i) = tr(\lambda^0)tr(\omega^0)\left[\prod \lambda^i\right]\left[\prod \omega^i\right] \quad . \qquad (45)$$

The twist operator would be

$$\Omega = (-1)^R\ \Omega_J\ \Omega_K \quad . \qquad (46)$$

The on-shell physical states would still have $\Omega=1$, so, in the covariant formulation where the massless bosons have $R=1$, the gauge boson states would satisfy

$$\Omega_J\Omega_K\ J(\lambda^i)K(\omega^i)|0\rangle = -\ J(\lambda^i)K(\omega^i)|0\rangle \quad . \qquad (47)$$

Since the gauge bosons have internal twist parity -1, consistency at the massless level requires that

$$J(\lambda^i)K(\omega^i)J(\lambda^j)K(\omega^j) - J(\lambda^j)K(\omega^j)J(\lambda^i)K(\omega^i) \qquad (48)$$

be a sum of group theory factors for massless states. But

$$J(\lambda^i\lambda^j)K(\omega^i\omega^j)-J(\lambda^j\lambda^i)K(\omega^j\omega^i) = \frac{1}{2} J([\lambda^i,\lambda^j])K(\{\omega^i,\omega^j\}) + \frac{1}{2} J(\{\lambda^i,\lambda^j\})K([\omega^i,\omega^j]) \qquad (49)$$

Consider the case where both $G_J$ and $G_K$ are orthogonal groups. Then regardless of whether $\lambda^i$ and $\omega^i$ are antisymmetric (i in adjoint rep, A) or symmetric (i in singlet or symmetric tensor rep, S), $[\lambda^i,\lambda^j]$ and $[\omega^i,\omega^j]$ are antisymmetric while $\{\lambda^i,\lambda^j\}$ and $\{\omega^i,\omega^j\}$ are symmetric. Consistency then requires that under the direct product group the massless bosons transform as

$$A \otimes S \oplus S \otimes A \qquad (50)$$

The same result is obtained if $G_J$ and $G_K$ are symplectic groups. On such a state the twist operator $\Omega_J\Omega_K$, given by a product of two operators of the form of eq. 20 with v=+1 , automatically satisfies eq. 47. The $\xi$ factor for this theory is a product of the $\xi$ factors for $G_J$ and $G_K$ and hence is +1 if $G_J$ and $G_K$ are either both orthogonal or both symplectic. We have, therefore, a finite type I string theory if

$$tr(\lambda^0)tr(\omega^0) = 32 \qquad (51)$$

The symmetry group is therefore of the form

$$G_J \otimes G_K = SO(N_1) \otimes SO(N_2) \quad \text{with } N_1N_2 = 32 \qquad (52)$$

or

$$G_J \otimes G_K = USp(N_1) \otimes USp(N_2) \quad \text{with } N_1N_2 = 32 \qquad (53)$$

If the symmetry is a direct product of an orthogonal group with a symplectic group, the $\xi$ parameter is -1 and no finite string theory results. The dimension of the representation (eq. 50) of either direct product group in eqs. 52 or 53 is:

$$N_1(N_1-1)N_2(N_2+1)/4 + N_1(N_1+1)N_2(N_2-1)/4 = N_1N_2(N_1N_2-1)/2 = 496$$

The theory has the same number of massless states as the SO(32) string. As presently formulated, the two theories, in fact, have the same internal symmetry since in addition to the direct product symmetry which is a subgroup of SO(32), there is a further symmetry in which a state of the first term of eq. 50 is taken into a state of the second term. The present attempt does not therefore lead to a distinct finite string theory although generalizations such as that discussed here could be useful in a complete theory of symmetry breakdown wherein mass degeneracy between the two terms of eq. 50 is somehow broken.

This article is largely a report of work[2] done in collaboration with Professors Paul Cox and Ben Harms. The consideration of direct product groups was a result of interaction with Professor J. Shapiro. Discussions with all of the above and with Dr. Allen Stern at the University of Alabama and with Professor Pierre Mathieu of Laval University are gratefully acknowledged. This work was supported in part by the Department of Energy under contract DE-FG05-84ER40141.

REFERENCES

1. N. Marcus and A. Sagnotti, Phys. Lett. 188B, 58 (1987)
2. L. Clavelli, P.H. Cox, and B. Harms, Alabama preprint UAHEP874, to be published
3. M.B. Green, J.H. Schwarz, and E. Witten, Superstring Theory (Cambridge University Press, 1987)
4. M.R. Douglas and B. Grinstein, Phys. Lett. 183B 52 (1987) Err. Phys. Lett. 187B 442 (1987)
   S. Weinberg, Phys. Lett. 187B, 278 (1987)
5. P. Mathieu and Q. Ho-Kim, Cancellation of the leading divergence in open bosonic string one-loop amplitudes, Laval University preprint (revised) (1987)
6. L. Clavelli, in Proceedings of the Lewes String Theory Workshop, Lewes Del, 6-27 July (1985), eds. L. Clavelli and A. Halprin, World Scientific Press (1986); Phys. Rev. D33, 1098 (1986); Prog. Theoret. Phys. Suppl. No 86, 135 (1986)
7. M.B. Green and J.H. Schwarz Phys. Lett. 151B, 21 (1985)

## PARTICIPANTS

ATICK, JOSEPH J., *Theory Group, SLAC, Bin 81, P.O. Box 4349, Stanford, CA 94303 USA*

BJORKMAN, JON, *Department of Physics, University of Colorado, Boulder, CO 80309 USA*

CAMPBELL, DAVID E., *Department of Mathematics, University of Texas, Austin, TX 78712 USA*

CARLIP, STEVEN, *Center for Relativity, Department of Physics, The University of Texas at Austin, Austin, TX 78712-1081 USA*

CHANG, DARWIN, *Physics Department, Northwestern University, 2145 Sheridan Road, Evanston, IL 60201 USA*

CLAVELLI, LOUIS J., *Physics Department, University of Alabama, P.O. Box 1921, University, AL 35486 USA*

CLEMENTS, MINOT, *Theory Group, RLM5.208, Physics Department, University of Texas at Austin, Austin, TX 78712-1081 USA*

DE ALWIS, S. PRASHANTA, *Theory Group, Physics Department, University of Texas, Austin, TX 78712 USA*

DELLA PIETRA, STEPHEN, *Theory Group, RLM5.208, Physics Department, University of Texas at Austin, Austin, TX 78712-1081 USA*

DOLAN, LOUISE, *Department of Physics, Rockefeller University New York, NY 10021 USA*

ENGLERT, FRANCOIS, *Service de Physique Théorique, Campus Plaine, c.p. 225, Bd du Triomphe, B-1050 Brussels, Belgium*

ENSIGN, PHILIP, *Shell Development Company, Box 481, Houston, TX 77001 USA*

FRAMPTON, PAUL, *Physics Department, University of North Carolina, Chapel Hill, NC 27514 USA*

FREUND, PETER G.O., *Enrico Fermi Institute, University of Chicago, 5460 Ellis, Chicago, IL 60637 USA*

GATES, S. JAMES, *Physics Department, University of Maryland, College Park, MD 20742 USA*

GEPNER, DORON, *Physics Department, Princeton University, Princeton, NJ 08544 USA*

GERVAIS, J.-L., *Laboratorie de Physique Theorique, Ecole Normale Superieure, 24 rue Lhomond, 75231 Paris Cedex 05, France*

GODDARD, PETER, *DAMTP, Cambridge University, Silver Street, Cambridge, CB3 9EW, Britain*

HOSOTANI, YUTAKA, *Physics Department, University of Minnesota, Minneapolis, MN 55455 USA*

JEVICKI, ANTAL, *Physics Department, Brown University, Providence, RI 02912 USA*

KAC, VICTOR G., *Department of Mathematics, Massachusetts Institute of Technology, Cambridge, MA 02139 USA*

KAKU, MICHIO, *Physics Department, City College of C.U.N.Y., New York, NY 10031 USA*

KAUFMAN, WILLIAM, *Department of Physics, University of Colorado, Boulder, CO 80309 USA*

KIKKAWA, KEIJI, *Department of Physics, Osaka University, Toyonaka, Osaka 560, Japan*

KIM, JIHN E., *Physics Department, Seoul National University, Seoul 151, Korea*

KLEBANOV, IGOR, *Theory Group, SLAC, Bin 81, P.O. Box 4349, Stanford, CA 94305 USA*

KOURES, V. BILL, *Department of Physics, University of Colorado, Boulder, CO 80309 USA*

LOFT, RICHARD, *Department of Physics, University of Colorado, Boulder, CO 80309 USA*

MATHANTHAPPA, K. T., *Department of Physics, University of Colorado, Boulder, CO 80309 USA*

MANSOURI, FREYDOON, *Mail Location 11, 210 Braunstein, University of Cincinnati, Cincinnati, OH 45221 USA*

MEZINCESCU, LUCA, *Physics Department, P.O. Box 248.046, University of Miami, Coral Gables, FL 33124 USA*

NEPOMECHIE, RAFAEL I., *Physics Department, P.O. Box 248.046, University of Miami, Coral Gables, FL 33124 USA*

OVRUT, BURT, *Physics Department, University of Pennsylvania, Philadelphia, PA 19104 USA*

PAL, PALASH, *Physics Department, University of Massachusetts, Amherst, MA 01003 USA*

PREITSCHOPF, CHRISTIAN, *Physics Department, University of Maryland, College Park, MD 20742 USA*

RAJEEV, S.G., *Center for Theoretical Physics, 6-408, MIT, Cambridge, MA 02139 USA*

RUBIN, MARK, *Physics Department, Box 272, Rockefeller University, New York, NY 10021 USA*

SAKITA, BUNJI, *Department of Physics, City College of C.U.N.Y., New York, NY 10031 USA*

SCHWARZ, JOHN, *Physics Department, Caltech, Pasadena, CA 91125 USA*

SEN, ASHOKE, *Theory Group, SLAC, Bin 81, P.O. Box 4349, Stanford, CA 94305 USA*

SHAPIRO, JOEL, *Physics Department, Rutgers University, P.O. Box 849, Piscataway, NJ 08854 USA*

STELLE, KELLY, *Department of Theoretical Physics, Imperial College, London, SW7 2AZ, Britain*

SUSSKIND, LEONARD, *Physics Department, Stanford University, Stanford, CA 94305 USA*

TAYLOR, JOHN G., *Department of Mathematics, King's College, Strand London, WC2R 2LS. Britain*

TOSA, YASUNARI, *Department of Physics, University of Colorado, Boulder, CO 80309 USA*

INDEX